Self-Assembled Nanostructures

Nanostructure Science and Technology

Series Editor: David J. Lockwood, FRSC
National Research Council of Canada
Ottawa, Ontario, Canada

Current volumes in this series:

Polyoxometalate Chemistry for Nano-Composite Design
Edited by Toshihiro Yamase and Michael T. Pope

Self-Assembled Nanostructures
Jin Z. Zhang, Zhong-lin Wang, Jun Liu, Shaowei Chen, and Gang-yu Liu

---

A Continuation Order Plan is available for this series. A continuation order will bring delivery of each new volume immediately upon publication. Volumes are billed only upon actual shipment. For further information please contact the publisher.

Self-Assembled Nanostructures

Jin Z. Zhang

University of California
Santa Cruz, California

Zhong-lin Wang

Georgia Institute of Technology
Atlanta, Georgia

Jun Liu

Sandia National Laboratories
Sandia, New Mexico

Shaowei Chen

Southern Illinois University
Carbondale, Illinois

and

Gang-yu Liu

University of California
Davis, California

Springer Science+Business Media, LLC

Library of Congress Cataloging-in-Publication Data

Self-assembled structures/Jin Zhang ... [et al.].
p. cm. — (Nanoscience science and technology)
Includes bibliographical references and index.
ISBN 978-1-4757-8724-5 ISBN 978-0-306-47941-0 (eBook)
DOI 10.1007/978-0-306-47941-0
1. Nanostructures. 2. Self-organizing systems. I. Zhang, Jin. II. Series.

QC176.8.N35 S45 2003
530.4'1—dc21

2002030089

Top cover illustrations:

Quantum dot: Created by Jordon Johnson and David Lockwood.
(Used by permission of David Lockwood, National Research Council of Canada.)

Carbon nanotube: Created by Professor J.-C. Charlier.
(Used by permission of Prof. J.-C. Charlier, Université Catholique de Louvain, Belgium.)

ISBN 978-1-4757-8724-5

Originally published by Kluwer Academic/Plenum Publishers, New York in 2003
Softcover reprint of the hardcover 1st edition 2003

http://www.wkap.nl

10 9 8 7 6 5 4 3 2 1

A C.I.P. record for this book is available from the Library of Congress

Scope of the Series "Nanostructure Science and Technology"

Nanostructure science and technology now forms a common thread that runs through all physical and materials sciences and is emerging in industrial applications as nanotechnology. The breadth of the subject material is demonstrated by the fact that it covers and intertwines many of the traditional areas of physics, chemistry, biology, and medicine. Within each main topic in this field there can be many subfields. For example, the electrical properties of nanostructured materials is a topic that can cover electron transport in semiconductor quantum dots, self-assembled molecular nanostructures, carbon nanotubes, chemically tailored hybrid magnetic-semiconductor nanostructures, colloidal quantum dots, nanostructured superconductors, nanocrystalline electronic junctions, etc. Obviously no one book can cope with such a diversity of subject matter. The nanostructured material system is, however, of increasing significance in our technology-dominated economy and this suggests the need for a series of books to cover recent developments.

The scope of the new series is designed to cover as much of the subject matter as possible—from physics and chemistry to biology and medicine, and from basic science to applications. At present, the most significant subject areas are concentrated in basic science and mainly within physics and chemistry, but as time goes by more importance will inevitably be given to subjects in applied science and will also include biology and medicine. The series will naturally accommodate this flow of developments in the science and technology of nanostructures and maintain its topicality by virtue of its broad emphasis. It is important that emerging areas in the biological and medical sciences, for example, not be ignored as, despite their diversity, developments in this field are often interlinked. The series will maintain the required cohesiveness from a judicious mix of edited volumes and monographs that, while covering subfields in depth, will also contain more general and interdisciplinary texts.

Thus the series is planned to cover in a coherent fashion the developments in basic research from the distinct viewpoints of physics, chemistry, biology, and materials science and also the engineering technologies emerging from this research. Each volume will also reflect this flow from science to technology. As time goes by, the earlier series volumes will then serve as reference texts to subsequent volumes.

David J. Lockwood
National Research Council
Ottawa, Canada

Foreword

Self Assembled Nanostructures by J. Z. Zhang, Z .L. Wang, J. Liu, S. Chen, and G.-Y. Liu is an exhaustive compilation of the different active areas in nanoscience today. The authors should be congratulated for their great efforts in writing such an extensive volume. They have amassed ten chapters on the different aspects of nanoscience covering topics such as synthesis, natural nanomaterials, nanocrystal self-assembly, structure characterization, lithographic techniques, chemical and photochemical reactivity, optical electronic dynamic properties of both semiconductor and metallic nanoparticles, and their electrochemical properties.

This book is valuable because of both its breadth and its focus. The chapters cover a wide span of topics relevant to the field of nanoscience with a focus on self-assembly. At the same time, the authors have commendably integrated their efforts and produced a very readable book.

The authors recognize that this important technique is crucial to furthering progress in nanoscience since our potential ability to employ nanoparticles in useful technological applications depends, in large part, on our ability to self-assemble them correctly.

Professor Mostafa A. El-Sayed
Julius Brown Chair and Regents Professor
Director, Laser Dynamics Laboratory
School of Chemistry and Biochemistry
Georgia Institute of Technology
Atlanta, Georgia

Preface

Nanostructured materials, or nanomaterials, refer to materials that have relevant dimensions on the nanometer length scales and reside in the mesoscopic regime between isolated atoms or molecules and bulk matter. These materials have unique physical and chemical properties that are distinctly different from bulk materials. Their size-dependent properties, their sensitivity to surface phenomena, and how they are spatially arranged present a significant challenge to our fundamental understanding of how these materials should behave. Intense interest in nanostructured materials is also fueled by tremendous economic and technological benefits anticipated from nanotechnology and nanodevices. Nanomaterials have demonstrated great potentials for applications in opto-electronics, photovoltaics, photocatalysis, microelectronics, sensors, and detectors. The advance in these areas will affect our daily life from how we design a faster computer, to how we use and conserve energy, preserve the environment, and how we diagnose and treat disease.

The research and development in nanomaterials involve three key aspects: assembly and synthesis of the nanomaterials, characterization of their properties, and exploration or implementation of their applications. Several books on the general subject of nanomaterials have appeared in the last few years. Most of these books have focused on the theoretical and physical aspects of nanomaterials and are written for professionals or experts in the field. To our best knowledge, there have been no introductory-level books that provide a systematic coverage of the basics of nanomaterials that covers the three important aspects of materials assembly and synthesis, characterization, and application. This book is designed to bridge that gap and offers several unique features. First, it is written as an introduction to and survey of nanomaterials with a careful balance between basics and advanced topics. Thus, it is suitable for both beginners and experts, including graduate and upper-level undergraduate students. Second, it strives to balance the chemistry aspects of nanomaterials with physical principles. Third, the book highlights nanomaterial based architectures including assembled or self-assembled systems rather than only isolated nanoparticles. Finally, the book provides an in-depth discussion of important examples of applications, or potential applications of nanoarchitectures, ranging from physical to chemical and biological systems.

The book is organized as follows. Chapter 1 provides an introduction to the field of nanomaterials. It covers some of the key historic developments, novel properties, and basic physical principles of nanomaterials. Chapter 2 deals primarily with molecular self-assembly or synthetic self-assembled nanostructures, including Langmuir–Blodgett films, surfactant directed nanoporous materials, and molecularly directed films and composites. Chapter 3 presents a survey of naturally self-assembled nanostructures important in biology and biomedical sciences, for example, bone tissues, natural laminated composites in sea shells, and cell membranes. Issues discussed include molecularly directed nucleation and growth, hierarchical ordering, and functional selectivity. A discussion of nanocrystal self-assembly is presented in Chapter 4, including different synthetic methods for making nanocrystals with emphasis on wet chemistry methods. Assembly and fabrication into desired two-dimensional and three-dimensional structures and magnetic properties are also covered. Chapter 5 deals with characterizations of structure and chemical functionality of nanoarchitectures using different microscopy techniques. Chapter 6 focuses on lithiography-engineered nanoarchitecture and nanofabrication, with emphasis on structure and chemical characterization of nanoarchitectures. Chemical reactivity of nanoarchitectures is the subject of Chapter 7, as illustrated with various chemical reactions conducted involving nanomaterials, including redox reactions involved in catalysis, photocatalysis, photoelectrochemistry, molecular recognition, and specific surface interaction. Chapter 8 deals specifically with optical and electronic functionality of nanoarchitectures, with focus on optical and electronic properties of semiconductor nanomaterials studied using frequency and time-resolved optical techniques. Optical and dynamic properties of metal nanomaterials are the subject of Chapter 9. Chapter 10 concentrates on electrochemical and transport properties of nanoarchitectures studied using electrochemical as well as electrical techniques. The scientific and technological impacts of nanoarchitectures are covered in many of the chapters. Several key application examples of nanoarchitectures, including nanoelectronics, solar cells, photocatalysis, photoelectrochemistry, light emitting diodes, lasers, detectors, sensors, and bioimaging are considered.

Acknowledgments

We are fortunate to have and grateful to many people who have supported and assisted us during the writing of this book.

First and foremost, it is the initiation, persistence, and camaraderie of Mr. Kenneth Howell, the book editor, that made the whole project possible. While organization and humor qualify a person to be a book editor, Ken's knowledge in nanoscience makes him a team member and a colleague. We are indebted to all other members at Kluwer Academic/Plenum Publishers for their assistance, especially Ms. Annette Triner, production editor, on whom one can always count.

Second, our sincere and deep appreciation go to many of our colleagues, associates, and collaborators. In particular, Jin Z. Zhang would like to thank his students, postdoctors, collaborators, and colleagues, including Julic Evans, Robert O'Neil, Donny Magana, Christian Grant, Thaddeus Norman, Adam Schwartzberg, Fanxin Wu, Bo Jiang, Drs. Trevor Robert, Archita Sengupta, Michael Brelle, Brian Smith, Nerine Cherepy, Greg Smestad, Michael Gratzel, Hongmei Deng, Alan Joly, Wei Chen, Anthony van Buuren, Frank Bridges, Jim Lewis, David Kliger, Greg Szulczewski, Tim Lian, Greg Hartland, Prashant Kamat, Arthor Nozik, Mostafa El-Sayed, Gunter Schmid, and Rajesh Mehra. Zhong Lin Wang thanks his colleagues, Drs. R. L. Whetten, M. A. El-Sayed, U. Landman, S. A. Harfenist, J. S. Yin, S. H. Sun, Z. R. Dai, and many students in the associated groups. Jun Liu would like to thank Mr. M. J. Mcdermott and Z. R. Tian for their help on manuscript preparation of Chapters 1, 2, and 3. Some of the research discussed in Chapters 1, 2, and 3 was funded by The Materials Division of Basic Energy Science Office of the U.S. Department of Energy. Shaowei Chen is grateful to his current and former student co-workers, Yiyun Yang, Renjun Pei, Kui Huang, Fengjun Deng, Siyuan Liu, Ivan Greene, Jennifer Sommers, for their contributions. Gang-yu Liu expresses her deep gratitude to her graduate students, Nabil A. Amro, Jayne C. Garno, Yile Qian, and Guohua Yang.

Finally, we would like to thank our family members for their understanding and support during the period of writing this book.

Contents

CHAPTER 4. Nanocrystal Self-Assembly

1

Introduction

Seldom in the history of scientific research has a single topic such as nanotechnology energized so much excitement and imagination in the scientific community and among the general public. The importance of nanotechnology has been identified by researchers from academic and industry worldwide. Potential applications of nanotechnology were indicated in President Clinton's speech at California Institute of Technology on January 21, 2000: "···shrinking all the information housed at the Library of Congress into a device the size of a sugar cube ··· detecting cancerous tumors when they are only a few cells in size." Nanotechnology involves manipulation of matter at the nanometer length scale ($1 \text{ nm} = 10^{-9}$ m). While methods allowing for the fabrication of smaller devices have always heralded new economic opportunities, fabrication on the nanometer scale is especially significant as this is the approximate size of the largest biological molecules: proteins and DNA. It is at this level where, over the next decade, the smallest length scale reachable by humans, and the largest molecular length scale of nature, will coincide. It is clear that nanotechnology brings opportunities for devices and materials capable of greatly improving the quality of human life.

Nanoscale materials, the foundation of nanoscience and nanotehcnology, have become one of the most popular research topics in a very short period of time. The intense interest in nanotehcnology and nanoscale materials is fueled by the tremendous economical, technological, and scientific impact anticipated in several areas: (1) The exponential growth of the capacity and speed of semiconducting chips, the key components that virtually enable all modern technology, is rapidly approaching their limit of art and demands new technology and new materials science on the nanometer scale; (2) novel nanoscale materials and devices hold great promise in energy, environmental, biomedical, and health sciences for more efficient use of energy sources, effective treatment of environmental hazards, rapid and accurate detection and diagnosis of human diseases, and improved treatment of such diseases; and (3) when a material is reduced to the dimension of the nanometer, only tens of the dimension of a hydrogen atom, its properties can be drastically different from those of either the bulk material that we can see and touch even though the composition is essentially the same, or the atoms or molecules that make up the materials. Therefore, nanoscale materials prove to be a very fertile ground for great scientific discoveries and explorations.

Nanoscale materials, or nanomaterials, can be defined as materials with at least one dimension that is on the nanometer scale, and with its properties influenced by this dimensional confinement. Materials with relevant physical dimensions on the nanometer scales possess unique chemical and physical properties compared to their corresponding bulk or isolated atoms and molecules. Their thermodynamic, dynamic, mechanical, optical, electronic, magnetic, and chemical properties can be significantly altered relative to their bulk counterparts. These properties are dependent not only on size but also on morphology and spatial arrangement. In addition, as the materials are reduced to nanometer scale, more and more atoms will be exposed to the surface. Therefore, surface phenomena, such as wetting, begin to play a critical role. Even with the recent surge of popularity of nanotehcnology, nanoscale materials are much more abundant than most people think. The soft and hard tissues in our body, such as muscles and bones, are superior nanoscale materials, whose mechanical properties cannot be matched by any man-made materials with the same chemical compositions due to the unique architecture on the nanometer scale. The ability for us to command and move our body parts, and for the cells to communicate, grow and divide, is closed related to movement of some molecular micormachines only a few nanometer in size (motor proteins). Nobody has been able to fabricate a functional motor machine with such small sizes, not to mention the incredible efficiency of the biological nanomachines (40–100%). In nature, bacteria actually use nanometer sized magnetic particles within the cell to sense the direction of their movement. Not realizing the importance of this particular length scale, mankind has been using nanoscale materials for a long time. Some good examples include the artistry to produce colorful glassware by dispersing nanoparticles in glass and the use of silver colloids in photography. In both cases, the color strongly depends on the size of the particles and how they are aggregated. Many of the materials we use in our daily life, such as metals and ceramics, are in fact nanoscale materials. These materials are usually treated in a special way so that one portion of the material is dispersed in another on a nanometer scale in order to improve the mechanical properties.

In the last decade, microelectronics and computers have penetrated every aspect of our life. The industry has been able to keep up with the demand by exponentially increasing the memory density and the speed of the semiconduting chips (Moore's law), namely increasing the memory capacity by a factor of 4 every three years. The main mechanism that has enabled this exponential growth is through the reduction in size of the individual chips. Currently, the relevant length scale in the fabrication of many new devices has reached the nanometer scale. For example, computer chip fabrication has reached the length scale of about 150 nm in commercial productions. It is anticipated that this length scale needs to reach 50 nm in the near future in order to keep up with the demand of the consumers, but microelectronic features on the nanometer scale presents serious challenges. First, the tools and methodologies for fabrication of such devices are not ready. Second, on such small length scale the electronic and other physical properties are not well understood. On these length scales, materials exhibit new properties such as quantum size confinement. Their surface properties become increasingly important. The unusual properties of nanoscale materials have tremendous implications on the performance and reliability of the devices. For example, at this length scale, the traditional insulation material (silica) between the interconnecting wires is not insulating enough to separate the signals. To make a Si-gate complementary metal oxide semiconductor (CMOS) less than 100 nm in dimension, the gate oxide layer separating the metal gate from the semiconducting layer needs to be as thin as 1.2 nm, only four silicon atoms across the layer. The question is how we can guarantee that this 1.2 nm oxide layer is uniform and does not leak current, and how we can guarantee that there is no tendency for this oxide layer to degrade and affect the long-term reliability. Even for something as simple as doping, which is crucial for semiconductors in device applications, becomes an intriguing one on the nanometer scale. For a nanoparticle with less than 100 atoms, not all the particles can contain one dopant atom for a 1% or less doping level. Then, how can one achieve homogeneous doping among all the particles and how would the doped particles behave differently from the undoped ones?

In the biomedical and human health area, the need to search for better detection and diagnosis, and for more effective treatment is insatiable. In the last few years, many technologies based on nanodevices and nanomaterials have emerged. Gold particles modified with specific DNA strands, antibody labeled magnetic particles, and antibody coded nanobars seem to promise easy and quick detection of protein molecules, DNA sequences, and other bio-organisms. Many techniques based on atomic force microscopy (AFM) have also been investigated for sensitive measurement of the interactions between biomolecules on the nanometer scale. Fluorescent nanoparticles have the potential to replace organic fluorescent dyes with high intensity and improved stability. Inorganic and organic nanoparticles, nanovesicles, and smart nanocomposite materials have also been widely investigated for drug formulation and for targeted and regulated delivery of drugs. In other areas, nanoparticles and nanoporous materials have been playing a pivotal role as catalysts and supports in chemical and energy industry for a long time. Reducing the scale and dispersion of the catalyst materials is the key to improve the catalytic efficiency. In environmental and health science, nanoparticles, nanophase materials, and nanocomposites have also demonstrated great potential for effective selection and separation of biomolecules, bio or industrial hazards. In all these areas, great challenges still remains. The challenges are related to more efficient fabrication and production of the nanomaterials so that they are more accessible, and have improved sensitivity and selectivity for fast and quantitative analysis, and better compatibility and stability in the environment (e.g., in biotissues). In addition, the interfacing (connection) of the nanomaterials with the system is not trivial.

Much has been learned about the unique properties of nanoscale materials in the last decades. For example, the color or absorption spectrum changes dramatically with size when the size is small compared to the de Broglie wavelength or the Bohr exciton radius of the electron. The emission wavelengths change with size as well. This could be useful for tunable lasers or light emitting diodes. The tunable properties of nanomaterials make them attractive for a variety of applications. The melting points of nanocrystals have been found to be much lower than those of bulk crystals. When a metal particle such as gold is smaller than 10 nm, it essentially exists in a state that is neither liquid nor solid. The shape of the particle is in a constant state of changing from one to another. When a common liquid such as water is confined to space that is only a few nanometer in dimension, its properties are significantly different from those of the liquid water and solid ice that we are familiar with. The unusual, not liquid-like and not solid-like properties of water play a very important role in the function of protein molecules in the body. When reduced to nanometer scale, ferroelectric materials and piezoelectric materials that are important in our daily electronic devices and appliances may no longer be ferroelectric or piezoelectric because they now have a different crystalline structure. Therefore, besides the tremendous potentials in improving our life, the fascinating properties of nanoscale materials will remain a strong motivation for scientific discovery and exploration.

From the above discussion, it is clear that in order for nanoscale materials to realize their full potential, evolutional progress and revolutional breakthrough are needed in the fabrication and synthesis of nanoscale materials, and in the understanding of the fundamental properties of such materials. The synthesis and fabrication of nanomaterials usually involve two drastically different approaches: the top-down approach and the bottom-up approach. Currently photolithography, which includes chemical vapor deposition (CVD) or metal oxide CVD (MOCVD), remains the only method accepted by the semiconductor industry, but this technique is rapidly reaching its size limit. To further reduce dimension, deep UV lithography, e-beam lithography, soft lithography, dip pen lithography, etc., are being investigated. The bottom up approach involving self-assembly of molecular species, with controlled chemical reactions, is much more efficient and flexible. Recent developments of new chemical synthesis techniques have reached the degree of sophistication that high quality samples can be produced. Samples with well-controlled surface properties and shape, for example, spherical particles vs. rods, can be conveniently prepared. Synthesis of particles with narrow size distributions has allowed the construction of superlattice or artificial atomic structures using different assembly techniques. Self-assembled superlattice structures have properties modified compared to isolated particles due to interparticle interactions. Very exciting results have been obtained with self-assembly using crystalline surfactants and polymer liquid crystalline structures, with proteins and other biomolecules. However, the self-assembly approach still needs to demonstrate the large scale control required for functional devices, and needs to find ways to make connections and interfaces between the devices and the systems.

Characterization of nanomaterials is a major part of nanomaterial research. Common techniques include optical spectroscopy (electronic, Raman, and infrared), X-ray techniques (diffraction, absorption, and photoemission) microscopy (transmission electron microscopy (TEM), scanning tunnelly microscopy (STM), and atomic force microscopy (AFM)), electrochemistry, and time-resolved laser techniques. For example, X-ray and microscopy techniques are often used to determine the structural properties, including particle size, shape, and crystal structure, whereas optical spectroscopy is sensitive to their electronic and optical properties. Each of these techniques serves to probe some specific aspects of the properties of the nanoparticles. One of the most exciting areas is the development of techniques for single molecular spectroscopy and single molecular microscopy, which shed new lights on the mechanisms involved in natural and synthetic nanoscale materials. The biggest challenges in characterization include preservation of the natural state of the materials while they are being analyzed, observation of real-time kinetic processes, quantitative analysis in addition to qualitative analysis, and sensitive detection of very minor components.

It is apparent that a complete coverage of all the topics is beyond the scope of any books or publications. Nor can anybody provide answers to all questions the challenges raised. A cursory survey of the market reveals numerous publications and books in the area of nanotechnology and nanoscale materials. However, a careful analysis of these publications indicates some important deficiencies. Many publications and books emphasize specific areas and topics. As a result, it is difficult to find books that provide comprehensive coverage of the important topics in materials production, characterization, and applications in a very concise manner for the general audience. The objective of this book is to provide a concise and comprehensive discussion of the latest research and development activities in the synthesis (including fabrication), characterization, and applications of nanomaterials. The book is also intended to illustrate dynamic relationships between materials synthesis, the understanding of the fundamental principles, and how the new understanding and new approaches lead to new applications. The applicability, the limitations, and the potentials of different approaches will be discussed. In addition to the coverage of new developments, the book will provide sufficient background information so that the readers can easily understand the most important scientific principles, whether it is related to synthesis, characterization, or application. Finally, the book is written with the hope that it will not only provide a ready reference about the general field of nanoscale materials, but also stimulate and challenge the readers to think about new things, to think about ways to break the boundaries between disciplines and to provide solutions from an angle that may not be familiar to material scientists.

In Chapters 2 and 3 of the book, the concept of self-assembly is introduced. Both synthetic and natural self-assembled materials are discussed. The principles of self-assembly and underlying molecular interactions are summarized. Natural self-assemblies in biologically important systems are discussed. Composite organic/inorganic and biological/inorganic nanomaterials are also presented. Chapters 2 and 3 are intended not only to provide updated information about the development of the very exciting field of self-assembly and the fundamental scientific principles behind it, but also to point out the inference between the "living" world and the synthetic world. The formation and the functions of living materials and living systems are fundamentally different from those of synthetic materials and devices. Living systems extract energy from the environment to generate and maintain highly ordered and functional structures on every length scale. Although the individual components, such as proteins molecules and cells, are in a constant state of evolution and regeneration, the collective functions and stability of the system is maintained. In addition, living materials and living systems are adaptable and responsive with respect to changes in environmental conditions. In contrast, in synthetic materials and devices, orders and functions are created by mechanical means (such as micromachining) or by the applications of external fields (as in thermally or electrically driven self-assembly). The stability, rigidity, and reliability of each individual component dictate the stability and life span of the whole system. Contrary to the adaptive and responsive properties of the living materials, the usefulness of synthetic materials and devices depend on their ability to stay unchanged regardless of the environment.

Chapter 4 focuses on self-assembly of inorganic semiconductor, insulator, and metal nanocrystals with an emphasis on shape and surface characterization and on understanding the growth mechanism of self-assembly. Chapter 5 concentrates on structural characterization of nanostructures using X-ray, TEM, and scanning microscopy techniques. Fabrication of nanostructures using nanolithographic techniques is covered in Chapter 6, with examples such as X-ray, electron, and ion-beam lithography, nanoparticle lithography, and scanning probe lithography. Both operating principles and some technical details are provided. Photochemical and photoelectochemical properties and related applications are the subject of Chapter 7. Application examples include air and water pollutant control using nanomaterials as photocatalysts. Molecular recognition and surface specific interactions are also presented. Optical, electronic, and dynamic properties and relevant applications of semiconductor and metal nanomaterials are covered in Chapters 8 and 9, respectively. These applications are illustrated with examples ranging from solar cells, light emitting diodes, and optical filters to biological labeling and imaging. Electrochemical properties and their applications are discussed in Chapter 10. Quantized charging behavior of metal nanoparticles is used to highlight the unique electronic properties of nanomaterials.

2

Synthetic Self-Assembled Materials: Principles and Practice

2.1. MICROSCOPIC AND MACROSCOPIC INTERACTIONS

Self-assembly has become a very effective and promising approach to synthesize a wide range of novel nanoscale materials. In the self-assembly processes, atoms, molecules, particles, and other building blocks organize themselves into functional structures as driven by the energetics of the system. Self-assembly also implies that if the system is taken apart into the appropriate subunits, these subunits can then be mixed to reform the whole structure under favorable conditions. The most important driving force for self-assembly is the interaction energies between the subunits, whether they are atoms, molecules, or particles.

2.1.1. Molecular Interaction Energies

On the most fundamental level, there are interactions among atoms, ions, and molecules. These interactions can be further classified into three categories, depending on if the species are charged:[1,2] Coulomb interactions due to the electrostatic effects from the permanent charges, van der Waals interactions due to instantaneous polarizations induced by the neighboring molecules (or atoms, ions), and short range strong repulsions. These interaction forces have been extensively discussed in Refs. 1–6, but a brief summary is provided below based on these references:

2.1.1.1. Coulomb Interactions

Coulomb interactions are caused by charged particles and can be positive or negative (attraction or repulsion) depending on the sign of the charges.

(i) Ion–ion pair interaction

$$E = (Z_1 e)(Z_2 e)/4\pi\epsilon_0 x \tag{2.1}$$

where E is the interaction energy, Z_1 and Z_2 are the valence (or number of charges) of the ions (species), e is the electronic charge, ϵ_0 the dielectric constant in vacuum, and x is the separation distance between the two ions.

(ii) Ion–permanent dipole interaction

$$E = (Ze)\mu \cos\theta/4\pi\epsilon_0 x^2 \tag{2.2}$$

where μ is the dipole moment, and θ the angle between the line of centers and the axis of the dipole.

(iii) Permanent dipole–permanent dipole interaction

$$E = (\text{constant})\mu_1\mu_2/4\pi\epsilon_0 x^3 \tag{2.3}$$

The constant depends on the relative orientation between the two dipoles: constant $= \sqrt{2}$ for average over all orientations; constant $= 2$ for parallel dipoles; constant $= -2$ for antiparallel dipoles.

2.1.1.2. van der Waals Interactions

The van der Waals interactions are due to instantaneous polarizations induced by the neighboring molecules (or atoms, ions) and are always negative (attraction). The potential usually has a power law exponent of 6 with respect to the separation distance.

(i) Permanent dipole-induced dipole interaction (Debye interaction)

$$E = (\alpha_1\mu_2^2 + \alpha_2\mu_1^2)/(4\pi\epsilon_0)^2 x^6 \tag{2.4}$$

where α_1 and α_2 are the polarizabilities of the two dipoles.

(ii) Permanent dipole–permanent dipole (Keesom interaction, due to the average effect of the rotational contribution of the polarizability)

$$E = -(2/3)\mu_1^2\mu_2^2/(4\pi\epsilon_0)^2 kTx^6 \tag{2.5}$$

where k is the Boltzmann constant and T is the temperature.

(iii) Induced dipole–induced dipole interaction (London interaction)

$$E = -(3h/3)\lfloor \nu_1\nu_2/(\nu_1 + \nu_2)\rfloor \alpha_1\alpha_2/(4\pi\epsilon_0)^2 x^6 \tag{2.6}$$

where h is Plank's constant, and ν is the characteristic vibration frequency of the electrons.

2.1.1.3. Short-Range, Strong Interaction

At a very close separation distance, a very strong repulsive force develops that rises sharply with respect to distance like two hard spheres approaching each other.

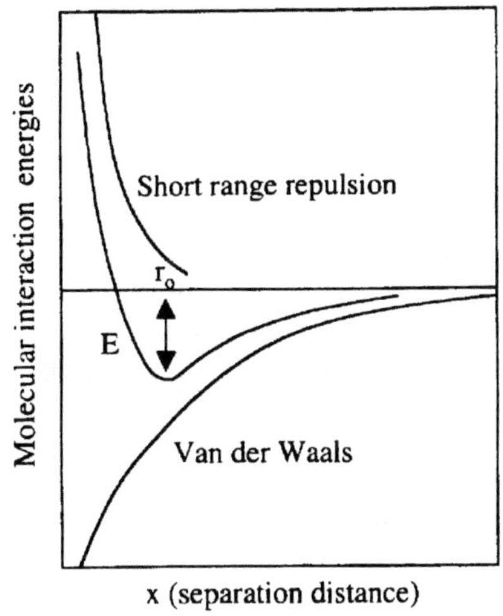

FIGURE 2.1. Schematic illustration of the inter-molecular energies as a function of distance. r_o indicates the equilibrium position (the equilibrium size) of the molecules.

The form of this interaction is not well defined but is usually treated as a power law with an exponent of 12, where ζ is a pre-exponent constant:

$$E = \zeta x^{-12} \tag{2.7}$$

2.1.1.4. *Total Interaction Potential*

There are other forms of intermolecular interactions, but at a close distance the most important interactions are van der Waals forces and the short-range repulsion. For convenience, these interaction energies are usually summarized in a power law equation (Lennard–Jones potential) including a repulsive term and an attractive term:

$$E = \zeta x^{-12} - \beta x^{-6} \tag{2.8}$$

The van der Waals six-power term reflects summation of the contribution from the random dipole–dipole interactions from (2.4) to (2.6) related to the polarizability of the molecules, and is always attractive (negative). The intermolecular energies as a function of distance are schematically illustrated in Fig. 2.1.

2.1.2. *Macroscopic Interaction Energies*

2.1.2.1. *van der Waals Attraction between Two Spherical Particles*

For macroscopic bodies, such as spherical particles, the van der Waals attraction energy can be assumed to be the addition of all the contributions from individual atoms (molecules).[2] If the number of atoms per unit volume is ρ, using the six-power expression for the van der Waals attraction in the Lennard–Jones equation, the pair wise interaction between the increment volumes (dV_1, dV_2) of the two objects is

$$dE_a = -(1/2)\rho^2\beta/x^6 dV_1 dV_2 \tag{2.9}$$

The total attraction energy is the integration over the volumes of the two objects:

$$E_a = -(1/2)\rho^2\beta \iint dV_1 dV_2/x^6 \tag{2.10}$$

With certain geometric assumptions, it is not difficult to derive the van der Waals attraction energies between a wide range of macroscopic bodies. Some of these expressions, under the approximate conditions, assume very simple and elegant forms, as shown below:[2]

Two identical spheres ($R \gg x$)

$$E_a = -AR/12x \tag{2.11}$$

Two spheres of the same composition but different size (R_1 and $R_2 \gg x$):

$$E_a = -AR_1R_2/6x(R_1 + R_2) \tag{2.12}$$

Two surfaces with indefinite thickness:

$$E_a = -A/12\pi x^2 \tag{2.13}$$

In these equations, R is the particle radius and A is the Hamaker constant as defined below:

$$A = (\rho\pi)^2\beta \tag{2.14}$$

2.1.2.2. Electrostatic Repulsive Energy

The electrostatic interactions between macroscopic bodies are more difficult to treat quantitatively than the van der Waals interactions. Usually, a charged surface is assumed to be composed of two regions (the so-called double layer structure): an inner region consisting of the charged surface itself and a layer of adsorbed species, and an outer diffuse region in which the charged species are distributed according to Boltzmann distribution functions as determined by the electric potential:[3]

$$n_i = n_{io} \exp[-z_i e\varphi/kT] \tag{2.15}$$

where n_i is the concentration of the charged species (number of charges per unit volume), z_i is the number of charges on the charged species, and φ is the electrical potential.

The net volume charge density ρ^* at any position is

$$\rho^* = \sum n_i = \sum n_{io} \exp[-z_i e\varphi/kT] \tag{2.16}$$

The variation of the surface potential as a function of distance from the surface and the charge distribution in space satisfies the Poisson equation:

$$\nabla^2\varphi = -\rho^*/\epsilon \tag{2.17}$$

where ∇^2 is the Laplacian operator, and ϵ is the permitivity.

These equations can be solved together with the proper boundary conditions ($\varphi = \varphi_0$ when $x = 0$; $\varphi = 0$ and $d\varphi/dx = 0$ when $x = \infty$) to give the electrical potential as a function of distance from the surface.

Normally, the solution of φ is complicated, but when the electric potential is low compared to the thermal effect, a simple solution can be derived (The Debye–Hückel approximation) for flat surfaces:[3]

$$\varphi = \varphi_0 \exp[-x/\kappa^{-1}] \tag{2.18}$$

where κ^{-1} is called the double layer thickness, and is defined by the electrolyte concentrations (ionic strength):

$$\kappa^2 = [c^2 N_A/\epsilon kT] \sum z_i^2 c_i \tag{2.19}$$

where N_A is the Avogadro's constant, and c is the molar concentration of the charged species.

When the electric potential overlaps, a repulsive potential is produced between two surfaces. A variety of analytical and numerical solutions have been obtained, depending on the approximations made, including whether a constant potential or constant charge is assumed. One of the widely used expressions for two identical spheres was derived by Verwey and Overbeek by considering the balance of the electrostatic repulsive force and the osmotic pressure developed between the surfaces at constant surface potential:[4,5]

$$E_r = [32 \cdot \epsilon R k^2 T^2 \gamma^2 / c^2 z^2] \exp[-\kappa x] \tag{2.20}$$

where z is the number of charges on the counter ion, and γ is a constant related to the surface potential φ_0:

$$\gamma = [\exp(ze\varphi_0/2kT) - 1)]/[\exp(ze\varphi_0/2kT) + 1)] \tag{2.21}$$

Although this equation is only applicable under a strict restriction of low surface potential ($\kappa x > 1$), acceptable results have been obtained for both large and low surface potential.

The total interaction energy is the summation of the electrostatic repulsion and the van der Waals attraction, as shown in Fig. 2.2. With a high particle surface potential, the repulsion is stronger, and the total interaction is positive. The particles will remain separated because an energy barrier needs to be overcome for the particles to approach each other. For a low surface potential, the repulsion is not strong and the total interaction energy is mostly attractive. Under these conditions, the particles will come together to form aggregated clusters.

2.1.3. Hydrogen Bonding, Hydrophobic, and Hydrophilic Interactions[6]

Hydrogen Bond. Water molecules are polar molecules and can be considered as a tetrahedral structure, with two positive charges (due to the two hydrogen atoms) pointing to one direction and two negative charges on the oxygen pointing

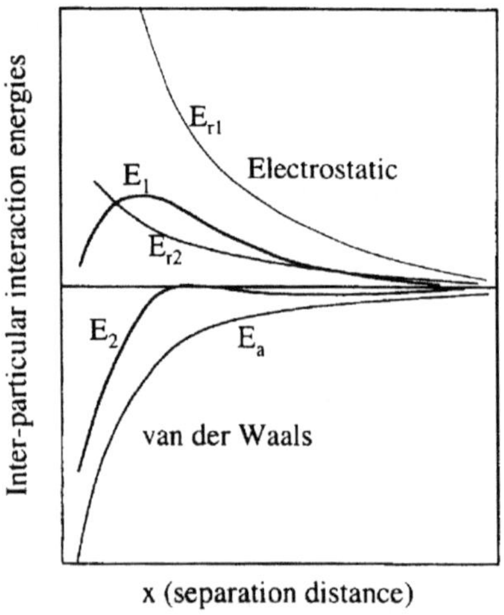

FIGURE 2.2. Schematic interaction energy curves as a function of separation between two particles. E_1 and E_2 represent the total interaction energies corresponding to a high repulsive energy E_{r1} and a lower repulsive energy E_{r2}.

in the opposite direction.[6] The positively polarized hydrogen atoms can interact strongly with neighboring electronegative atoms (such as the negatively polarized oxygen in water) and form an electrostatic bond (the hydrogen bond). Although the hydrogen bond is electrostatic in nature, it is directional. Hydrogen bonding is also commonly encountered between water and many other molecules containing electronegative atoms such as O, N, F, and Cl. It can occur intramolecularly (within one molecule) as well as intermolecularly (between two different molecules).

Hydrophobic Interaction. When water molecules approach an inert surface that cannot form a hydrogen bond such as alkanes, hydrocarbons, and fluorocarbons. The molecules need to reorient themselves so that the four charges on the water molecules will point away from the surface. The tetrahedral water molecules rearrange themselves so that the polarized groups on the molecules are still available for hydrogen bonding with the rest of the molecules, while also minimizing contact with the inert surface. As a result, the water molecules near the surface become more ordered as compared to free water molecules, producing a hydrophobic hydration layer of interconnected water molecules with open cage structures. A consequence of the hydrophobic hydration is the so-called hydrophobic attractive interaction between nonpolar molecules and the surfaces. The hydrophobic attraction occurs in water between hydrophobic molecules and surfaces because of the rearrangement of the hydrogen bond configurations in the overlapping solvation zone as the two hydrophobic surfaces come closer. The hydrophobic attraction is strong and has a long range that cannot be accounted for by van der Waals interaction as discussed earlier.

Hydrophilic Effect. Certain hydrophilic molecules and groups (charged ions, molecules, polar molecule and groups, molecules and groups capable of hydrogen bonds) are water soluble and repel each other in water. Contrary to hydrophobic molecules, hydrophilic molecules, and groups prefer to be in contact with water. Sometimes, the water molecules can be associated with these hydrophilic molecules. When dispersed in water, hydrophilic molecules tend to decrease the ordering of the water network, rather than increase the ordering.

2.2. SURFACTANTS AND AMPHIPHILIC MOLECULES

Amphiphilic molecules (molecules containing both hydrophilic and hydrophobic groups), such as surfactants, copolymers, and proteins, play a critical role in a wide range of self-assembly phenomena. The properties of these amphiphilic molecules are determined by unique intermolecular interactions: the hydrophobic interactions between the tail groups, and the hydrophilic and/or the electrostatic interactions between the head groups. Among these molecules, surfactants are mostly encountered in our daily life as detergents, soaps, shampoos and conditioners, dispersants for paints, food processing aids, foaming agents, emulsifiers, antistatic agents, fabric softeners, cosmetic ingredients, etc.

Surfactants, also called surface active agents, are molecules containing at least one hydrophilic head group and one hydrophobic tail group.[7] When present even at a low concentration, these molecules are able to absorb to the surface or interface to significantly reduce the surface energy. In general, surfactants can be classified into cationic surfactants, anionic surfactants, zwitterionic and nonionic surfactants. The cationic surfactants are molecules with positively charged head groups. These surfactants are usually made of long-chain amines and long-chain ammonium salts. Anionic surfactants are molecules with negatively charged head groups. Examples include carboxylic acid salts and sulfonic acid salts (sulfonates). Zwitterionic surfactants are molecules with head groups containing both a positive group (ammonium) and a negative group (carboxylic or sulfonate). Nonionic surfactants are molecules with neutral head groups, such as polyethylene oxides.

Figure 2.3 shows some commonly encountered cationic, anionic, zwitterions and neutral surfactants. The most important characteristics of the surfactants are the head group charge, the chain length, and the head group size. The long-chain amines are positively charged below pH 7. At high pH (>7) long-chain amines are no longer charged and lose activity. On the other hand, quaternary ammonium salts remain charged at all pH conditions and retain their activity all the time. The anionic surfactants are usually negatively charged at $\text{pH}>7$. At $\text{pH}<7$, anionic surfactants are not ionized and lose their activity. Beside pH, the activity of the ionic surfactants is also affected by the ionic strength of the solutions, and by the counter ions. The existence of counter ions can quickly neutralize the charges on the head group, and even cause the surfactant to precipitate. Nonionic surfactants have the advantage in that pH, counter ions, or solvents do not affect their activity.

2.3. TRANSITION FROM DISPERSED STATE TO CONDENSED STATE: THE BEGINNING OF SELF-ASSEMBLY

In general, materials can be regarded as existing in two states: the dispersed state and the condensed state. The transition from a dispersed state to a condensed state is a universal phenomenon and is the beginning of self-assembly, whether the system is made of molecular species, polymers, or microscopic objects such as particles. Several examples are given below to illustrate the

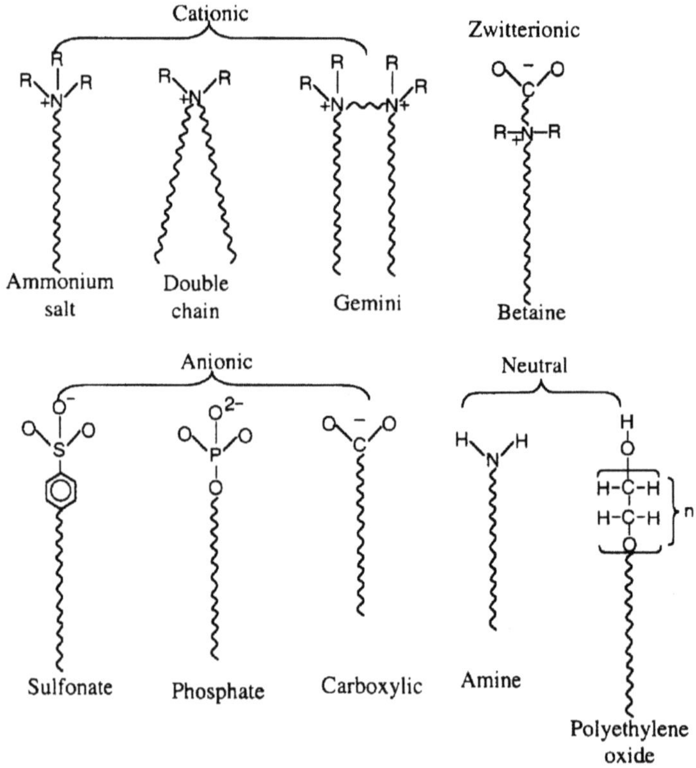

FIGURE 2.3. Commonly encountered surfactants.

similarity and role of intermolecular and interparticular interactions in the phase transition.

The behavior of an ideal gas is governed by the equation of state

$$PV = RT \tag{2.22}$$

where P is the pressure, V, the volume, R, the ideal gas constant, and T the absolute temperature.

In the ideal gas equation, it is assumed that there is no interaction between the molecules and that the molecules do not have a physical size. In reality, as discussed in the previous sections, all uncharged gaseous molecules interact with one another through van der Waals attractions. As a result of these molecular interactions, the molecules have a finite size. Therefore, the ideal gas equation needs to be replaced by the van der Waals equation of state:[8]

$$(P + a/V^2)(V - b) = RT \tag{2.23}$$

where a is the correction term reflecting the reduced pressure due to the van der Waals attraction force between the molecules, and b is the constant to count for the finite size of the molecules.

The van der Waals equation can be rewritten as

$$V^3 - (b + RT/P)V^2 + (a/P)V - ab/P = 0 \tag{2.24}$$

This is a cubic equation. For any given pressure and temperature, there will be three solutions for the volume V. The relationships between the temperature and volume are plotted in Fig. 2.4(a) at fixed temperature. Within a certain range of pressure, the three solutions for the volume, for example, B, F, C, in curve I are

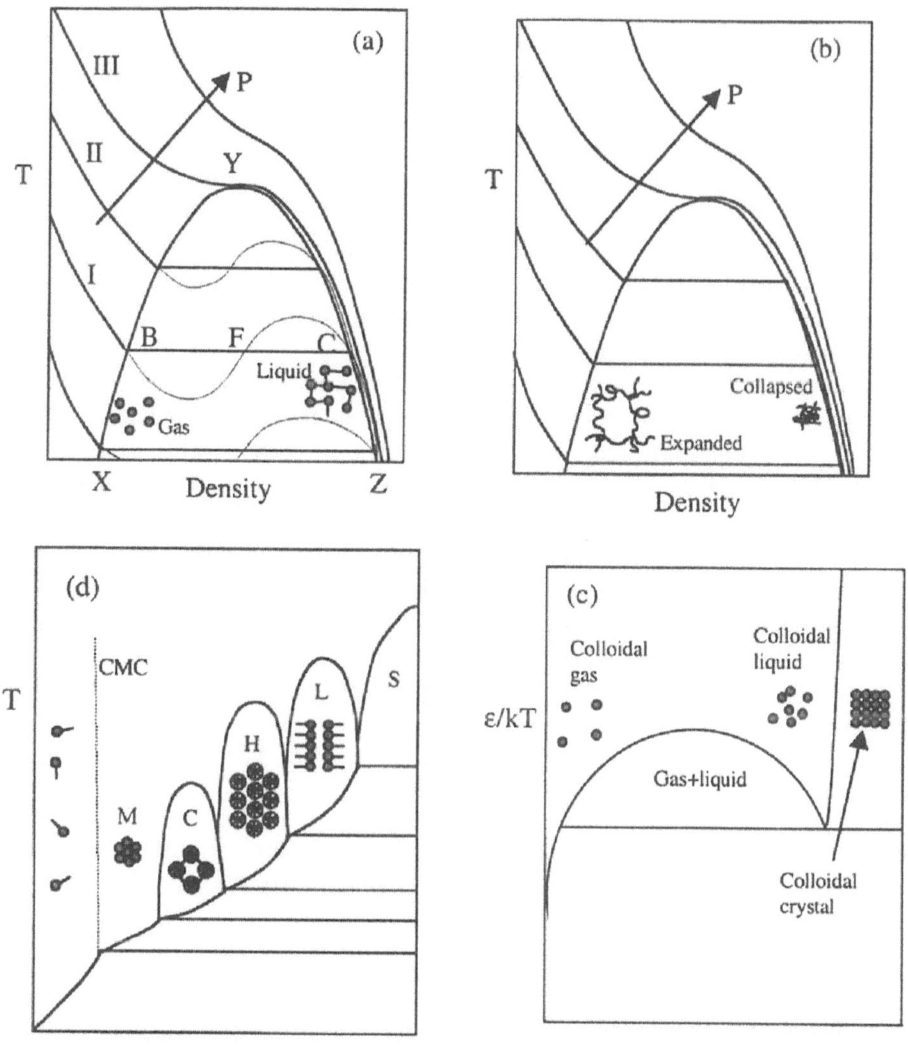

FIGURE 2.4. Phase diagrams of several systems showing the transition from dispersed states to condensed states. (a) Gas-liquid phase transition. (b) Swelling-deswelling in polymer gels. (c) Consolidation of colloidal crystals. (d) Surfactant phase diagrams. In Fig. 2.4(d) M, C, H, L, S represent micelles, cubic phase, hexagonal phase, lamellar phase, and solid phase respectively. Figures 2.4(a) and (b) are redrawn after Refs. 9, 10. Figure 2.4(c) is redrawn after Ref. 13. Figure 2.4(d) is redrawn after Ref. 16.

real. Therefore, B to C corresponds to the gas to liquid condensation when the temperature is reduced, or liquid to gas transition when the temperature is increased. Similar situations are encountered in curve II except that the transition occurs over a narrower volume range. At a much higher temperature, only one real solution exists and liquid condensation does not occur any more. In curve III, the three solutions for the volume become one and the gas–liquid transition is condensed to one point Y. This point is called the critical point because at this point the difference between the gas and the liquid disappears. The curve XYZ is called the coexisting curve, and the area under the curve is called the coexisting region.

Similar transitions from dispersed state to condensed state can occur in polymeric gels.[9,10] As in the gaseous materials, the behavior of a polymer gel can be described by an equation of state derived from a modified Flory–Huggins mean-field theory.[11,12] This modified equation considers the contribution from rubber elasticity (entropy change), the effect of counter ion osmotic pressure, and electrostatic repulsive forces.

Without going into details of the calculation, minimization of the free energy leads to a solution of the volume as a function of reduced temperature $(1-\Delta F/kT)$ (or other environmental parameters) as shown in Fig. 2.4(b). Figure 2.4(b) represents the transition from the dispersed state (expanded polymer gel) to the condensed state (collapsed polymer gel) similar to the gas–liquid transition shown in Fig. 2.4(a).

The transition from a dispersed state to a condensed state is not limited to molecular systems. Microscopic objects, such as colloidal particles also exhibit this kind of transition, and this phenomenon is widely used to prepare self-assembled colloidal crystals and supper-lattice structures. Although more detailed discussion on the self-assembly phenomena with colloidal particles or nanoparticles can be found in the next few chapters, here we will briefly describe the fundamental driving force. A simple solution to the colloidal phase diagram[13] involves the use of a two-dimensional (2D) square lattice simulation and the calculation of the pair-wise interactions between the particles, and their contribution to the entropy.[14,15] The total free energy of the system can be written as

$$F = E - TS \tag{2.25}$$

The internal energy E is the summation of the total pair-wise interaction energies, as discussed earlier, and the entropy S is related to the number of ways the particles and the pairs can be arranged on the lattice Ω:

$$S = k\,ln\,\Omega \tag{2.26}$$

With the use of the simplest form of a finite repulsion energy E between the particles, a colloidal phase diagrams can be derived as shown in Fig. 2.4(c).[13] In the colloidal phase diagram, the colloidal gas corresponds to a dilute colloidal suspension that exhibits Newtonian flow behavior. As the particle concentration is increased, non-Newtonian flow is observed corresponding to a continuous transition to a colloidal liquid phase. Once the critical particle concentration is reached, a sharp transition from dispersed colloids to a condensed crystalline-like behavior is observed. The transition from the colloidal liquid (dispersed state) to the colloidal crystalline phase (condensed state) is first order in nature, similar to the phase transitions observed in molecular systems.

Phase transitions in molecular systems, in polymers, and in colloidal particle systems are very similar in nature. Thus, it is reasonable to expect the transition from a dispersed state to a condensed state to be a universal phenomenon that can be used for self-assembled material. It is not surprising, for example, that surfactant systems commonly encountered also have similar phase transitions as shown in Fig. 2.4(d).[16] The difference in Fig. 2.4(d) is that within the condensed region, there may exist many condensed phases of a different structure. If an envelop is drawn along the outer boundaries of the condensed phases and the detailed structures under the envelop is ignored, the surfactant phase diagram is in fact very similar to the colloidal phase diagrams.

2.4. PACKING GEOMETRY: ATTAINING THE DESIRED SELF-ASSEMBLED STRUCTURES

Just knowing the transition from a dispersed state to a condensed state does not necessarily provides information on the microstructure of the phases that are formed. Amphiphilic molecules exist in a wide range of ordered structures in the condensed states. These structures can also transform from one to another when the solution conditions, pH, temperature, or electrolyte concentrations are changed. The equilibrium structures are determined by the thermodynamics of the self-assembly process and the inter- and intra-aggregate forces. The major driving forces for the amphiphiles to form well-defined aggregates are the hydrophobic attractions at the hydrocarbon–water interfaces and the hydrophilic ionic or steric repulsion between the head groups. A simple way to describe this kind of interaction is to use a geometric packing parameter (shape factor), $R = v/a_o l_c$, where v is the volume of the hydrocarbon chains, a_o, the optimal head-group area, and l_c the critical chain length.[17] As shown in Fig. 2.5, a small critical packing parameter (<0.5) favors the formation of a highly curved interface (spherical micelles and rod-like micelles), and a larger critical packing parameter (>0.5) favors the formation of flat interfaces (flexible bilayers and planar bilayers). A critical packing parameter greater than 1 will produce inverse micelles. In addition, the surfactant geometry is related to the experimental conditions.

Phase behavior can be explained based on the critical packing parameters and the amphiphile concentration. Figure 2.6 is a simplified diagram of a surfactant–water–oil system.[18,19] Depending on the solution compositions, spherical micelles, rod-like micelles, hexagonally ordered crystals, cubic crystals, lamellar phases, inverse micelles, and inverse micellar liquid crystals can be formed.

Clearly, as shown in Fig. 2.6, many experimental parameters affect which phase is formed. However, the change from one phase to another is not random, but rather follows a consistent pattern. For example, as the surfactant

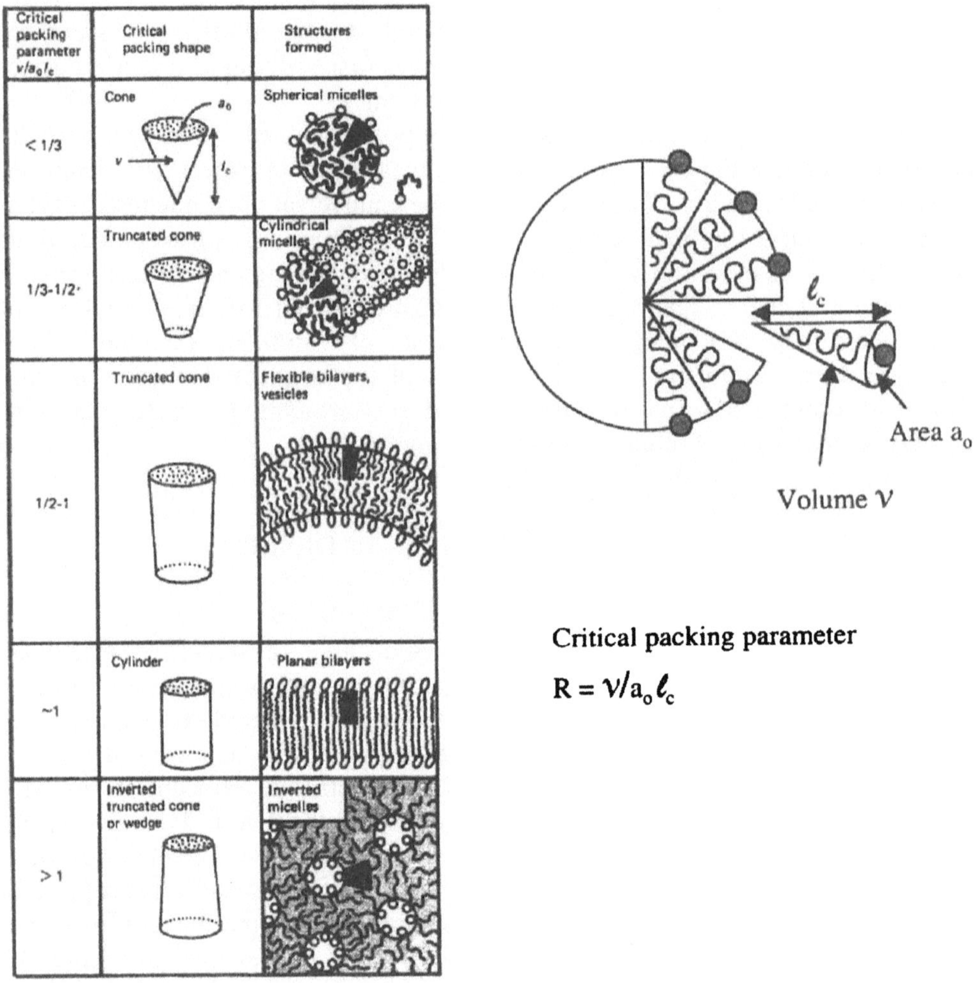

FIGURE 2.5. Relationship between the bilayer structures and the packing parameters (shape factors) (reproduced and adapted from Ref. 17 with permission from Academic Press).

concentrations are increased, the phases go through spherical micelles, rod-like micelles, hexagonal, cubic, and lamellar phases (Fig. 2.7).[20] These changes are consistent with the change of packing geometry when the experimental conditions are changed.

2.4.1. Effect of Surfactant Concentration

As the surfactant concentration increases, the amount of water available for association with the surfactant head group decreases. As a result, the degree of hydration of the surfactant head group decreases. The reduced hydration of the head groups leads to a decrease in the effective head group area. Based on the critical packing parameters, for fixed surfactant tail length, a reduction of

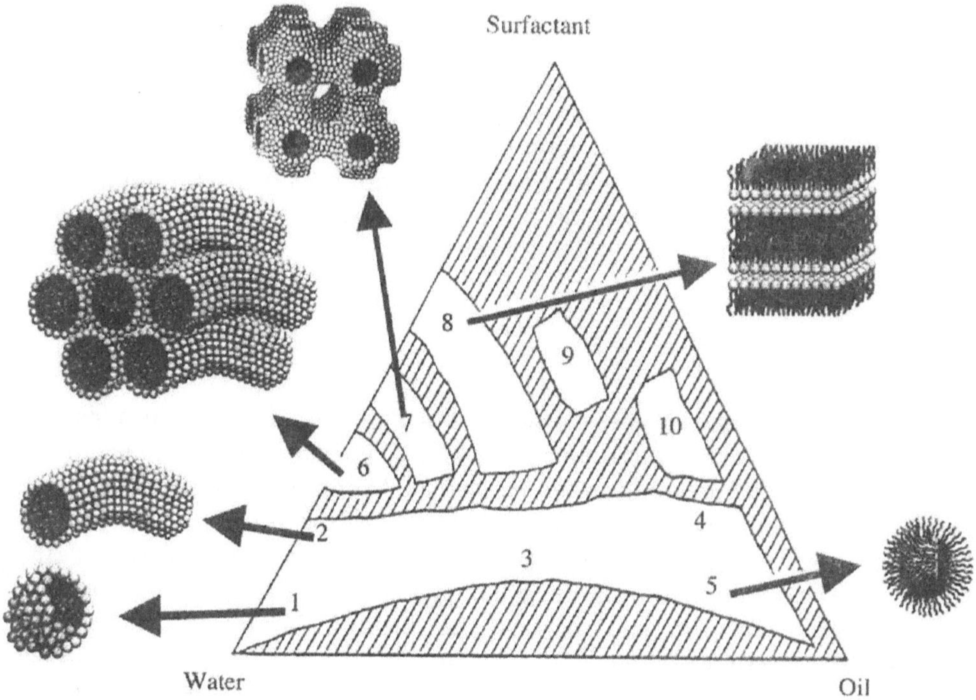

FIGURE 2.6. Schematic phase diagrams of surfactant-water-oil systems. 1. Spherical micelles. 2. Rod-like micelles. 3. Irregular bicontinuous phase. 4. Reverse cylindrical micelles. 5. Reverse micelles. 6. Hexagonal phase. 7. Cubic phase. 8. Lamellar phase. 9. Reverse cubic phase. 10. Reverse hexagonal phase (Derived from Refs. 18 and 19 with permissions from the publishers).

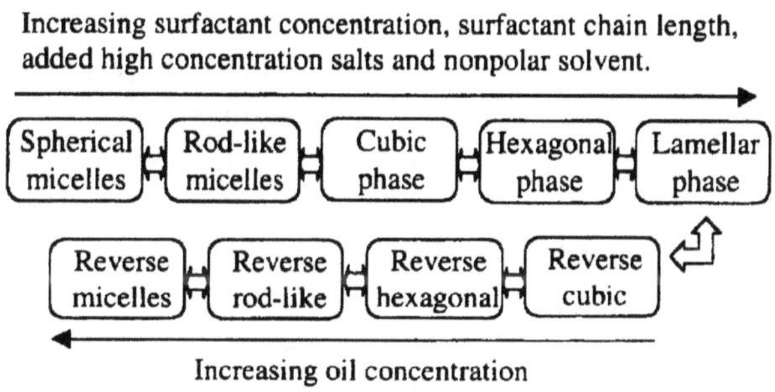

FIGURE 2.7. Phase structures as a function of external conditions.

the head group area increases the critical packing parameters. A larger critical packing parameter favors a less curved geometry. The transition from the more curved spherical micelles to flat lamellar phase observed in Fig. 2.7 agrees with the packing geometry illustrated in Fig. 2.5.

2.4.2. *Effect of Chain Length*

Increasing the surfactant chain length has a similar effect as increasing the surfactant concentrations. In this case, the head group area is fixed. Both the volume and the chain length increases, but their effects on the packing geometry do not cancel out. From Fig. 2.5, it can be seen that for a fixed number of surfactants within a fixed head group area in one spherical micelle, that if the chain length is increased beyond certain limit, the packing of the surfactants is no longer space filling and the spherical geometry will not be stable. This implies that when the chain length increases, the packing parameter is also increasing, and phase transition from spherical to less curved hexagonal and lamellar structures is favored.

2.4.3. *Effect of Cosolvents*

Polar solvents, like alcohols, tend to associate with the head groups and reduce the tendency for the surfactant molecules to associate. In fact, the addition of alcohols in a surfactant solution can make the CM completely disappear, which means the surfactants will not aggregate at all. On the other hand, nonpolar solvent molecules tend to associate with the hydrophobic chains of the surfactants. The addition of the nonpolar groups will, therefore, increase the volume of the surfactant, and increase the packing parameter. For normal aggregates, there will be a tendency for the transition from a more curved structure to a less curved structure. If the packing parameter is further increased, reverse micellar structures will form with an increased tendency for reverse curved structure.

2.4.4. *Effect of Salts and Ionic Species*

The effect of adding salts and ionic species to ionic surfactant systems is not difficult to understand. For ions that do not specifically bind to the charged head group, the increased ionic strength has a screening effect on the charged head group, as discussed in the previous sections. The screening effect reduces the repulsive energy between the head groups, therefore, reducing the head group area. This leads to an increase in the packing parameter, and a similar phase transition from a more curved structure to less curved structure. For ionic species that strongly bind to the head group, significant increase in the packing parameters can lead to the formation of larger vesicles, bilayer structures, and even reverse micelles. For nonionic surfactants, the addition of salts will have less effect. However, the inorganic species can interact with the micellar structures in a more complicated fashion and alter the phase diagrams. More discussions of these effects will be continued in later sections.

2.5. SELF-ASSEMBLED BLOCK COPOLYMER NANOSTRUCTURES

Block copolymers are amphiphilic molecules containing distinctively different polymer segments (blocks).[21] The block copolymers can contain two, three, or more blocks. Some common block copolymers include polystyrene ($H(C_6H_5–CH_2–CH)–$) and polyisoprene ($(–CH_2–H_3(CH{=}CH_2)C–)$) block copolymer [Fig. 2.8(a)], polystyrene and polybutadiene (–CH–CH=CH–CH)–) block copolymer, polyethylene (–(CH_2–CH_2)–) polypropylene (–(CH_2–CH_3CH)–) block copolymer, polyurethane (–(CH_2–CH_2–O_2C–NH–$C_6H_{10}CH_2$–C_6H_{10}–NH–OC)–) polyurethane block copolymers, etc. As shown in the figure, block copolymers can contain, two, three, or more blocks, and can have either linear or branched configurations [Fig. 2.8(b) and (c)]. The variety of the molecular architecture and compositions is endless. In general, they can be classified by the number of the blocks, and whether they are linear or branched. Depending on the number of blocks, they are called AB diblock copolymers, ABC triblock copolymers, ABC star block copolymers, etc.

Like in any other multicomponent polymer systems or multicomponent solutions, phase separation is commonly encountered in block copolymers because of the difference in the physical and chemical properties between the different blocks. However, unlike in ordinary polymers, the different blocks in

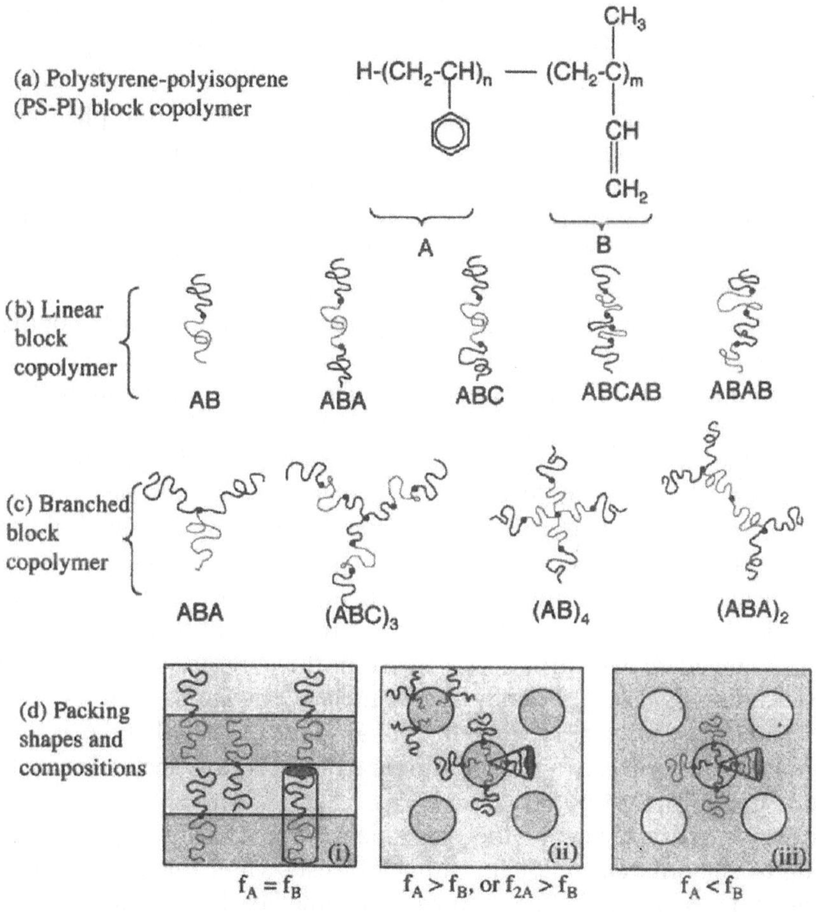

FIGURE 2.8. Variations of block copolymer configurations and the consequences on the packing geometry of the nanostructural ordering. (a) A common block copolymer, polystyrene-polyisoprene (PS-PI). (b) Linear block copolymers. (c) Branched block copolymers. (d) Packing shapes as a function of the compositions or molecular compositions (redrawn from Ref. 22).

a single polymer chain are covalently bonded. The covalent linkage between the different blocks makes macroscopic phase separation impossible. Instead, the phase separation occurs on the nanometer scale, as determined by the dimension of the blocks. If all the polymer chains have narrow size distribution, the phase separation will produce ordered nanostructures similar to surfactant systems.

Several factors control the phase separation and morphology of the phases: the choice of monomers, the composition and molecular size, and the molecular configuration. Whether or not phase separation will occur depends on a thermodynamic parameter called the Flory–Huggins segment–segment interaction parameter χ:[22]

$$\chi = [E_{AB} - 1/2(E_{AA} + E_{BB})]/k_B T \tag{2.27}$$

where E_{AB}, E_{AA}, and E_{BB} are the interaction energies between block A and B, between A and A, between B and B, respectively.

If χ is negative, the polymer will mix into a homogeneous system without phase separation. If χ is positive, phase separation is favored.

In the last two decades, sophisticated theories have been developed that allow the calculation and prediction of the microphase separation phenomena in block copolymer systems. These theories can, to a large extent, qualitatively and even quantitatively account for the rich domain shapes, dimensions, and symmetries observed in an experimental system. Discussions of these theories are not within the scope of this chapter. To be consistent with the discussions in the surfactant systems, simple packing geometries will be used to qualitatively illustrate the formation of the rich phases encountered in block copolymers, depending on the molecular architecture, the compositions, and the chain length.

The compositions, f_A and f_B, usually refer to the relative volume fractions of block A and block B for a linear diblock copolymer. As shown in Fig. 2.8(d), if f_A equals f_B, the two blocks have similar dimensions, and the packing geometry can be considered as a straight cylinder. Under these conditions, the lamellar phase will form. If f_A is greater than f_B, segment A takes up a larger volume than segment B. The packing geometry can be regarded as a cone structure, and spherical micelles will form along with the B phase dispersed in the A phase as spherical micelles. If f_B is greater than f_A, the situation will be reversed. Spherical micelles of segment A will form and disperse in the continuous phase formed by B. Similar arguments can be made for branched polymers, such as ABA star polymer. Because of the branching in the A segment, it will take up a larger volume than segment B. As a result, segment B will form spherical micelles in a continuous phase formed by A.

A very cursory review of the theoretical work and the experimental data reviews up to 280 different phase assemblages as shown in Fig. 2.9. Similar to the surfactant systems, many different forms of lamellar, hexagonal, cubic phases have been reported. Because of the advance in the polymer chemistry to synthesize block copolymers with very tight molecular weight distribution, and the flexibility to modify the compositions and molecular architecture, the rich phase behavior in block copolymers far exceeds what is observed in surfactants.

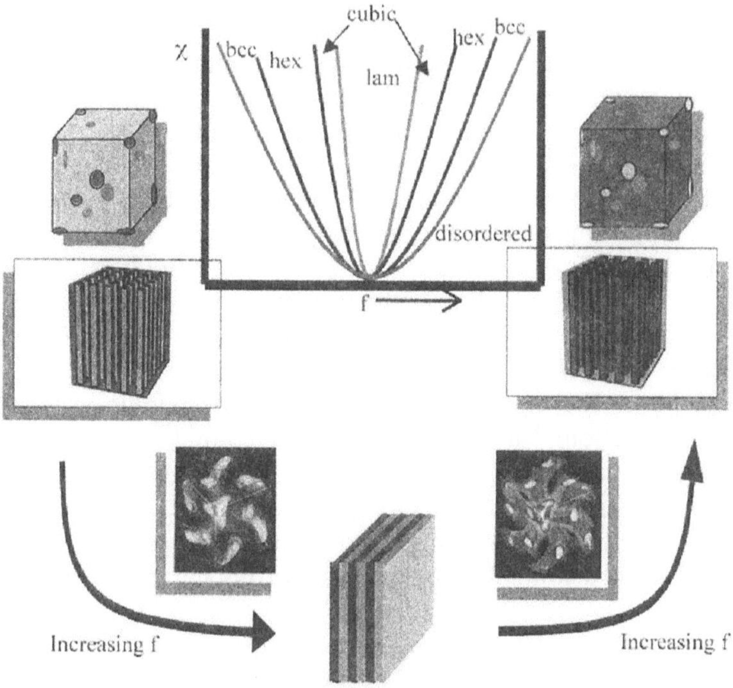

FIGURE 2.9. A schematic phase diagrams of block polymers as a function of the polymer composition and interaction parameters. This figure was provided by Dr. Gregory Exarhos of the Pacific Northwest National Laboratory, Richland, Washington.

2.6. CO-ASSEMBLY OF LIQUID CRYSTALLINE STRUCTURES AND INORGANIC MATERIALS

In the early 1990s, a new class of ordered nanostructured ceramics were reported based on self-assembly principles in surfactant solutions.[23–25] These materials are characterized by their well-ordered structure, tunable pore size from 2 to 50 nm, and simple preparation methods. The synthesis of nanoporous materials closely resembles the self-assembly process in surfactant systems. In principle, the self-assembled liquid crystals from the surfactant act as a template to support the growth of the ceramic materials. These ordered structures are cross-linked together through the condensation of the aluminosilicate ions. Subsequently, the surfactant molecules can be removed by thermal or chemical treatment without collapsing the ordered structure. Besides the immediate applications of such materials as new catalyst and catalyst supports, ion exchange media and separation media, and novel semiconductors and optical devices, synthesizing such three-dimensional (3D) ordered structures provides important lessons for making ceramic materials and composites.

The amazing ordering of the self-assembled periodic nanostructures can be illustrated by two 3D cubic structures reported in the literature, SBA-6 and SBA-16.[26,27] The structures of SBA-6 and SBA-16 have been resolved by high re-solution transmission electron microscopy (HRTEM) and image reconstruction.[28]

The high resolution images are shown in Fig. 2.10. SBA-6 is made of globular arrays packed in a cubic structure and has a pm3n symmetry. It contains a large cage (8.5 nm) and a small cage (7.3 nm). SBA-16 has an Im3m structure and is made of spherical cavities packed into body-centered cubic arrays.

Since 1992, the surfactant self-assembly approach has become a very active area because of its great potential. Numerous papers have been published on the preparation of nanoporous materials of novel chemical compositions and on the fundamental understanding of the reaction processes, and it is impossible to summarize all the research published in the last ten years or so. The publications in this area are numerous and a complete review does not serve the purpose of this book. Many review articles are already available. [29–33]

In terms of how the materials are prepared, direct ionic interaction,[23,24] mediated ionic interaction,[34] and neutral hydrogen bonding,[35,36] can be used. Almost all different kinds of surfactants illustrated in Fig. 2.3 have been used. Besides traditional surfactants, block copolymers have also been used as templates.[26,27,37] The block copolymer directed synthesis greatly expanded the pore size and compositional range of mesoporous materials. In addition to

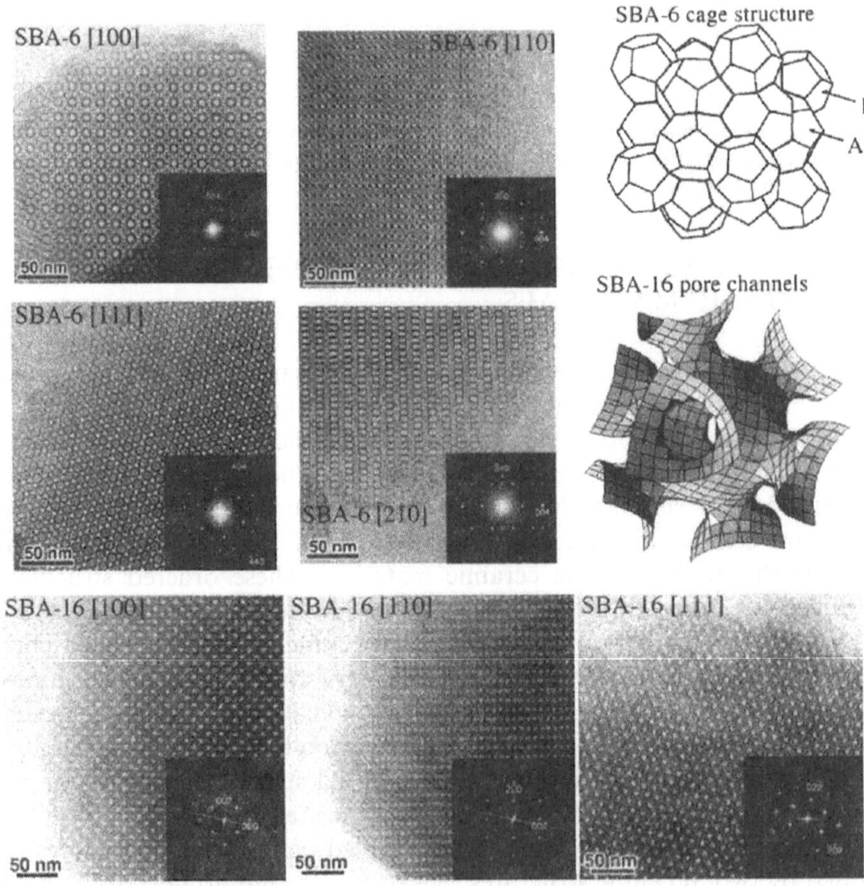

FIGURE 2.10. High resolution images of two cubic structures, SBA-6 and SBA-16, and the reconstruction of the pore structures (reproduced from Ref.28 with permission from Nature).

the aqueous phase synthesis, which is widely used, non-aqueous phase synthesis has been investigated.[38] In terms of the range of compositions, besides the widely investigated silica or non-silica based oxides, semiconducting and conducting materials (manganese oxide,[39] tungsten oxide,[40] germanium sulfides,[34] selenide,[41] and other semiconducting superlattices[42]), periodic carbons,[43] and nanostructured metals,[44,45] have attracted wide attention recently.

The self-assembled periodic structures can be further used as a structural framework to develop new materials. Several approaches have been investigated: (1) by incorporating functional molecules and active sites into the porous channels, (2) by using the periodic nanoporous structures as a template, and (3) by physically confining a new material in the organized nanostructures. The first approach has been extensively investigated for catalytic and other chemical applications. Nanoporous silica has been doped with elements possessing catalytic properties.[46–48] Optimum functionality can be achieved by placing a close paçked monolayer within the pore channels.[49] The chemistry of the molecular monolayer can be tailored for a wide range of applications. In the second approach, the periodic silica pore structure is used as a template and filled with carbon or metals.[39,41] The removal of the silica templates leaves a ordered nanostructured carbon or metal material, a very difficult material to make with other approaches. In addition, framework nanoporous structures containing bridge-bonded ethene (and other) organic groups were reported.[50–52] These types of materials raise the possibility of generating integrated organic–inorganic hybrid materials. Ordered ruthenium clusters were also introduced to the porous channels.[53] Finally, the unique nanostructures can be used as physical confinement for other materials with novel properties. Wu and Bein[54] first introduced conducting polymers into ordered nanoporous silica. Later, the energy transfer of conjugated polymers within and outside the pore channels were investigated by polarized femtosecond spectroscopy.[55] The porous channels also find novel applications as a nano-extruder to produce liner polyethylene nanofibers of high molecular weight.[56] By incorporating conjugated groups (diacetylene groups) into the surfactant, Lu *et al.*[57] prepared nanostructured chromic films that can change color in response to thermal, mechanical, and chemical stimuli.

Microstructural and morphological controls have also attracted wide attention. Great progress has been made in fabricating oriented nanophase materials on various substrates[58,31] and in making free-standing films, [59] spheres,[60,61] fibers,[62,63] and single crystalline mesoporous materials in which all the pore channels are aligned.[64] Nanostructured silica films using a rapid evaporation approach (dip coating[65,66] or spin coating[67]) have been considered good candidates for ultralow dielectric interlayer structures in next generation of microelectronic devices. A magnetic field has been used to align the pore channels before the silicate was polymerized.[68] In addition, a wide range of intriguing morphologies and patterns has been reported, including spiral and toroidal shapes,[69] disks,[70] and silica vesicles.[71] Hierarchically ordered nanoscale patterns were generated by combining multiscale templating and micromolding, or lithography.[72,73] The microscopic pattern can also be defined by optical beams through photosensitive sol–gel chemistry.[74] These patterns will have potential as novel optical waveguides.[75] Biogenic or biomimetic directing agents (copolypeptides) can also be used to control the morphologies and shapes.[76]

2.6.1. Interactions between Cationic Surfactants and Anionic Silicates

These new mesoporous materials were prepared by mixing aluminosilicate precursors such as sodium aluminate, tetramethyl ammonium silicate, and silica in a surfactant solution under mild hydrothermal conditions. Then, surfactant molecules such as cetyltrimethylammonium chloride (CTAC) and aluminosilicate organize into ordered lyotropic liquid crystalline structures.

A liquid crystalline templating mechanism was discussed, as shown in Fig. 2.11.[23,24] This figure suggests two pathways: (1) Ordered surfactant structure formed first followed by the intercalation of the soluble silicate species into the hydrophilic region between the head groups (route B). (2) The addition of the soluble silicate species caused the spherical micelles to transform into high order aggregates (route A). In either case, a highly ordered inorganic–organic nanocomposite was formed. After the silicate was condensed into 3D network ceramics through further hydrothermal treatment, the inorganic phase (the surfactant template) was burned out by calcinations at more than 500°C. The removal of the inorganic phase produced a well-ordered honey cone type porous ceramic material.

In the liquid crystalline templating mechanism, the surfactant liquid crystals are formed first, and the ceramic materials are introduced to the pre-existing liquid crystals to replicate the ordered structure. This mechanism was first illustrated in lamellar mesophase by introducing silicic acid monomers into a lamellar structure, followed by *in situ* polymerization.[77,78] The polymerization process of the silicic acid caused some distortion of the lamellar structure. Attard *et al.* later prepared mesoporous silica and other metallic nanostructures similar to MCM41 via the same mechanism by first forming the liquid crystalline template with a nonionic surfactant, followed by adding the silicate species.[43,79–81]

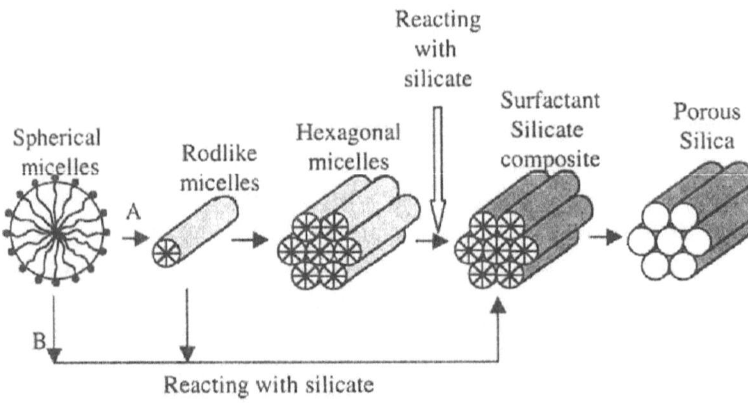

FIGURE 2.11. Liquid crystaline templating mechanisms initially suggested for the formation of ordered nanoscale materials (redrawn after Ref. 23).

The condensed silicate essentially replicates the pre-assembled liquid crystalline structure. Braun *et al.* synthesized non-oxide semiconducting superlattices using preordered mesophase.[41,82] The mesophase template was formed by nonionic amphiphiles, followed by precipitation of cadmium sulfide and cadmium selenite. An ordered inorganic–organic superlattice compound was formed in which the crystalline structures of all inorganic nanoparticles were correlated.

Although it has been established in the last few years that the ordered nanophase materials can be formed through many different pathways,[83] the principle of self-assembly was well illustrated by a cooperative mechanism, as shown in Fig. 2.12.[84] The co-assembly of surfactant and silicate into liquid crystalline structures can be carried out under conditions in which the silicate alone would not condense, and the surfactant alone would not form a liquid crystalline phase. Nanoporous materials were formed with cetyltrimethylammonium bromide (CTABr) surfactant in solution as dilute as 1 wt.%, which, according to the phase diagram, only spherical micelles would be expected by surfactants alone. The phase diagrams derived with and without the silicate species are drastically different. Therefore, the inorganic ions played an important role in the self-assembly process. Based on the observation of continuous-phase transition from lamellar to hexagonal phase, it is suggested that the driving forces for self-assembly are multidentate binding of silicate ions, polymerization of silicate at silicate-surfactant interfaces, and charge–density matching across the interfaces. It was observed that when a cationic surfactant was mixed with hydrolyzed silicate under basic conditions, lamellar precipitates formed immediately. The lamellar phase later transformed into the hexagonal phase. The formation of the transient phase can be explained by the change of the packing

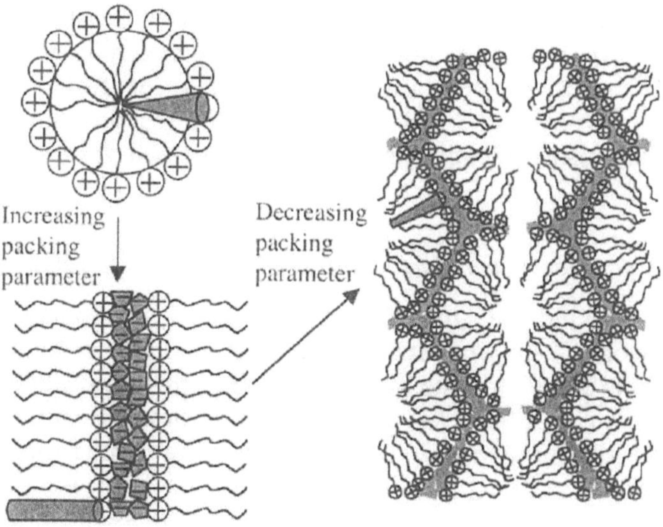

FIGURE 2.12. The cooperative mechanism is consistent with the change of the packing parameters during the reaction between silicate and surfactants. The phases formed as a function of time are: spherical micelles, lamellar composite, and hexagonal composite. The representative packing shapes are highlighted by the shaded cone structures (redrawn after Ref. 84).

parameter during the reaction between the surfactant and the silicate. Under the experimental conditions with dilute surfactant, the surfactant molecules only exist as dispersed spherical micelles. No high order aggregates were expected. After the addition of the hydrolyzed silicate, negatively charged silicates, whether as monomers ($Si(OH)_3O^-$) or more cross-linked oligomers ($SiO_x(OH)_{4-x}{}^{x-}$), bond strongly to the positively charged surfactant head groups. This strong binding immediately led to the formation of precipitates of solid materials. At the same time, the neutralization of the head group charge significantly reduced the repulsive energy between the head groups, and the head group area. Based on the definition of the packing parameter, the reduction of the head group area led to an increase of the packing parameter. As a result, the lamellar phase was formed within the solid precipitates. This observation is consistent with the formation of lamellar phases when charged surfactants are mixed with oppositely charged ions.

As the reaction proceeded, the silicate anions continued to condense with one another. For example, the silicate monomers could nominally follow the following condensation reaction:

$$2Si(OH)_3O^- = 2SiO_2 + 2H_2O + 2OH^-$$

The net result of the condensation reaction was the elimination of the negative charge on the inorganic species that was bonded to the positively charged surfactant head groups. The reduction of the number of negative charge in the head group region implied an increase in the repulsive interaction between the head groups, and thus an increase of the head group area. As a result, the packing parameter was decreased, and a transition form the lamellar phase to the hexagonal phase occurred. It should also be noted that in the initial phase of the reaction, the silicate was far from being completely condensed. During this period, the inorganic phase still had enough flexibility to allow the phase transition to occur. In fact, even after the silicate is mostly condensed, some flexibility for the silica to rearrange still remains. Several groups have demonstrated that post-hydrothermal treatment of the reaction nanostructured silica/surfactant composites in alkylamine solutions caused the silica frame to expand considerably.[85,86]

2.6.2. Interactions between Other Surfactants, Polymers, and Ceramics

Many different kinds of surfactants,[87] cationic,[23,24] anionic,[85] neutral,[34] nonionic,[35] zwitterionic,[88] polymerizable surfactants, and single chain or double chain,[89] have been explored for the synthesis of ordered nanoscale ceramics or composites, as shown in Fig. 2.13. The different approaches all rely on the interaction between the surfactants and the solubilized inorganic species, and on the modification of the packing geometry resulting from such interactions and from the effect of other additives. Apparently, self-assembly of ordered nanostructures can occur under a variety of experimental conditions, although the detailed reactions differ greatly from system to system. In Fig. 2.13, the double chained Gemini surfactant is used as an example on how the packing geometry can be modified. In this case, a wide range of experimental parameters can be

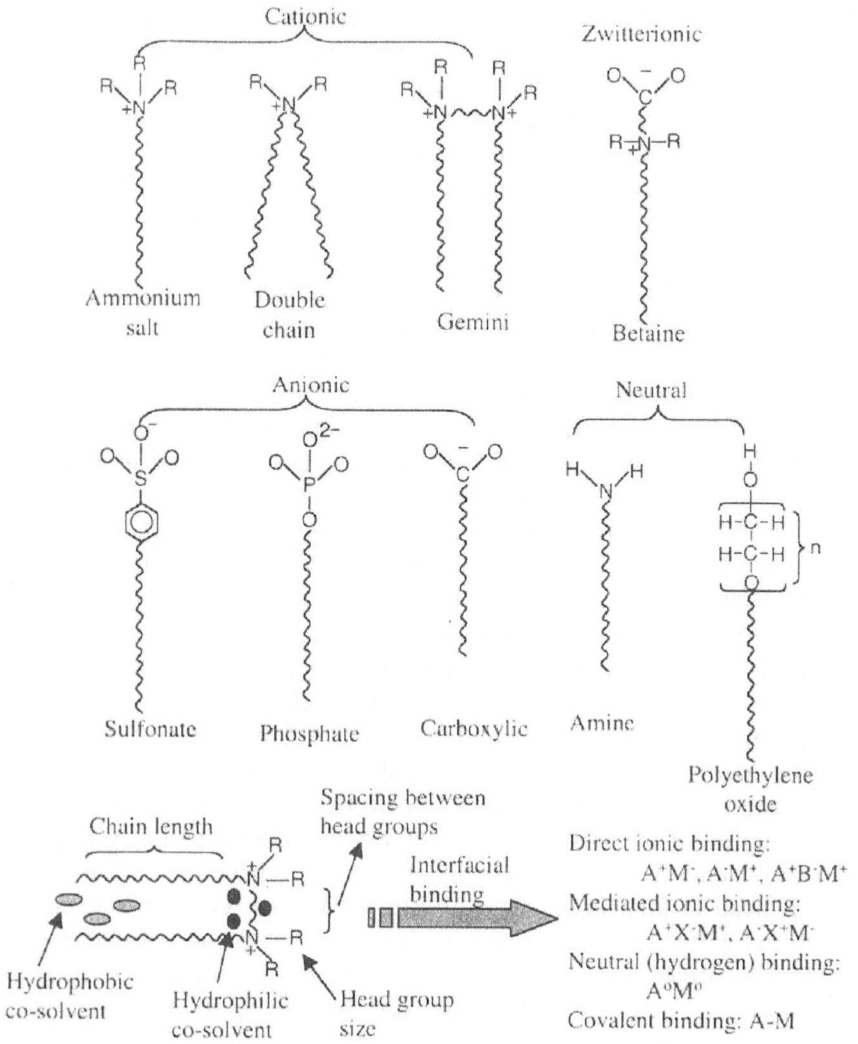

FIGURE 2.13. Different surfactants that have been used for the self-assembly of ordered nanostructured ceramics, and the factors that can influence the packing geometry of the micelles. A and B represents the head group of the surfactant. M is the metal oxide precursor, and X is the mediating counter ions.

explored. First, the chain length can be systematically varied, the effect of which was already discussed in previous sections. Second, the spacing between the two head groups, as well as the head group size, as determined by the R groups, can be increased. The increase of these parameters increases the head group area, therefore, decreasing the packing parameter, which favors the formation of the more curvatured structures. This is why a 3D hexagonal structure with close packed spherical micelles ($P6_3/mmc$), instead of the regular 2D hexagonal structure with close packed rod-like micelles (P6m), was formed when the Gemini surfactant was used. As indicated in Figure 2.13, co-solvents can also have a significant effect on the packing geometry. While a nonpolar solvent swells the hydrophobic region of the tail groups, a polar solvent can penetrate the regions close to the head groups. As discussed earlier, the swelling of head groups is usually not favorable for formatting high ordered liquid crystalline structures. Huo *et al.*[87] also demonstrated that the addition of *t*-amyl alcohol to the Gemini surfactant system caused an increase of the surfactant volume. This led to a transition from the spherical hexagonal phase to cylindrical hexagonal phase due to an increase of the packing parameter.

The interaction between the head groups and the solubilzed inorganics has the largest effect on forming the ordered liquid crystalline nanostructure. Normally, the interfacial binding causes a decrease of the head group area, and, therefore, an increase of the packing parameter. This is why ordered nanostructures are usually formed under conditions when the liquid crystalline structures are not expected. The MCM41 series of silicate-based materials were prepared using direct ionic bonding with a cationic surfactant.[23,24] The phase transition caused by the direct ionic binding of opposite charges were already discussed. The direct binding encountered with anionic surfactants, and zwitterionic surfactant has a similar effect. A more generalized synthesis route was developed to include ion-mediated ionic bonding.[33,37] Here, it is assumed that the surfactant head groups and the silicates had similar positive charges, and that the head groups would first bond to the negatively charged counter ions such as chlorine and bromine, and then bond to the positively charged silicate. The binding of the counter ions, or even the screening effect of the ionic species, would change the packing parameters enough (by decreasing the head group area) to cause the formation of the higher order hexagonal and cubic phase. When neutral (dodecyl amine) and nonionic (polyethylene oxide) surfactants were used, the change in the head groups' area was caused by hydrogen bonding. The change in the head group regions was also expected when the head group covalently bonded to the inorganic species, followed by further condensation of the inorganics.

The similarity between surfactants and block copolymers were already discussed. Therefore, it is natural that block copolymers have also been used to prepare self-assembled nanoscale composites.[90-92] The advantage with block copolymers is in the flexibility of the polymer chemistry, and the wide range of length scale available when compared to normal surfactants. Zhao *et al.* used block copolymer surfactants to prepare ordered nanoporous ceramics to expand the pore size from 50 to 300 Å.[26,27] Other new structures, such as the "Plumber's Nightmare" has been recently reported with block copolymer-silica nanocomposites.[93]

2.7. INTELLIGENT NANOSCALE MATERIALS

In the first part of this Chapter, the principles and the methods available to build nanoscale architectures through self-assembly approaches have been discussed. One of the main deficiencies in these materials is the lack of functionality. For many potential applications, two important concepts need to be introduced: molecular recognition and responsive properties. As will be discussed in Chapter 3, these two concepts form the foundations of the functions with any biological material. In the next few sections, we will discuss several synthetic approaches to introduce intelligent functionalities into self-assembled nanoscale materials.

2.7.1. Molecular Recognition

2.7.1.1. Functionalized Nanoscale Materials

2.7.1.1.1. Molecular Recognition in Functionalized Nanoparticles. Nanoparticles made of metals (gold, silver, etc.), semiconductors (ZnO, CdS, Si, etc.), and other oxides (SiO_2, TiO_2, etc.) form an ideal platform for functionalized nanoscale materials. Nanoparticles of uniform size can be prepared from a variety of aqueous and non-aqueous media. The particle size can be varied from 2 to > 10 nm. In particular, a wide range of functional molecules and functional groups can be attached to the particle surfaces. The small particle size also implies high sensitivity and selectivity.

Usually, the first step in functionalizing nanoparticles involves building a self-assembled monolayer (SAM) on the particle surface. In this approach, bifunctional molecules containing a hydrophilic head group and a hydrophobic tail group adsorb onto a substrate or an interface as closely packed monolayers, mostly through the attraction between the hydrophobic chains. The tail group and the head group can be chemically modified to contain functional groups. In the literature, the assembly of molecular structures and monolayers on flat substrates has been studied for a long time. SAMs are widely explored for engineering the surface and interfacial properties of materials such as wetting, adhesion, and friction.[94,95] These monolayers are also used to mediate the molecular recognition processes and to direct oriented crystal growth.[96] Chemical and physical modification of solid surfaces is extensively used in chromatography, chemical analysis, catalysis, electrochemistry, and electronic industry. Molecular monolayers are also widely investigated as an effective means to fabricate novel sensing and electronic devices. The molecular arrangement and the chain conformation on flat substrates have been extensively studied by AFM, contact angle measurement, small angle scattering, and many other techniques.[1,2]

Depending on the substrate surface, two kinds of SAMs are commonly encountered. On metal surfaces such a gold and silver thiolated alkyl chains are often used. The SAM molecules are attached to the substrate through the sulfide group, but there is no chemical reaction between the substrate and the SAM molecules. The monolayer structure is highly ordered, and the crystalline ordering in the monolayer is dependent on the atomic structure of the substrate. On oxide surfaces such as silica, alkyl silanes are usually used. The silane end groups hydrolyze first, and then become covalently bound to the substrate through a condensation reaction. The SAMs molecules may also be covalently linked with one another through siloxane groups as a result of the hydrolysis of the silane groups and their subsequent condensation. Because of the disordered nature of the substrate surfaces, SAMs on the oxide surfaces are usually disordered.

One novel application of gold–SAMs complex is in the detection of DNA sequencing (Fig. 2.14).[97] Instead of regular alkanethiols, DNA molecules capped with thiol groups were used. Two different, non-complementary DNA oligonucleotides capped with thiol groups were attached to two batches of 13 nm gold particles. The two non-complementary DNA gold batches were mixed. This mixture is stable and shows a typical wine red color of colloidal gold. When another kind of DNA molecules that have a complementary sequence to the grafted oligonucleotides were introduced into the solution, hybridization occurs, which causes the aggregation of the gold particles. The aggregation caused the wine red solution to turn bluish. If the aggregates were heated, denaturing of the DNA causes the bonds formed from hybridization between the particles to break down, and the gold particles became dispersed again. The solution restored the typical colloidal gold color. In fact, the color change during heating can be used as an indication of the closeness of the DNA sequence matching. Therefore, this technique is very sensitive for DNA sequence detection.

The SAM molecules serve two purposes: (1) protecting the particles and (2) introducing functionality to the surface. Without the protection of SAMs, most nanoparticle systems would aggregate and precipitate very quickly. In order to introduce functionality, mostly different functional groups such as sulfate, phosphate, and carboxylic acid, can be introduced to the head groups of the SAM molecules via proper chemical treatment. However, in many cases homogeneous SAMs are not sufficient for molecular recognition of proteins

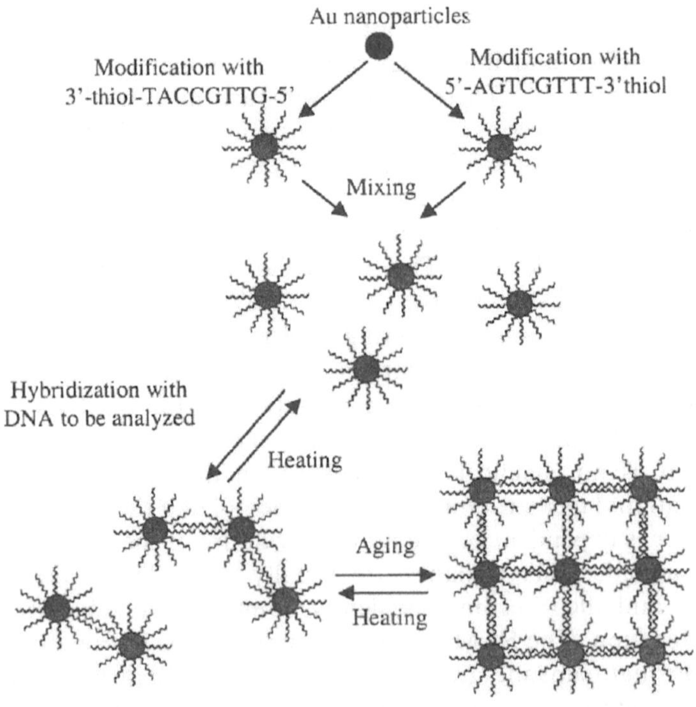

FIGURE 2.14. DNA capped gold nanoparticles for DNA sequence detection (redrawn after Ref. 97).

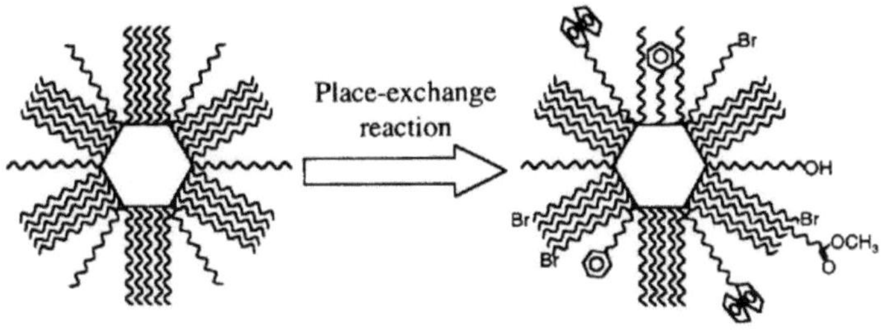

FIGURE 2.15. Place-exchange reactions to build multiligand binding sites on nanoparticles (redrawn after Ref. 98).

and other complex molecules. Multiligand binding is usually required. To solve the multiligand problem, Ingram *et al.*[98] developed a place-exchange reaction, in which the SAMs molecules on gold are replaced by other functional groups in a solvent. In this approach, as illustrated in Fig. 2.15, gold nanoparticles protected by alkanethiol SAMs were produced in toluene mixed with alkanethiol molecules. ω-Functionalized alkanethiol molecules and the thiol protected gold particles were codissolved in methylene chloride. After stirring the mixture for some time, the original simple thiol molecules were partially replaced by functionalized thiols. Using this approach, more than one type of functional molecules can be introduced either by simultaneous place-exchange, or by a multi-step exchange reaction. In the simultaneous exchange reaction, up to five different ligands were simultaneously reacted with the thiol protected gold. In the multistep exchange reaction, ligands with progressively long-chains were incorporated step by step.

Boal and Rotello[99] used the place-exchange reaction to construct multi-ligands receptors for molecular recognition. Three different types of receptors were compared as shown in Fig. 2.16: (1) monoligand bonding involving hydrogen bonding between flavin (target) and diaminopyridine (receptor); (2) bi-ligand bonding of flavin with diaminopyridine and pyrene; (3) target induced reorganization and bi-ligand bonding. In the bi-ligand bonding, the target (flavin) molecules interact with diaminopyridine groups through hydrogen bonding, and with the pyrene groups through aromatic stacking. As a result, there is a twofold increase in the binding constant as compared with the mono-ligand bonding. In the target induced reorganization and multi-ligand bonding, dilute hydrogen bonding ligands and aromatic recognition elements were placed in the SAMs. The interaction with the target causes the binding ligands to rearrange and relocate to the nearest neighbor of the targets. A 71% increase in the binding constant was observed due to the rearrangement of the ligands.

2.7.1.1.2. Functionalized Nanoscale Composites. By combining the ordered nanoporous materials formed through surfactant self-assembly and SAMs, a new class of hybrid nanoscale materials has been developed (Fig. 2.17).[49,100] This approach allows us to systematically modify the surface chemistry and tailor the molecular recognition process of nanoporous materials toward the targets. In this approach, high quality oriented molecular monolayers are spontaneously grown

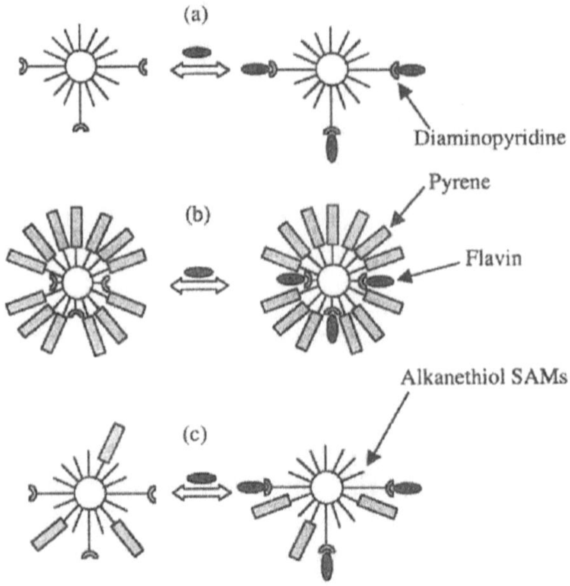

FIGURE 2.16. Multiligand recognition receptors on nanoparticles. (a) Monoligand receptors. (b) Bi-ligand receptors. (c) Target induced bi-ligand receptors (redrawn after Ref. 99).

on ordered nanoporous ceramic substrates with controlled pore shape and pore size. The functional molecules are closely packed and cross-linked with one another. The terminal functional groups on the monolayer can be easily modified, thereby allowing rational design and layer-by-layer construction of host sites on the nanoporous substrates that match the shape, size, or chemical properties of heavy metals, transition metals, or organic molecules. This makes these materials extremely efficient scavengers of these species, or effective catalysts for reactions involving these species.

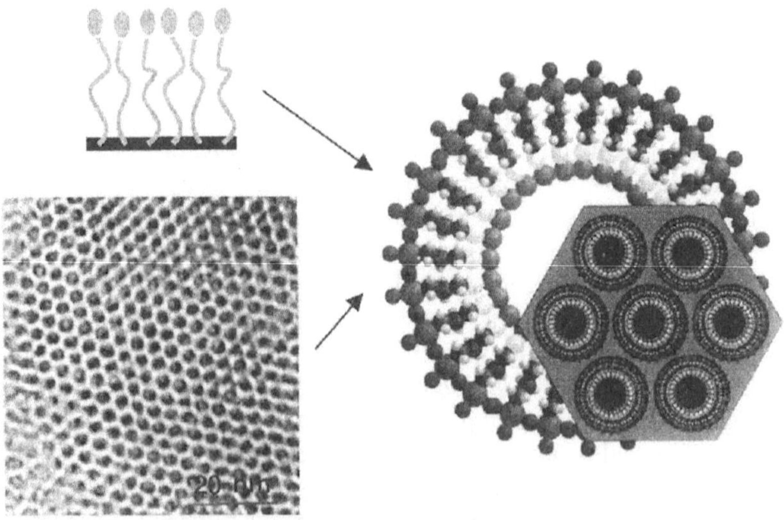

FIGURE 2.17. Formation of self-assembled monolayers in ordered nanoporous silica.

The quality of the molecular monolayer is reflected on the interfacial binding between the monolayer and the silica substrate. High quality, closely packed monolayers can be formed on the nanoporous supports through the introduction of several layers of physically adsorbed waters on the nanoporous surface. The role of the water molecules is to physically confine all the hydrolysis and condensation reactions of the organosilanes at the interface. Currently, the population densities of functional groups on the nanoporous materials can be systematically varied from 10% to 100% of the full surface coverage.

The interfacial chemistry can be easily tuned for different applications. For example, Fig. 2.18 shows two different types of functional molecules on the silica substrate. Figures 2.18(a) and (b) involve a direct chemical bonding between heavy metals and alkyl thiols [*tris*(methoxy)mercaptopropylsilane, TMMPS] as the functional molecules for heavy metal adsorption. This material has exceptional selectivity and capability of hybrid materials to bind mercury and other heavy metals. A loading capacity of 600 mg (Hg)/g (absorbing materials) has been obtained. By using a different monolayer chemistry, highly selective binding for arsenate removal was demonstrated.[101] The nanoporous silica was functionalized with an ethylenediamine (EDA) terminated silane [(2 aminoethyl)-3-aminopropyl trimethyl silane]. Cu(II) ions were bound to the EDA monolayer with a 3:1 EDA to Cu ratio, forming an approximate octahedral $Cu(EDA)_3$

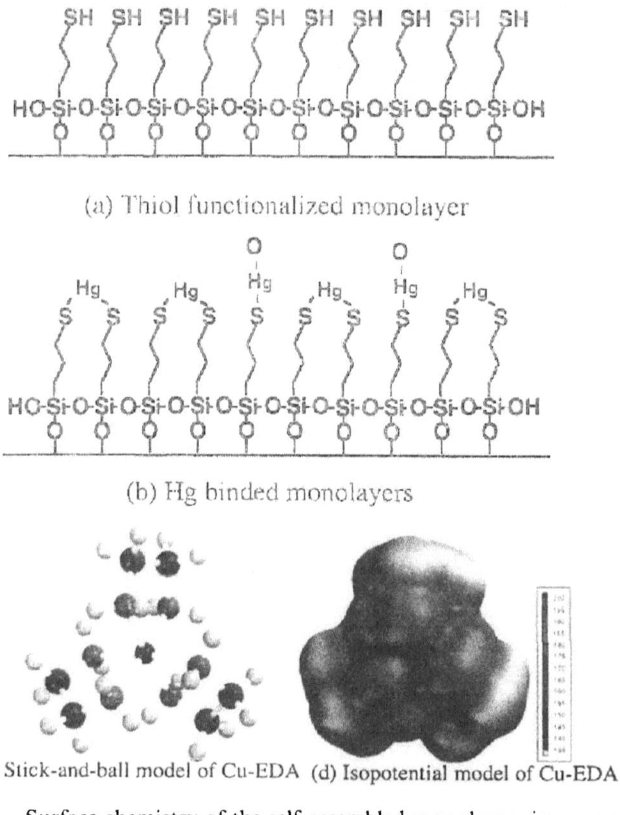

FIGURE 2.18. Surface chemistry of the self-assembled monolayers in nanoporous silica. (c) and (d) are reproduced from Ref. 101 with permission from ACS.

complex structure. The cationic octahedral complex contains an electrophilic basket with C3 symmetry that forms an ideal host for a tetrahedral anion. A binding capacity of more than 120 mg (anion)/g of adsorption materials was observed. This approach is especially promising considering the rich chemistry that can be explored with monolayers, nanoporous silica, and the possibility of designing better anion recognition ligands.

2.7.1.2. Building Molecular Architectures by Molecular Imprinting

2.7.1.2.1. Principles and Applications of Molecular Imprinting. In order to mimic the molecular mechanisms found in nature, a molecular imprinting technique has been investigated for decades.[102] The principle of molecular imprinting is quite simple, as shown in Fig. 2.19. First, functional groups, ligands that contain polymerizable sites are bound to suitable target (or template) molecules. Second, a highly cross-linked polymer matrix is formed around the ligand bound target. At this stage, the position and orientation of the ligands and the location and the geometry of the interfaces between the target and the polymer are fixed. Finally, the target molecules are removed by washing with a good solvent. This process leads to a microporous polymer matrix that contains a predetermined cavity shape, a predetermined position, and orientation of the binding ligands. The binding between the target and the ligands can be either covalent or non-covalent. In selecting the polymer matrix, one needs to consider both the mechanical stability and the accessibility of the microporosity. Normally, macroporous polymers are used with a high content of a cross-linking agent in the presence of an inert solvent. During the polymerization, phase separation occurs. The removal of the solvent leaves permanent porosity in the matrix. In addition to polymer matrix, inorganic matrix can also be used. For example, molecular imprinting was carried out in silica gels, and on solid surfaces.

Molecular imprinting has been investigated on many length scales (Fig. 2.20). On the smallest length scale, a single small molecule was used as the target.[103] This almost resembles the antibody, antigen binding in biology. The antibody mimicking polymer was used to accurately measure drug levels in human serum with results comparable to those using more conventional immunoassay techniques. Molecular imprinting can also be used with large protein molecules. Shi *et al.*[104] patterned surfaces with proteins that show good recognition of a variety of proteins (albumin, immunoglobulin G, lysozyme, ribonuclease, and

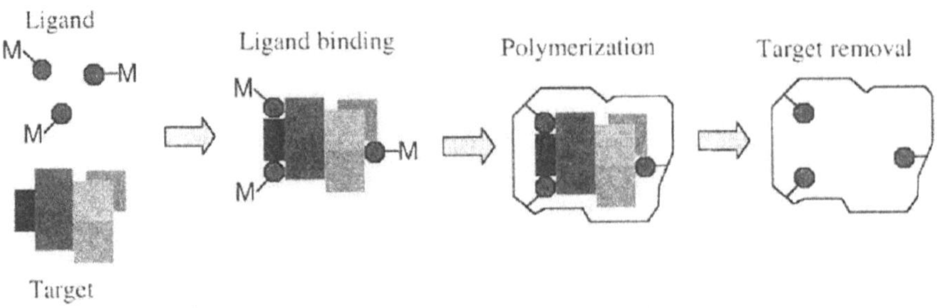

FIGURE 2.19. Principle of molecular imprinting.

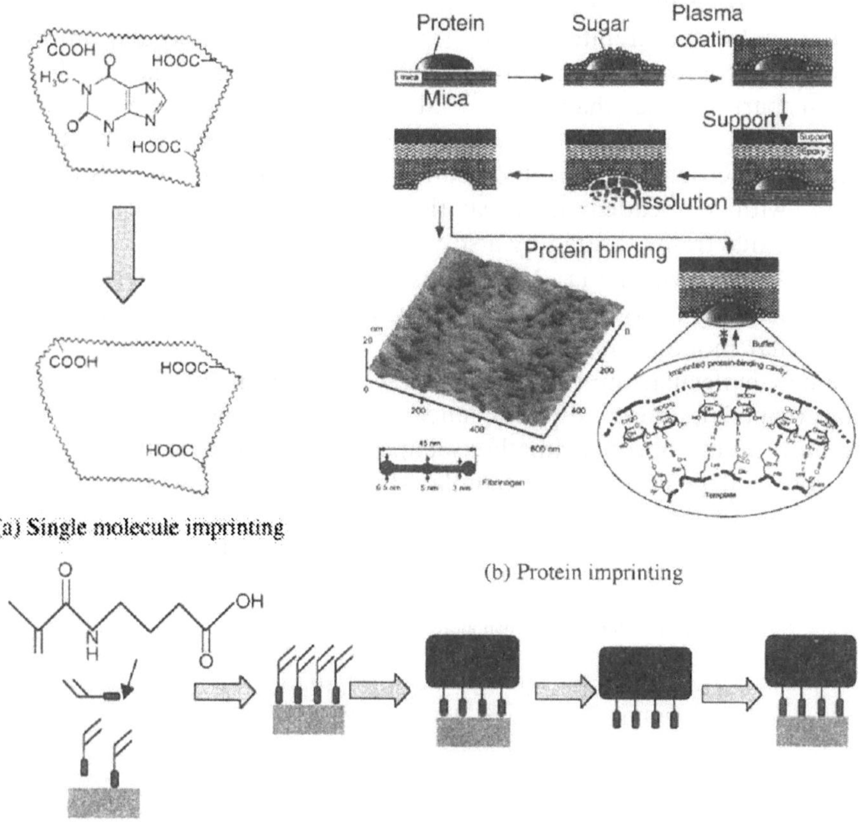

FIGURE 2.20. Molecular imprinting on different length scales. (a) Single molecule molecular imprinting. (b) Protein imprinting. (c) Mineral imprinting. (a) is reproduced from Ref. 103 with permission from Nature. (b) is reproduced from Ref. 104 with permission from Nature. (c) is redrawn after Ref. 105.

streptavidin). The imprinting was done by first depositing protein molecules on to a mica surface in buffer solutions. The mica surfaces are atomically flat and are hydrophilic to prevent denaturing of proteins. Disaccharide was coated onto the protein bound mica surfaces to provide multiple hydrogen binding sites for proteins and also to protect the protein from damage during further polymerization. A polymer coating was applied by plasma deposition to yield a smooth and robust overlayer that is covalently bound to the sugar layer. The mica substrate and the proteins were removed by elution with basic solutions.

Molecular imprinting was also investigated for controlled nucleation and growth of inorganic minerals.[105] Here, polymerizable ligands were bound to the mineral ($CaCO_3$) surfaces. After the polymer was formed on the mineral surfaces, the mineral was dissolved. A comparison of the imprinted and non-imprinted regions revealed that the imprinted region significantly enhanced the nucleation rate, indicating the ligands were arranged with a favorable position and orientation to stimulate the nucleation of the mineral.

2.7.1.2.2. Molecular Imprinting in Self-Assembled Nanostructures. To control the molecular architecture in three-dimensions, molecular imprinting techniques were combined with self-assembled nanoscale materials as discussed in the first part of the chapter.[106,107] This approach also created a novel hierarchical porous material with tunable size and shape selective pore structures to mimic the microchannels in biomembranes. Soft molecular coatings were introduced with embedded microcavities. This material contains a rigid mesoporous oxide frame, coated by a soft, "microporous" molecular monolayer, as schematically shown in Fig. 2.13. Unlike in zeolite-based materials, the size and shape selectivity is not determined by the oxide framework, but by the long-chain molecular monolayers.

The new molecular imprinting process was shown in Fig. 2.21. Nominally, triangular (tripods), linear (dipods), and point cavities were created by first bonding short 3-aminopropyltrimethoxysilanes (APS) to the template molecules, followed by the deposition of the template-bound silanes onto the nanoporous substrate. The tripod chemistry on silica substrates was first reported by Tahmassebi *et al.*[108,109] Subsequently, long-chain molecules were deposited on

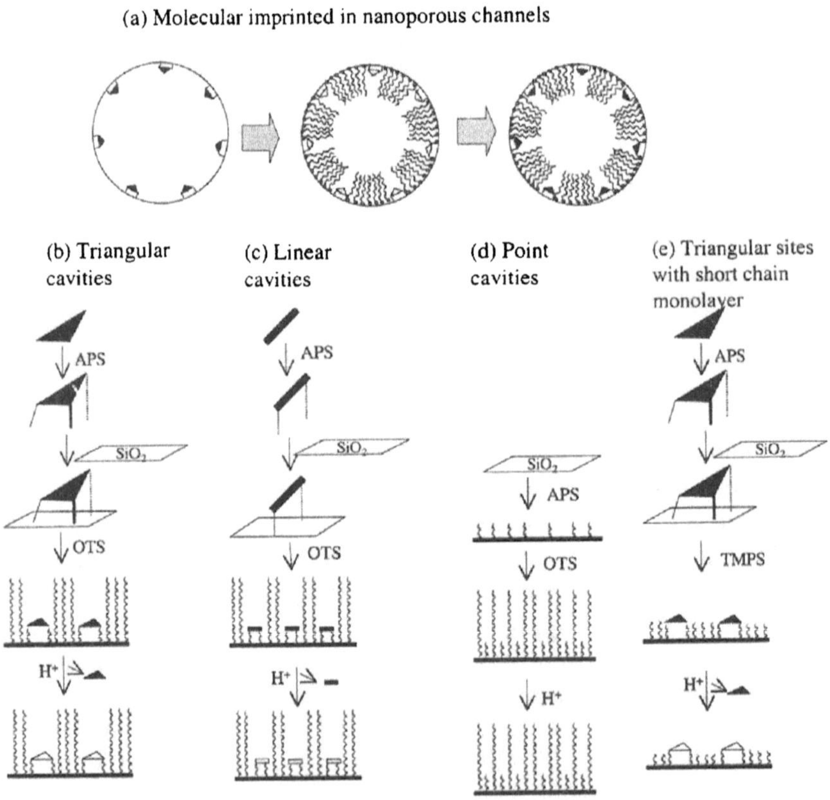

FIGURE 2.21. Hierarchical nanoporous materials with organized microcavities through molecular imprinting and ordered nanoporosity through surfactant templating. (a) Molecular imprinting in nanoporosity. (b) to (e) Different microcavity shapes generated by different imprinting molecules. (b) to (d) were reproduced from Ref. 106 with permission from Wiley-VCH.

the unoccupied surface of the silica substrates. The template molecules were selectively removed from the substrates with a mild acid wash to leave the desired cavities in the monolayer coating.

The molecular chain conformations have been studied using N_2 adsorption and NMR techniques during each step of the imprinting process. The layer thickness derived from the pore-size change is consistent with the expected chain length and chain conformation. The surface area in each step was reduced, depending on the amount of organic silanes added and on the final pore size. Ideally, the imprinting procedure should produce triangular cavities (10 Å wide, based on the dimension of the template) and linear cavities (10 Å long). For point cavities, APS molecules were first deposited on the porous substrate, followed by the deposition of octadecyltrimethoxysilane (OTS). The dimension of the point cavities was determined from the area an APS molecule would occupy (about 4.5 Å for one point cavity) on the surface.

The microcavities on the monolayer coatings showed considerable selectivity between the tripod and dipod molecules. Considering the fact that the difference between the tripod and dipod molecules is not great, the capability of the substrate to differentiate between these two molecules is remarkable. A tripod molecule has three benzaldehyde arms, and a dipod has two. The rest of the molecular structures are similar. In principle, the dipod molecules can easily fit into the triangular cavities and conform to the majority of the cavities nicely. The two aldehyde groups can also bind with the two amine groups at the corner of the cavity (hydrogen bonding or forming Schiff base). We would expect the dipod molecules to adsorb on triangular cavities as well as the linear cavities. The large difference in the adsorption behavior of dipod molecules on triangular and linear cavities demonstrated that the shapes of the cavities are very important.

The selectivity of the microcavities is directly related to the 3D nature of the long-chain molecule coatings. One-to-one correspondence of the binding sites does not give any selectivity. Materials have been prepared using the same procedure without the long-chain molecule monolayer coating or with a coating layer made of small molecules (hexamethyl disilazane or trimethoxypropylsilane (TMPS)). These materials contain exactly the same bindings sites (amine groups from the APS) arranged the same way as in the long-chain monolayers. No selectivity was observed on any of these materials without the long-chain capping. If the selectivity were solely attributed to the matching of the binding sites, the materials without capping or with the short-chain capping layer should have similar selectivity as the long-chain materials. Therefore, it can be concluded that the steric effect of the cavities in the long-chain monolayers largely contributed to the selectivity of such materials.

2.7.2. *Responsive Nanoscale Materials*

2.7.2.1. *Environmentally Sensitive Polymers*

Many synthetic polymers exhibit a unique phase transition at a certain critical temperature.[9,10,110–112] As discussed in the beginning of the chapter, the phase transition, as reflected in a transition from the dispersed state to condensed state, is a universal phenomenon. For example, poly(*N*-isopropylacrylamide) (PNIPAAm) exhibits a soluble–insoluble change at its lower critical solution temperature (LCST).[113] PNIPAAm chains contain both hydrophilic and hydrophobic groups. At a temperature below the LCST, the hydrophilic groups are surrounded by hydrogen bonded water molecules and the hydrophobic groups are surrounded by structured water. The PNIPAAm chains assume an expanded conformation, thus forming a homogenous system. At a temperature above the LCST of PNIPAAm, the structured water molecules are dissociated and the PNIPAAm chains assume a collapsed conformation and phase separate from water.[114] If PNIPAAm chains are cross-linked, the thermal transition of the polymer chains will be reflected in a remarkable swelling–deswelling transition.[115,116] The unusually large volume change and the changes in other physical properties associated with these environmentally sensitive polymers as when they go through the phase transition make them good candidate materials for many applications. These applications include artificial muscles,[117,118] drug delivery systems,[119–121] reversible surfaces,[122–124] separation membranes,[125,126] enzyme immobilization,[127–129] catalysis substrates,[130,131] and actuator and chemical valves.[132,133]

One of the most important applications of environmentally sensitive polymers is drug delivery. Many mechanisms associated with environmentally sensitive polymers have been investigated for drug release (Fig. 2.22). Copolymers of NIPAAm containing hydrophobic co-monomers are usually used for negative temperature release[134,118] via a skin layer regulated diffusion mechanism. However, intelligent delivery of antipyretics (or antibiotics) that responds to

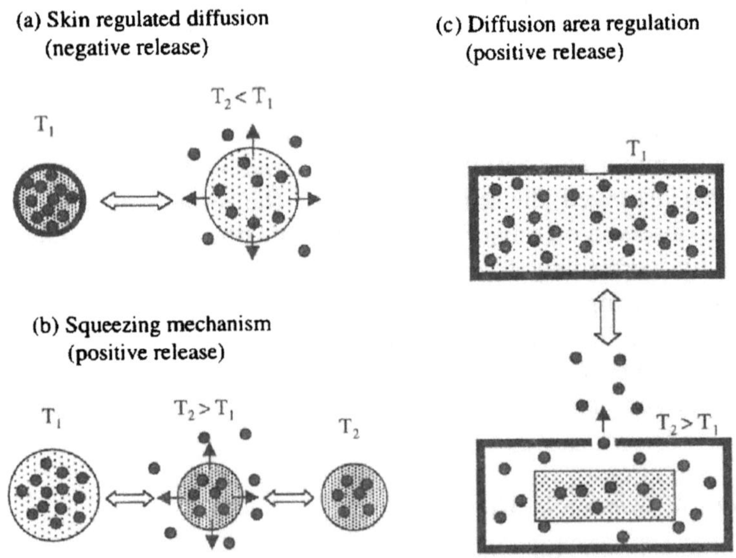

FIGURE 2.22. Common drug delivery mechanism involving thermal sensitive polymers (Redrawn after Ref. 134, 135, 136).

a rise in body temperature requires positive thermo-responsive delivery. One approach to achieve a positive thermosensitive release is to use a squeezing mechanism.[135] In this mechanism, drugs loaded into PNIPAAm gels are squeezed out of the polymer when the temperature is raised above, the LCSTs. Yoshida *et al.*[136] introduced a diffusion area-regulating mechanism by encapsulating PNIPAAm gels in an impermeable capsule equipped with a rate-determining release orifice. At a high temperature, the drug is squeezed out of the polymer gel, but the release of the drug is controlled by the diffusion through the orifice. Other concepts have been proposed for positive thermosensitive release.[137,138] For example, Ichikawa and Fukomori recently reported a microcapsule (MC) containing a membrane of composites of nano-sized polymer gel dispersed in cellulose matrix.[65] The composite MCs showed an enhanced release at 50°C and reduced release at 30°C.

2.7.2.2. Hybrid Responsive Nanoscale Materials

The environmentally sensitive polymers are mostly considered as soft materials, and in many cases, these soft materials alone cannot provide the desired function either for the lack of mechanical robustness and for other reasons. In biological materials, the soft phase is often integrated with or supported by hard tissues that may contain inorganic minerals and high strength polymers. Recently, a wide range of natural and synthetic nanoscale ceramic and composite materials have been investigated. For example, dispersing nanoscale ceramic minerals in a polymer matrix has been shown to improve the thermomechanical stability.[139,140] Such nanocomposites were first reported by a research group from Toyota.[141–143] Subsequently, many polymer systems have been studied.[45,46] The nanoscale aluminosilicate platelets are derived from natural clay minerals (montmorillonite, hectorite, vermiculite, etc.), which are made of 2D, 1 nm layers of an aluminate sheet sandwiched between two silicate sheets.[144] These layers are stacked together by weak ionic and van der Waals forces. Under favorable conditions, the layers can be partially or completely separated.[145] In the intercalated composites, the distance between the aluminosilicate layers is expanded, but they retain the periodic coherence. In the exfoliated composites, the ceramic layers are totally separated and homogeneously distributed in the matrix. This high degree of dispersion of ceramic nanoplates in these composite materials alters the physical and chemical properties of the polymer materials, and improves the mechanical property and thermal stability.

2.7.2.2.1. Responsive Colloidal Crystals. Unique thermally sensitive nanocomposites with tunable periodicity was created by combining the thermosensitive polymer, PNIPAAm, with ordered colloidal crystals.[146] This material was prepared by dispersing monodispersed 100 nm polystyrene (PS) particles in a solution containing NIPAAm monomers and a cross-linking agent. The chemicals and the solutions were carefully purified to remove any impurities, especially any charged species, which would screen the electrical potential of the particles and reduce the repulsive forces between them. Polymerization was induced by UV curing. This procedure produced a nanocomposite in which the PS particles were

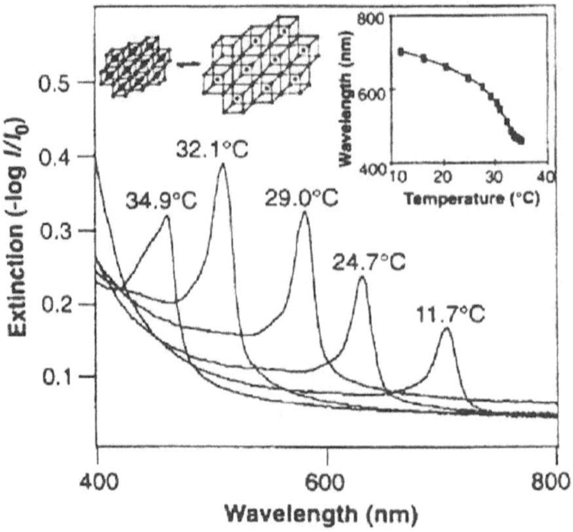

FIGURE 2.23. Bragg diffraction patterns of PS PNIPAAm nanocomposites as a function of temperature using UV–visible–near IR spectrometer (Reproduced from Re. 146 with permission from Science).

periodical, and gave strong Bragg diffractions for infrared, visible, and ultraviolet light sources. The colloidal crystals had either body centered symmetry (bcc) or faced centered symmetry (fcc). In addition, when the temperature was changed around the LCSTs, the polymer matrix in which the colloidal PS particles were embedded expanded and shrank, causing the periodicity of the colloidal crystal to change in accordance with the temperature change. The periodicity was directly correlated with temperature and could be directly monitored through Bragg diffraction techniques (Fig. 2.23).

The environmentally sensitive colloidal crystals have potential for sensing and detection of a wide range of chemicals and biomolecules.[147] For example, crown molecules which can selectively bind to Pb^{2+}, Ba^{2+}, K^+ was attached to the thermosensitive polymer matrix containing the colloidal crystals. The binding of the metal ions increased the osmotic pressure in the polymer and caused the polymer to swell, which led to an increase in the lattice spacing between the particles. The lattice spacing could be directly monitored through the Bragg diffraction, and was used as an indication of the metal ion concentration. Similarly, the enzyme glucose oxidase was incorporated into the polymer–PS composites. The exposure of this material to a glucose solution caused the gel to swell and resulted in an expansion of the lattice spacing. Therefore, glucose sensor devices could be conceived based on this principle.

2.7.2.2.2. Responsive Clay-Polymer Nanocomposites. The preparation of responsive polymer-clay nanocomposites involves several steps.[148] The reaction steps involve the intercalation of the clay with alkylammonium ions through ion exchange followed by polymerization of NIPAAm inside and outside the clay galleries. First, montmorillonite clay was ion exchanged to produce

Na^+-montmorillonite. The Na^+-montmorillonite was mixed with *N*-tetradecyl-trimethylammonium chloride (TDTAC) to produce quaternary alkylammonium-exchanged montmorillonite. The resulted organo-clay dispersed in *N*-isopropy-lacrylamide (NIPAAm) and *N,N'*-methylenebisacrylamide (BisAAm) in dioxane. Polymerization was carried out at 65°C. By dispersing the clay in the PNIPAAm gel, a nanocomposite PNIPAAm–clay gel was generated.

The polymers in the clay composite are able to go through the swelling–deswelling phase transition, causing further expansion of the clay galleries (Fig. 2.24). One of the most important, and unusual, observations in this system is that for the first time the thermal response (swelling ratio) is significantly enhanced in the nanocomposites as compared with the pure polymer hydrogels. On would normally expect the addition of an "inert" ceramic phase to have a negative effect on the phase behavior of such polymers, as previously reported.[149,150] The TEM images of the thermosensitive polymer–clay nanocomposites and the corresponding thermal transition, are shown in Fig. 2.22.

In the intercalated region, the clay platelets are parallel to one another, and separated by the same distance. The optical Fourier transformation of the TEM image gives distinct periodic spots, corresponding to the expanded basal spacing, 26 Å. In the partially exfoliated region, the clay platelets are further separated, and mostly no longer retain the regular periodic basal spacing. The optical

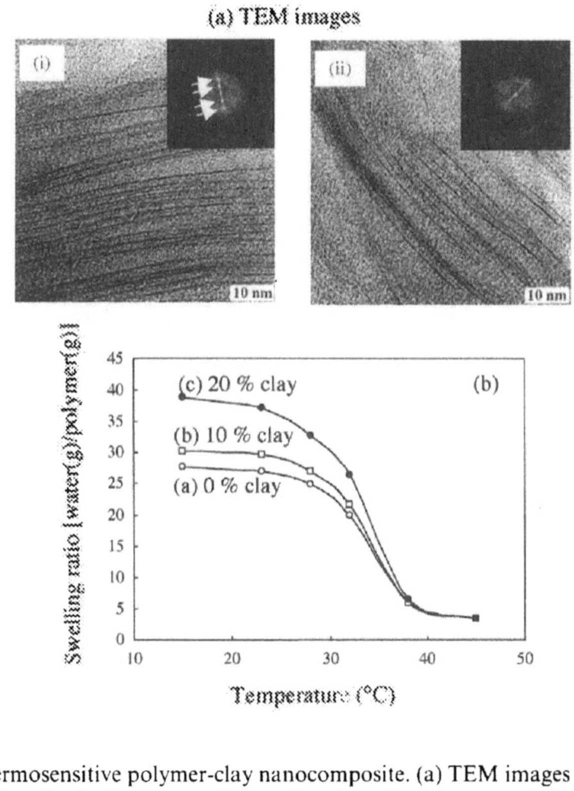

FIGURE 2.24. Thermosensitive polymer-clay nanocomposite. (a) TEM images of intercalated region (i) and exfoliated region (2). (b) Swelling ratio as a function of temperature. Reproduced from Re. 148 with permission from ACS.

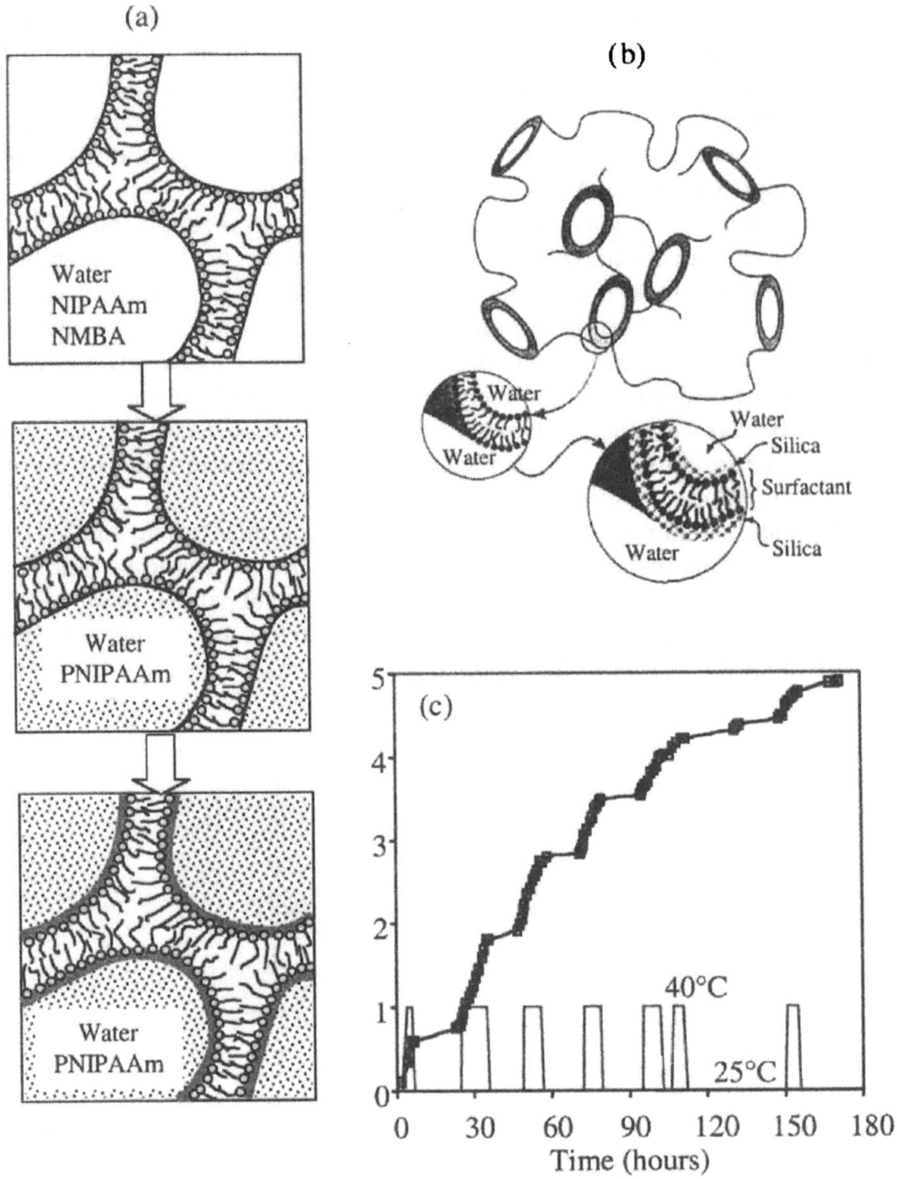

FIGURE 2.25. Bicontinuous thermal sensitive nanocomposites and the drug delivery application. (a) Steps to make the nanocomposite. (b) L3 phase. (c) Amount of drug released as a function of time with cyclic temperature change between 25°C and 40°C. (b) is reproduced from Ref. 152 with permission from ACS. (c) is reproduced from Ref. 153 with permission from J. Controlled Release.

Fourier transformation only gives streaking perpendicular to the platelet orientation due to the 2D nature of the platelets. Thermal response of the composite in water was tested by typical equilibrium swelling experiments. Both conventional gel and PNIPAAm–clay nanocomposite exhibit typical swelling–shrinking transition around 20–46°C. What is remarkable is that the degree of swelling (as measured by the water uptake) and shrinking (as measured by the water release) in the composite is much higher than in the pure polymer gel. For a composite containing 20 wt.% clay (20% of the dry polymer weight), 38 g water can be released when the temperature is increased from 20°C to 45°C. For the pure polymer gel, only 27 g water was released. Therefore, the swelling ratio of the 20 wt.% composite is 50% more than the pure polymer gel.

2.7.2.2.3. Bicontinuous Thermosensitive Nanocomposite. To prepare the bicontinuous hydrogel–silicate, a recipe originally developed to prepare a bicontinuous nanoporous silicate materials (L_3 phase)[151,152] was adapted.[153]

The formation of bicontinuous hydrogel–silicate nanocomposites is schematically illustrated in Fig. 2.25a. First, the surfactant–water system formed the surfactant L_3 phase. The PNIPAAm polymer gel was formed in the hydrophilic region of the surfactant L_3 phase. Finally, the condensation of the silicate network at the surfactant–water interface completed the preparation of the interpenetrating nano-network of silicate and polymer gels. The L_3 phase contains pore channels that are uniformly and homogeneously distributed, and are connected in three dimensions (Fig. 2.25b). This microstructure is ideal for controlling molecular diffusion.

The unique thermosensitive properties of the bicontinuous nanocomposite are illustrated with sustained positive thermosensitive drug release experiments. Indomethacin release from the nanocomposites was investigated during stepwise temperature changes between 25°C and 40°C (Fig. 2.25c). At a low temperature, the drug is trapped in the polymer and in the porous structure. The polymers are expanded, preventing significant release into the media. When the temperature is increased above the LCST of PNIPAAm, the polymer shrinks, squeezing the drug into the porous channels, and opening the pore structure. The silica network structure is stable and does not change significantly within this temperature range. The overall delivery of the drug into the media is controlled by diffusion through the porous channels. The advantage of the current nanocomposite approach is the uniform release rate when the temperature is maintained above the LCSTs. In addition, sustained release over a long period of time can be achieved by simply changing the nanocomposite composition. Based on previous studies,[61,62] reducing the water content from 95 to 55 wt.% can reduce the pore dimension from about 30 to 8 nm. By reducing the amount of water from 95% to 55% composite, the release is extended to a much longer period of time. The release rate did not decrease significantly after six cycles (a duration of more than 150 h).

REFERENCES

1. J. N. Israelachvili, in: *Intermolecular and Surface Forces*, 2nd edition (Academic Press, San Diego, 1992), Chapter 2–7.
2. P. C. Hiemenz, *Principles of Colloid and Surface Chemistry*, 2nd edition (Marcel Dekker Inc., 1986), Chapter 11, pp. 616–622.
3. D. J. Shaw, *Introduction to Colloidal and Surface Chemistry*, 3rd edition (Butterworths, Boston, 1980), Chapter 7, pp. 148–182.

4. E. J. W. Verwey and J. Th. G. Overbeek, *Theory of the Stability of Lyophobic Colloids* (Elsevier, Amsterdam, 1948).
5. J. Th. G. Overbeek, Recent development in the understanding of colloidal stability, *J. Colloid. Interface Sci.* **58**, 408–22 (1977).
6. J. N. Israelachvili, *Intermolecular and Surface Forces*, 2nd edition (Academic Press, San Diego, 1992), pp. 122–127.
7. M. J. Rosen, *Surfactants and Interfacial Phenomena* (John Wiley & Sons, New York, 1978), Chapter 1, pp. 1–25.
8. S. Glasstone and D. Lewis, *Elements of Physical Chemistry*, 2nd edition (D. Van Nostrand Company, New Jersey, 1960), Chapter 5, pp. 118–153.
9. T. Tanaka, Phase transition of gels, *ACS Symp. Ser.* **480**, 1–21 (1992).
10. Y. Li and T. Tanaka, Phase transition of gels, *Annu. Rev. Mat. Sci.* **22**, 247–277 (1992).
11. P. J. Flory, *Principles of Polymer Chemistry* (Cornell University Press, Ithaca, NewYork, 1957).
12. P. G. DeGennes, *Scaling Concepts in Polymer Physics* (Cornell University, Ithaca, New York, 1979).
13. I. A. Aksay and R. Kikuchi, in: *Science of Ceramic Chemical Processing*, edited by L. L. Hench and D. R. Ulrich, (Wiley, New York, 1987), pp. 513.
14. R. Kikuchi, Ternary phase diagrams calculation-I: General theory, *Acta Metallurgica* **25**, 195–205 (1977).
15. R. Kikuchi, Crystal-growth approach of lattice-order simulation and the cluster-variation method, *Phys. Rev. B* **22**, 3784–3789 (1980).
16. N. Boden, P. J. B. Edwards, and K. W. Jolley, in: *Structure and Dynamics of Strongly Interacting Colloids and Supermolecular Aggregates in Solutions*, edited by S. H. Chen, J. S. Huang, and P. Tartaglia (Kluwer Academic Publishers, Dordrecht, Netherlands, 1992), p. 433.
17. J. N. Israelachvili, *Intermolecular and Surface Forces*, 2nd edition (Academic Press, San Diego, 1992), Chapter 17, pp. 366–394.
18. P. K. Vinson, J. R. Bellare, H. T. Davis, W. G. Miller, and L. E. Scriven, Direct imaging of surfactant micelles, vesicles, discs, and ripple phase structures by cryo-transmission electron microscopy, *J. Colloid. Interface Sci.* **142,** 74–91 (1991).
19. D. F. Evans and H. Wennerström, *The Colloidal Domain, Where Physics, Chemistry, Biology, and Technology Meet* (VCH Publishers Inc., New York, 1994), pp. 14–15.
20. G. J. T. Tiddy, Surfactant-water liquid crystal phases, *Phys. Rep.* **57**, 1–46 (1980).
21. F. S. Bates and G. H. Fredrickson, Block copolymer-designer soft materials, *Phys. Today*, 32–28 (1999).
22. F. S. Bates, Polymer–polymer phase behavior, *Science* **251**, 898–905 (1991).
23. J. S. Beck, J. C. Vartuli, W. J. Roth, M. E. Leonowicz, C. T. Kresge, K. D. Schmitt, C. T.-W Chu, D. H. Olson, E. W. Sheppard, S. B. McCullen, J. B. Higgins, and J. L. Schlenker, A new family of mesoporous molecular sieves prepared with liquid crystalline templates, *J. Am. Chem. Soc.* **114**, 10834–10843 (1992).
24. C. T. Kresge, M. E. Leonowicz, W. J. Roth, J. C. Vartuli, and J. S. Beck, Ordered mesoporous molecular-sieves synthesized by a liquid-crystal template mechanism, *Nature* **359**, 710–712 (1992).
25. S. Inagaki, Y. Fukushima, and K. Kuroda, Synthesis of highly ordered mesoporous materials from layered polysilicate, J. Chem. Soc. *Chem. Commun.* 680–682 (1993).
26. D. Zhao, J. Feng, Q. Huo, N. Melosh, G. H. Fredrickson, B. F. Chmelka, and G. D. Stucky, Triblock copolymer synthesis of mesoporous silica with periodic 50 to 300 Å pores, *Science* **279**, 548–552 (1998).
27. D. Zhao, Q. Huo, J. Feng, B. F. Chmelka, and G. D. Stucky, Nonionic triblock and star diblock copolymer and oligomeric surfactant synthesis of highly ordered, hydrothermal stable, mesoporous silica structures, *J. Am. Chem. Soc.* **120**, 6024–6036 (1998).
28. Y. Sakamoto, M. Kaneda, O. Terasaki, D. Y. Zhao, J. M. Kim, G. Stucky, H. J. Shin, and R. Ryoo, Direct imaging of the pores and cages of three-dimensional mesoporous materials, *Nature* **408**, 449–453, (2000).
29. A. Sayari, *Periodic Mesoporous Materials: Synthesis, Characterization, and Potential Applications*, in: *Recent Advances and New Horizons in Zeolite Science and Technology*, edited by H. Chon, S. I. Woo, and S.-E. Park (Elsevier, Amsterdam, 1996), pp. 1–46.

30. J. Liu, A. Y. Kim, L. Q. Wang, B. J. Plamer, Y. L. Chen, P. Bruinsma, B. C. Bunker, g. J. Exarhos, G. L. Graff, P. C. Rieke, G. E. Fryxell, J. W. Virden, B. J. Tarasevich, and L. A. Chick, Self-assembly in the synthesis of ceramic materials and composites, *Adv. Colloids Interface Sci.* **69**, 131–180 (1996).
31. I. A. Aksay, M. Trau, S. Manne, I. Honma, N. Yao, L. Zhou, P. Fenter, P. M. Eisenberger, and S. M. Gruner, Biomimetic pathways for self-assembling inorganic thin films, *Science* **273**, 892 (1996).
32. J. S. Beck and J. C. Vartuli, Recent advances in the synthesis, characterization and application of mesoporous molecular sieves, *Curr. Opin. Solid State Mater. Sci.* **1**, 76–87, (1996).
33. D. Dabbs and I. Aksay, Self-assembled ceramics produced by complex-fluid templation. *Annu. Rev. Phys. Chem.* **51**, 601–622 (2000).
34. Q. Huo, D. I. Margolese, U. Ciesla, P. Feng, T. E. Gier, P. Sieger, R. Leon, P. M. Petroff, F. Schuth, and G. D. Stucky, Generalized synthesis of periodic surfactant inorganic composite-materials, *Nature* **368**, 317–321, (1994).
35. P. T. Tanev and T. J. Pinnavaia, A neutral templating route to mesoporous molecular sieves, *Science* **267**, 865–867 (1995).
36. S. A. Bagshaw, E. Prouzet, and T. J. Pinnavaia, Templating of mesoporous molecular sieves by nonionic polyethylene oxide surfactants, *Science* **269**, 1242–1244 (1995).
37. P. Yang, D. Zhao, D. I. Margolese, B. F. Chmelka, and G. D. Stucky, Generalized synthesis of large pore mesoporous metal oxides with semicrystalline framework, *Nature* **396**, 152–155 (1998).
38. M. J. MacLachlan, N. Coombs, and G. A. Ozin, Non-aqueous supermolecular assembly of mesostructured metal germanium sulphides from $(Ge_4S_{10})^{4-}$ clusters, *Nature* **397**, 681–684, (1999).
39. Z.-R. Tian, W. Tong, J.-Y. Wang, N.-G. Duan, V. V. Krishnan, and S. L. Suib, Manganese oxide mesoporous structures: mixed-valent semiconducting catalysts, *Science* **276**, 926–930 (1997).
40. C. Santato, M. Odzeiemkowski, M. Ulmann, and J. Augustynski, Crystallographically oriented mesoporous WO_3 films: synthesis, characterization, and application, *J. Am. Chem. Soc.* **123**, 10639–10649 (2001).
41. P. N. Trikalitis, K. K. Rangan, T. Bakas, and M. G. Kanatzidis, Varied pore organization in mesostructured semiconductors based on the $[SnSe_4]^{(4-)}$ anion, *Nature* **410**, 671–675 (2001).
42. P. V. Braun, P. Osenar, V. Tohver, S. B. Kennedy, and S. I. Stupp, Nanostructured templating in inorganic solids with organic lyotropic liquid crystals, *J. Am. Chem. Soc.* **121**, 7302–7309 (1999).
43. S. H. Joo, S. J. Choi, I. Oh, J. Kwak, Z. Liu, O. Terasaki, and R. Ryoo, Ordered nanoporous arrays of carbon supporting high dispersions of platinum nanoparticles, *Nature* **412**, 169–172 (2001).
44. G. S. Attard, P. N. Barttlet, N. R. B. Coleman, J. M. Elliott, J. R. Own, and J. H. Wang, Mesoporous platinum films from lyotropic crystalline phases, *Science* **278**, 838–840 (1997).
45. H. J. Shin, C. H. Ko, and R. Ryoo, Synthesis of platinum networks with nanoscopic periodicity using mesoporous silica as template, *J. Mater. Chem.* **11**, 260–261 (2001).
46. P. T. Tanev, M. Chibwe, and T. J. Pinnavaia, Titanium containing mesoporous molecular-sieve for catalytic-oxidation of aromatic-compounds, *Nature* **368**, 321–323 (1994).
47. A. Sayari, Catalysis by crystalline mesoporous molecular sieves, *Chem. Mater.* **8**, 1840–1852 (1996).
48. A. Stein, B. J. Melde, and R. C. Schroden, Hybrid inorganic-organic mesoporous silicate: nanoscopic reactors coming of age, *Adv. Mater.* **12**, 1403–1419 (2000).
49. X. Feng, G. E. Fryxell, L.-Q. Wang, A. Y. Kim, J. Liu, and K. M. Kemner, Functionalized monolayers on ordered mesoporous supports (FAMMS), *Science* **276**, 923–926 (1997).
50. S. Inagaki, S. Guan, Y. Fukushima, T. Ohsuna, and O. Terasaki, Novel mesoporous materials with a uniform distribution of organic groups and inorganic oxide in their frameworks, *J. Am. Chem. Soc.* **121**, 9611–9614 (1999).
51. M. H. Lim and A. Stein, Comparative studies of grafting and direct synthesis of inorganic–organic hybrid mesoporous materials, *Chem. Mater.* **11**, 3285–3290 (1999).
52. T. Asefa, M. J. MacLachlan, N. Coombs, and G. A. Ozin, Periodic mesoporous organosilica with organic groups inside the channels walls, *Nature* **402**, 867–871 (1999).
53. W. Zhou, J. M. Thomas, D. S. Shephard, B. F. G. Johnson, D. Ozkaya, T. Maschmeyer, R. G. Bell, and Q. Ge, Ordering of ruthenium cluster carbonyls in mesoporous silica, *Science* **280**, 705–708 (1998).

54. C. G. Wu and T. Bein, Conducting polyaniline filaments in a mesoporous channel host, *Science* **264**, 1757–1759 (1994).
55. T. Q. Nguyen, J. J. Wu, V. Doan, B. J. Schwartz, and S. H. Tolbert, Control of energy transfer in oriented conjugated polymer-mesoporous silica, *Science* **288**, 652–656 (2000).
56. K. Kageyama, J.-I. Tamazawa, and T. Aida, Extrusion polymerization: catalyzed synthesis of crystalline linear polyethylene nanofibers within a mesoporous silica, *Science* **285**, 2113–2115 (1999).
57. Y. F. Lu, A. Stellinger, M. C. Lu, J. M. Huang, H. Y. Fan, R. Haddad, G. Lopez, A. R. Burns, D. Y. Sasaki, J. Shelnutt, and C. J. Brinker, Self-assembly of mesoscopically ordered chromic polydiacetylene silica nanocomposites, *Nature* **410**, 913–917 (2001).
58. H. Yang, A. Kuperman, N. Coombs, S. Mamiche-Afrara, and G. A. Ozin, Synthesis of oriented films of mesoporous silica on mica, Nature **379**, 703–705 (1996).
59. H. Yang, N. Coombs, I. Sokolov, and G. A. Ozin, Free standing and oriented mesoporous silica films grown at the air–water interfaces, *Nature* **381**, 589–592 (1996).
60. Q. Huo, J. Feng, F. Scheth, and G. D. Stucky, Preparation of hard mesoporous silica spheres, *Chem. Mater*. **9**, 14–17 (1997).
61. Y. Lu, H. Fan, A. Stump, T. L. Ward, T. Ricker, and C. F. Brinker, Aerosol-assisted self-assembly of mesostructured spherical nanoparticles, *Nature* **398**, 223–226 (1999).
62. P. J. Bruinsma, A. Y. Kim, J. Liu, and S. Baskaran, Mesoporous silica synthesized by solvent evaporation: Spin fibers and spray-dried hollow spheres, *Chem. Mater*. **10**, 2507–2512 (1998).
63. P. Yang, D. Zhao, B. F. Chmelka, and G. D. Stucky, Triblock-copolymer-directed synthesis of large-pore mesoporous silica fibers, *Chem. Mater*. **10**, 2033–2036 (1998).
64. D. M. Antonelli, A. Nakahira, and J. Y. Ying, Ligand-assisted liquid crystal templating in mesoporous niobium oxide molecular sieves, *Inorg. Chem*. **35**, 3126–3136 (1996).
65. Y. Lu, R. Gangull, C. A. Drewien, M. T. Anderson, C. F. Brinker, W. Gong, Y. Guo, H. Soyez, B. Dunn, M. H. Huang, and J. I. Zink, Continuous formation of supported cubic and hexagonal mesoporous films by sol–gel dip coating, *Nature* **389**, 364–368 (1997).
66. D. Zhao, P. Yang, N. Melosh, J. Feng, B. F. Chmelka, and G. D. Stucky, Continuous mesoporous silica films with highly ordered large pore structures, *Adv. Mater*. **9**, 1380 (1997).
67. S. Baskaran, J. Liu, K. Domansky, N. Kohler, X. Li, C. Coyle, G. E. Fryxell, S. Thevathasan, and R. E. Williford, Low dielectric constant mesoporous silica films through molecular templated synthesis, *Adv. Mater*. **12**, 291–294 (2000).
68. S. H. Tolbert, A. Firouzi, G. D. Stucky, and B. F. Chmelka, Magnetic field alignment of ordered silicate-surfactant composites and mesoporous silica, *Science* **278**, 264–268 (1997).
69. H. Yang, N. Coombs, and G. A. Ozin, Morphologenesis of shapes and surface patterns in mesoporous silica, *Nature* **38**, 692–695 (1997).
70. T. Zemb, M. Dubois, B. Demé, and T. Gulik-Krzywicki, Self-assembly of flat nanodisks in sla-free cationic surfactant solutions, *Science* **283**, 816–819 (1999).
71. S. S. Kim, W. Zhang, and T. J. Pinnavaia, Ultrastable mesostructured silica vesicles, *Science* **282**, 1302 (1998).
72. P. Yang, T. Deng, D. Zhao, J. Feng, D. Pine, B. F. Chmelka, G. M. Whitesides, and G. D. Stucky, Hierarchically ordered oxides, *Science* **282**, 2244–2246 (1998).
73. M. Trau, N. Yao, E. Kim, Y. Xia, G. M. Whitesides, and I. A. Aksay, Microscopic patterning of oriented mesoscopic silica through guided growth, *Nature* **390**, 674–676 (1997).
74. D. A. Doshi, N. K. Huesting, M. Lu, H. Fan, Y. Lu, K. Simomons-Potter, B. G. Potter Jr., A. J. Hurd, and C. J. Brinker, Optically defined multifunctional patterning of photosensitive thin-film silica mesophase, *Science* **290**, 107–111 (2000).
75. P. Yang, G. Wienberger, H. C. Huang, S. R. Cordero, M D. McGehee, B. Scott, T. Deng, G. M. Whitesides, B. F. Chmelka, S. K. Buratto, and G. D. Stucky, Mirrorless lasing from mesostructured waveguides patterned by soft lithography, *Science* **287**, 465–467 (2000).
76. J. N. Cha, G. D. Stucky, D. E. Morse, and T. J. Deming, Biomimetic synthesis of ordered silica structures mediated by block copolypeptides, *Nature* **403**, 289–292 (2000).
77. M. Duboise and B. Cabane, Polymerization of silicic acid in a collapsed lamellar phase, *Langmuir*, **10**, 1615–1617 (1994).
78. M. Duboise, Th. Gulik-Krzywicki, and B. Cabane, Growth of silica polymers in a lamellar mesophase, *Langmuir* **9**, 673–680 (1993).

79. G. S. Attard, J. C. Glyde, and C. G. Göltner, Liquid-crystalline phases as the templates for the synthesis of mesoporous silica, *Nature* **378**, 366–368 (1995).
80. G. S. Attard, C. G. Göltner, J. M. Corker, S. Henke, and R. H. Templer, Liquid-crystalline templates for nanostructured metals, Angew. Chim. *Int. Ed. Engl.* **36**(2), 1315–1317 (1997).
81. C. G. Göltner and M. Antonietti, Mesoporous materials by templating of liquid crystalline phases, *Adv. Mater.* **9**(5), 431–436 (1997).
82. P. V. Braun, P. Osennar, and S. I. Stupp, Semiconducting superlattice by molecular assemblies, *Nature* **380**, 325–328 (1996).
83. M. Davis, C.-Y. Chen, S. L. Burkett, and R. F. Lobo, Synthesis of (alumino)silicate materials using organic molecules and self-assembled organic aggregates as structure-directing agents, *MRS Symp. Proc.* **346**, 831–842 (1994).
84. A. Monnier, F. Schuth, Q. Hua, D. Kumar, D. Margolese, R. S. Maxell, G. D. Stucky, M. Krishnamurty, P. Petroff, A. Firouzi, M. Janicke, and B. F. Chmelka, Cooperative formation of inorganic–organic interfaces in the synthesis of silicate mesostructures, *Science* **261**, 1299–1303 (1993).
85. A. Sayari, Unprecedented expansion of the pore size and volume of periodic mesoporous silica, *Angew. Chim. Int. Ed.* **39**(16), 2920–2922 (2000).
86. M. Kruk and M. Jaroniec, Application of large pore MCM-41 molecular sieves to improve pore size analysis using nitrogen adsorption measurements, *Langmuir* **13**(23), 6267–6273 (1997).
87. Q. Huo, D. I. Margolese, U. Ciesla, D. G. Demuth, P. Feng, T. E. Gier, P. Sieger, Firouzi, B. F. Chmelka, F. Schüth, and G. D. Stucky, Organization of organic molecules with inorganic molecular species into nanocomposite biphase arrays, *Chem. Mater.* **6**, 1176–1191 (1994).
88. A. Y. Kim, P. Bruinsma, Y. L. Chen, and J. Liu, Amphoteric surfactant templating route for mesoporous zirconia, *Chem. Commun.* 161–162 (1997).
89. Q. Huo, R. Leon, P. M. Petroff, and G. D. Stucky, Mesostructure design with Gemini surfactants: supercage formation in a three-dimensional hexagonal arrays, *Science* **268**, 1324–1327 (1995).
90. M. Templin, A. Franck, A. Du Chesne, H. Leist, Y. Zhang, R. Ulrich, V. Schädler, and U. Wiesner, Organically modified aluminosilicate mesostructures from block copolymer phases, *Science* **278**, 1795–1798 (1997).
91. C. G. Göltner, S. Henke, M. C. Wiessenberger, and M. Antonietti, Mesoporous silica from lyotropic liquid crystalline polymer templates, *Angew. Chim. Int. Ed.* **37**(5), 613–616 (1998).
92. L. Zhang, C. Bartels, Y. Yu, H. Shen, and A. Eisenberg, Mesosized crystal-like structure of hexagonally packed hollow hoops by solution self-assembly of diblock copolymers, *Phys. Rev. Lett.* **79**(25), 5034–5037 (1997).
93. A. C. Finnefrock, R. Ulrich, A. Du Chesne, C. C. Honeker, K. Schumacher, K. K. Unger, S. M. Gruner, and U. Wiesner, Metal oxide containing mesoporous silica with bicontinuous plumber's nightmare: morphology from block copolymer-hybrid mesophase, *Angew. Chem. Int. Ed.* **40**, 1208–1211, (2001).
94. G. M. Whitesides, Self-assembling materials, *Sci. Am.* **273**, 146–149 (1995).
95. A. Ulman, Formation and structure of self-assembled monolayers, *Chem. Rev.* **96**, 1533–1554 (1996).
96. B. C. Bunker, P. C. Rieke, B. J. Tarasevich, A. A. Campbell, G. E. Fryxell, G. L. Graff, L. Song, J. Liu, W. Virden, and G. L. McVay, Ceramic thin film formation on functionalized interfaces through biomimetic processing, *Science* **264**, 48–55 (1994).
97. C. A. Mirkin, R. L. Letsinger, R. C. Mucic, and J. J. Storhoff, A DNA based method for rationally assembly of nanoparticles into macroscopic materials, *Nature* **382**, 607–609 (1996).
98. R. S. Ingram, M. J. Hostetler, and R. W. Murray, Poly-hetero-w-functional alkanethiol-stabilized gold cluster compounds, *J. Am. Chem. Soc.* **119**, 9175–9178 (1997).
99. A. K. Boal and V. M. Rotello, Fabrication and self-optimization of multivalent receptors on nanoparticle scaffolds, *J. Am. Chem. Soc.* **122**, 734–735 (2000).
100. J. Liu, X. Feng, G. E. Fryxell, L- Q. Wang, A. Y. Kim, and M. Gong, Hybrid mesoporous materials with functionalized monolayers, *Adv. Mater.* **10**, 161–165 (1998).
101. G. E. Fryxell, J. Liu, T. A. Hauser, Z. Nie, K. F. Ferris, S. Mattigod, M. Gong, and R. T. Hallen, Design and synthesis of selective mesoporous anion traps, *Chem. Mater.* **11**, 2184–2154 (1999).

102. G. Wulff. Molecular imprinting in cross-linked materials with the aid of molecular templates: A way towards artificial antibodies. *Angew. Chem. Int. Ed. Engl.* **34**, 1812–1832 (1995).
103. G. Vlatakis, L. I. Anderson, R. Müller, and K. Mosbach, Drug assay using antibody mimics made by molecular imprinting, *Nature* **361**, 645–647 (1993).
104. H. Shi, W-B. Tsai, M. D. Garrison, S. Ferrari, and B. D. Ratner, Template-imprinted nanostructured surfaces for protein recognition, *Nature* **398**, 593–597 (1999).
105. S. M. DSouza, C. Alexander, S. W. Carr, A. M. Waller, M. J. Whitcombe, and E. N. Vulfson, Directed nucleation of calcite at a crystal-imprinted polymer surfaces, *Nature* **398**, 312–316 (1999).
106. Y. Shin, J. Liu, L.-Q. Wang, Z. Nie, W. D. Samuels, G. E. Fryxell, and G. J. Exarhos, Ordered hierarchical porous materials: Towards tunable size and shape selective microcavities in nanoporous channel, Angew. *Chim. Intl. Ed.* **39**, 2702–2707 (2000).
107. J. Liu, Y. Shin, G. E. Fryxell, L. Q. Wang, Z. Nie, J. H. Chang, G. E. Fryxell, W. D. Samuels, and G. J. Exarhos, Molecular assembly in ordered mesoporosity: A new class of highly functional nanoscale materials, *J. Phys. Chem. A* **104**, 8328–8339 (2000).
108. D. C. Tahmassebi and T. Sasaki, Synthesis of a new trialdehyde template for molecular imprinting, *J. Org. Chem.* **59**, 679–681 (1994).
109. K. O. Hwang, Y. Yakura, F. S. Ohuchi, and T. Sasaki, Template-assisted assembly of metal-binding sites on a silica surfaces, *Mater. Sci. Eng. C* **3**, 137–141 (1995).
110. Y. Osada and S. B. Ross-Murphy, Intelligent gels, *Sci. Am.* 82–87 (1993).
111. D. W. Urry, Elastic biomolecular machines, *Sci. Am.* 64–69 (1995).
112. D. W. Urry, Physical chemistry of biological free energy transduction as demonstrated by elastic protein-based polymers, *J. Phys. Chem. B* **101**, 11007–11028 (1997).
113. H.G. Schild, Poly(*N*-isopropylacrylamide): Experiment, theory and applications, *Prog. Polym. Sci.* **17**, 163–249 (1992).
114. M. Q. Chen, A. Kishida, and M. Akashi, Graft copolymer having hydrophobic backbone and hydrophilic branches. II. Preparation and thermosensitive properties of polystyrene micro spheres having poly(*N*-isopropylacrylamide) branches on their surfaces, *J. Polym. Sci., Part A: Polym. Chem.* **34**, 2213–2220 (1996).
115. E. S. Matsuo and T. Tanaka, Kinetics of discontinuous phase-transition of gels, *J. Chem. Phys.* **89**, 1695 (1988).
116. L. C. Dong and A. S. Hoffman, Synthesis and applications of thermally reversible heterogels for drug delivery, *J. Control. Release* **13**, 21–31 (1990).
117. Y. Osada, H. Okuzaki, and H. Hori, A polymer gel with electrically driven motility, *Nature* **355**, 242–244 (1992).
118. R. Yoshida, T. Takahashi, T. Yamaguchi, and H. Ichijo, Self-oscillating gels, *Adv. Mater.* **9**, 175–178 (1997).
119. T. Okano, Y. H. Bae, H. Jacobs, and S. W. Kim, Thermally on off switching polymer for drug permeation and release, *J. Control. Release* **11**, 255–265 (1990).
120. H. Kurahashi and S. Furusaki, Preparation and properties of a new temperature-sensitive ionized gel, *J. Chem. Eng. Jpn.* **26**, 89–93 (1993).
121. S. Shoemaker, A. S. Hoffman, and J. H. Priest, Synthesis and properties of vinyl monomer enzyme conjugates: Conjugation of L-asparaginase with n-succinimidyl acrylate, *Appl. Biochem. Biotechnol.* **15** 11–14 (1987).
122. T. Takushiji, K. Sakai, A. Kikuchi, T. Aoyagi, Y. Sakurai, and T. Okano, Graft architectural effects on thermoresponsive wettabiliy changes of poly(*N*-isopropylacrylamide)-modified surfaces, *Langmuir* **14**, 4657–4662 (1998).
123. C. DellaVolpe, C. Cassinelli, and M. Morra, Wilhelmy plate measurements on poly-(*N*-isopropylacrylamide)-grafted surfaces, *Langmuir* **14**, 4650 (1998).
124. L. Liang, X. D. Feng, J. Liu, P. Rieke, and G. E. Fryxell, Reversible surface properties of glass plate and capillary tube grafted by photopolymerization of *N*-isopropylacrylamide, *Macromolecules* **31**, 7834–7850 (1998).
125. Y. M. Sun and T. L. Huang, Prevaporation of ethanol-water mixture through temperature-sensitive poly(vinyl alcohol-g-*N*-isopropylacrylamide) membranes, *J. Membrane Sci.* **118**, 304–304 (1996).

126. L. Liang, X. D. Feng, L. Peurrung, and V. Viswanatham, Temperature-sensitive membranes prepared by UV photopolymerization of *N*-isopropylacrylamide on a surface of porous hydrophilic polystyrene, *J. Membrane Sci.* **162**, 235–246 (1999).
127. T. Nozaki, Y. Maeda, K. Ito, and H. Kitano, Cyclodextrins modified with polymer-chain which are responsive to external stimuli, *Macromolecules* **28**, 522–524 (1995).
128. S. Takeuchi, I. Omodaka, K. Hasegawa, Y. Maeda, and H. Kitano, Temperature-responsive graft-copolymers for immobilization of enzymes, *Makromol. Chem.* **194**, 1991–1999 (1993).
129. T. Shiroya, M. Yasui, K. Fujimoto, and H. Kawaguchi, Control of enzymatic-activity using thermosensitive polymers, *Coll. Sci. B: Biointerfaces* **4**, 275–285 (1995).
130. D. E. Bergreiter, B. L. Case, Y. S. Liu, and J. W. Caraway, Poly(*N*-isopropylacrylamide) soluble polymer supports in catalysis and synthesis, *Macromolecules* **31**, 6053–6062 (1998).
131. C. W. Chen and M. Akashi, Synthesis, characterization, and catalysis of colloidal platinum nanoparticles protected by poly(*N*-isopropylacrylamide), *Langmuir* **13**, 6465–6472 (1997).
132. R. Kishi, H. Hara, K. Sawahata, and Y. Osada, Conversion of chemical energy to mechanical energy by synthetic polymer gels, in: *Polymer Gels*, edited by D. K. DeRoss, K. Kajiwara, Y. Osada, and Y. Yamauch, (Elsevier, Plenum Press, New York, 1991) pp. 205–220.
133. L. Liang, X. D. Feng, and L. Peurrung, Temperature-sensitive switching from composite poly(*N*-isopropylacrylamide) sponge gels, *J. Appl. Polym. Sci.* **75**, 730–1739 (2000).
134. X. S. Wu, A. S. Hoffman, and P. Yager, Synthesis and characterization of thermally reversible macroporous poly(*N*-isopropylacrylamide) hydrogels, *J. Polym. Sci. Part A: Polym. Chem.* **30**, 2121–2129 (1992).
135. A. S. Hoffman, A. Afrassiabi, and L. C. Dong, Thermally reversible hydrogels: II. Delivery and selective removal of substances from aqueous solutions, *J. Control. Release* **4**, 213–222 (1986).
136. R. Yoshida, Y. Kaneko, K. Sakai, T. Okano, Y. Sakurai, Y. H. Bae, and S. W. Kim, Positive thermosensitive pulsatile drug-release using negative thermosensitive hydrogels, *J. Control. Release* **32**, 97–102 (1994).
137. S. W. Chun and J. D. Kim, A novel hydrogel-dispersed composite membrane of poly(*N*-isopropylacrylamide) in a gelatin matrix and its thermally actuated permeation of 4-acetamidophen, *J. Control. Release* **38**, 39–47 (1996).
138. H. Ichigawa and Y. A. Fukomori, A novel positively thermosensitive controlled-release microcapsule with membrane of nano-sized poly(*N*-isopropylacrylamide) gel dispersed in ethylcellulose matrix, *J. Control. Release* **63**, 107–119 (2000).
139. E. P. Giannelis, Nanoscale, Two-Dimensional Organic-Inorganic Materials, in: *Advances in Chemistry Series, Materials Chemistry*, edited by L. V. Interrante, L. A. Caspar, and A. B. Ellis (Elsevier, American Chemical Society, WA DC, 1995) pp. 259–281.
140. T. Lan, P. D. Kaviratna, and T. J. Pinnavaia, Mechanism of clay tactoid exfoliation in epoxy-clay nanocomposites, *Chem. Mater.* **7**, 2144–2150 (1995).
141. Y. Kojima, A. Okada, A. Usuki, M. Kawasumi, T. Kuraduchi, and O. Kamigaito, One-pot synthesis of nylon-6 clay hybrid, *J. Polym. Sci., Part A: Polym. Chem.* **31**, 1755–1758 (1993).
142. A. Usuki, M. Kawasumi, Y. Kojima, A. Okada, T. Kurauchi, and O. Kamigaito, Swelling behavior of montmorillonite cation exchanged for omega-amino acids by epsilon-carrolactam, *J. Mater. Res.* **8**, 1174–1178 (1993).
143. A. Usuki and A. Okakda, The chemistry of polymer-clay hybrids, *Mater. Sci. Eng. C* **3**, 109–115 (1995).
144. R. W. Grimshaw, in: *The Chemistry and Physics of Clays and Advanced Ceramic Materials* (Tech Books, Fairfax, VA, 1971), pp. 124–157.
145. M. V. Smalley, Electrical theory of clay swelling, *Langmuir* **10**, 2884 (1994).
146. J. M. Weissman, H. B. Sunkara, A. S. Tse, and S. A. Asher, Thermally switchable periodicities and diffraction from mesoscopically ordered materials, *Science* **274**, 959–960 (1996).
147. J. H. Holtz and S. A. Ahser, Polymerized colloidal crystal hydrogel films as intelligent chemical sensing materials, *Nature* **389**, 829–832 (1997).
148. L. Liang, J. Liu, and X. Gong, Thermosensitive Poly(*N*-isopropylacrylamide)-clay nanocomposites with enhanced temperature response, *Langmuir* **16**, 9895 (2000).

149. P. B. Messersmith and F. Znidarsich, Synthesis and LCST behavior of thermally responsive poly(N-isopropylacrylamide) layered silicate nanocomposites, *MRS Symp. Proc.* **457**, 507–512 (1997).
150. N. A. Churochkina, S. G. Starodoubtsev, and A. R. Khokhlov, Swelling and collapse of the gel composite based on neutral and slightly charged poly(acrylamide) gel containing Na-montmorillonite. *Polym. Gels Network* **6**, 205–215 (1998).
151. K. M. McGrath, D. M. Dabbs, N. Yao, I. A. Aksay, and S. M. Gruner, Formation of a silicate L-3 phase with continuous adjustable pore sizes, *Science* **277**, 552–556 (1997).
152. K. M. McGrath, D. M. Dabbs, N. Yao, K. J. Elder, I. A. Aksay, and S. M. Gruner, Silica gel with tunable nanoporous through templating of the L-3 phase, *Langmuir*, **16**, 398–406 (2000).
153. Y. Shin, J. H. Chang, J. Liu, R. Williford, Y.-K. Shin, and G. J. Exarhos, Hybrid nanogels for sustainable positive thermosensitive drug release, *J. Control. Release* **73**, 1–6 (2001).

3

Examples of Nanoscale Materials in Nature

3.1. MULTISCALE ORDERING AND FUNCTION IN BIOLOGICAL NANOSCALE MATERIALS

In Chapter 2, self-assembly approaches to synthesize nanoscale materials were discussed. However, the range of structures and the functions of the materials derived from self-assembly are limited compared to many examples of highly ordered, highly functional nanoscale materials in biological systems. A superficial comparison of synthetic nanoscale materials and the biological nanoscale material reveals some key differences. First, biological materials are ordered on different length scales, not just on the nanometer scale. The hierarchical order is not commonly encountered in man-made materials, and can be exemplified by the structures found in seashells, bone tissues, and muscle tissues. Second, the assembly and the ordering of biological materials are purpose and function driven. For example, the filament structure that is responsible for cell repair only appears when the cell is damaged and when a defect in the cell membrane causes an influx of calcium ions. The calcium ions trigger the assembly of nanotubes that function as bridges for materials to be transported to the defect site. Once the defect is repaired and the calcium ions no longer flow, the bridges cease to have any further function and disappear by themselves. This is in sharp contrast to man-made materials that have to be prepared in many steps, one step involving the formation of the desired microstructure and another involving the introduction of functionality into the system. Third, biological materials are constantly going through dynamic changes. During cell division, many microtubules are formed, and these microtubules self-organize into three-dimensional (3D) structures such as asters and spindles. The microtubules themselves also go through dynamic growth and shrinkage. During this process, the microtubules with the desired conformation and interfacial binding survive and continue to evolve, while those without the correct interfacial binding disappear. Last, but not least, biological materials have amazing capabilities to adapt and to respond. A good example is the directed response of muscle tissues. Other examples include the ability of tissues to sense, adapt, and repair. All these critical functions are mostly absent from man-made materials.

In this chapter, examples will be given to illustrate how natural materials order on different length scales, and how the different components interact with one another, and how these materials function.

3.2. HIERARCHICAL ORDERING IN NATURAL NANOSCALE MATERIALS

3.2.1. Proteins as the Basic Building Materials of the Biological World

Proteins, composed of amino acids, are the basic building materials in the biological world.[1-3] Each amino acid contains two important functional groups, the amine group ($-NH_2$) and the carboxylic acid group (–COOH), and another functional group called the R group. The general molecular formula for the amino acid is H_2N–CHR–COOH. The R group can be a protein, a simple hydrocarbon chain, or hydrocarbon chains containing a variety of functional groups such as an amine group, a carboxylic group, and a thiol group (–SH). Depending on the specific R groups, there are about 20 different amino acids. The amino acids are connected with one another through peptide bonds (–CO–NH–) to form long-chain polypeptides.

However, in nature, proteins do not exist as simple linear polypeptide chains. They form highly organized aggregates. Not only the molecular structure, but also the high level ordering of the polypeptides plays a critical role in the properties and functions of proteins. The hierarchical structures of proteins on different length scales are shown in Fig. 3.1. The molecular structure of the

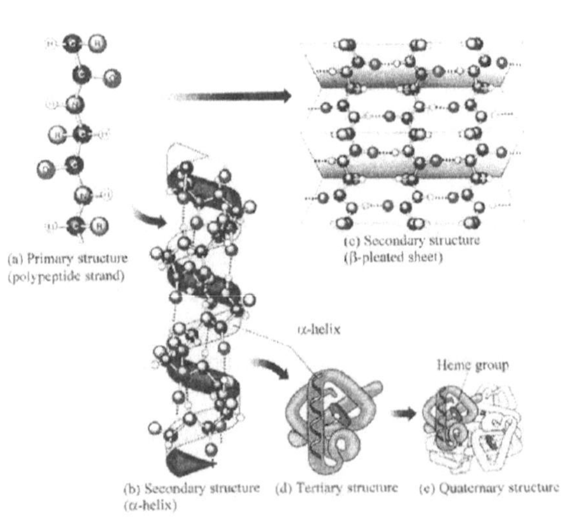

FIGURE 3.1. Hierarchical structure levels in proteins: (a) Primary structure (polypeptide strand). (b) Secondary structure of α-helix. (c) Secondary structure of β-pleated sheet. (d) Tertiary structure. (e) Quaternary structure. Reproduced Ref. 1 with permission from Benjamin/Cummings Science Publishing.

polypeptide is called the *primary structure*. The single polypeptide chain aggregates with another chain, and folds with itself, to form into a more complex *secondary structure*. The most common secondary structures are the alpha (α) helix and the beta (β) sheets. For the α-helix secondary structure, the chain is coiled by itself, and stabilized with hydrogen bonds between the NH– and CO– groups of the amino acid within the same chain. For the β-sheet, the polypeptide chains are parallel to one another, and are linked side-by-side through the hydrogen bonds between different chains forming a pleated ribbon-like structure. The α- and β-regions can fold upon one another to produce tertiary structures superimposed on the secondary structures. The tertiary structure is stabilized by both hydrogen bonds and covalent bonds in the same primary chain. When more than one polypeptide chains are aggregated together, they produce quaternary structures.

Proteins are not only the basic building materials in biology, but they also perform many different functions that are critical to the biological systems. Several of these functions will be discussed in this chapter. Even a single protein, such as a transmembrane protein, can perform several functions.[4] The transmembrane protein can regulate material transport, provide enzymatic catalytic function, recognize other proteins or cells, and attach to other surfaces or substrates. Many of these functions are related to molecular conformations. Moreover, the molecular conformations at different levels are very sensitive to the environment. Therefore, the protein molecules can respond to external and internal stimuli, and perform the required functions of the biological systems.

3.2.2. Hierarchical Structures in Bone Tissue

Bone is one of the most important mineralized connective hard tissues in higher vertebrates.[5] Bone tissue provides remarkable rigidity and strength, and at the same time retains some degree of structural elasticity and flexibility. In addition, the dynamic equilibrium of bone growth and resorbing regulates the calcium activities in the body.

Bone is made of an organic matrix that is strengthened by the deposition of a special form of calcium phosphate mineral, hydroxyapatite (HAP). The liquid phase, namely water containing various solubilized organic and inorganic species, also plays an important role in bone properties. The organic matrix primarily consists of type I collagen (95%).

Structural wise, bone tissues are remarkably diversified depending on the individual species, development age, anatomical sites, and functional needs. Although one cannot easily identify a single generalized bone structure, morphologically two forms of bones are encountered: cortical (compact) and cancellous (spongy). Cortical bones are made of organized and densely packed collagen fibers and minerals. Cancellous bones are made of a loosely packed, and disordered porous matrix. The microstructural difference between cortical bones and cancellous bones are related to their functions in the body because cortical bones provide structural support and protection, whereas cancellous bones only have metabolic functions.

Four different types of cells are involved in regulating bone growth: osteoblasts, osteoclasts, osteocytes, and bone lining cells. Osteoblasts are responsible for the production of the bone matrix. They not only produce the type I collagen and the noncollagenous proteins, but they also regulate the mineralization of the matrix, although this mechanism is not fully understood. Osteocytes are mature osteoblasts in the bone matrix, and are used for tissue maintenance. They maintain the connectiveness between different layers, and have some ability to synthesize and resorb the matrix. Osteoclasts are mainly responsible for resorbing bone minerals. The combined action of osteoblasts and osteoclasts helps maintain a healthy balance between growth and overgrowth. The role of bone lining cells is not understood yet. The bone growth activity depends on the life cycle of the individual. In the early stage (bone molding stage), bone formation exceeds bone resorbing, and these two activities are not coupled. In middle age, adult bone tissue is continuously maintained by removing and replacing (remolding). In the remolding process, the bone resorbed at a particular location will be precisely replaced by bone growth at the same location. Thus, bone growth and resorbing is coupled. In the later life cycle, bone resorbing exceeds bone growth, resulting in decreasing tissue density and strength. Bone growth and bone resorbing become uncoupled again.

Through extensive characterization using electron microscopy, X-ray diffraction-(XRD) and small angle diffraction, a large body of information on the ultrastructures of bone tissues has been obtained. Bone tissue usually contains remarkably ordered organic and inorganic materials extending through all length scales. Figure 3.2 schematically illustrates the hierarchical ordering of cortical bones. On the macroscopic scale, cortical bones are made of closely packed concentric lamellae around the vascular channels. These perimuscular structure units are called osteons or Haversian system. These lamellar structures are made of fibrils about 0.1 μm in diameter. Within each layer, all the fibrils are parallel with one another, but fibrils in adjacent layers are arranged in the perpendicular directions, as in plywood. The fibrils are made of type I collagen molecules.[6] Type I collagen molecules are defined as "structural proteins of the extracellular matrix that contain one or more domains harboring the conformation of a collagen helix."[6] Each molecule contains three polypeptide chains. These collagen molecules are spontaneously assembled into parallelly staggered fibrils due to hydrophobic interactions and static interactions through the charged amino acids on the molecules. Each molecule is 300 nm ($4.4D$, $D = 67$ nm) long. In the fibril, the molecules overlap with one another by a multiple of distance D. The ends of the molecules are also separated by a distance of about 40 nm. It is suggested that HAP crystals are initially deposited in the gap between the heads and ends of the collagen molecules.[7] The HAP are initially plate-like particles with a depth of 4–6 nm, a width of 30–45 nm, and a length of about 50 nm. They are all separated by 5 nm and are parallel to one another.[8] As the mineral grows, the particles begin to fuse with the neighboring particles forming long and broad plates. Further growth may even disturb the arrangement of the collagen fibers. In the highly calcified tissue, the minerals may have deposited both in between the collagen fibrils and outside the collagen fibrils.

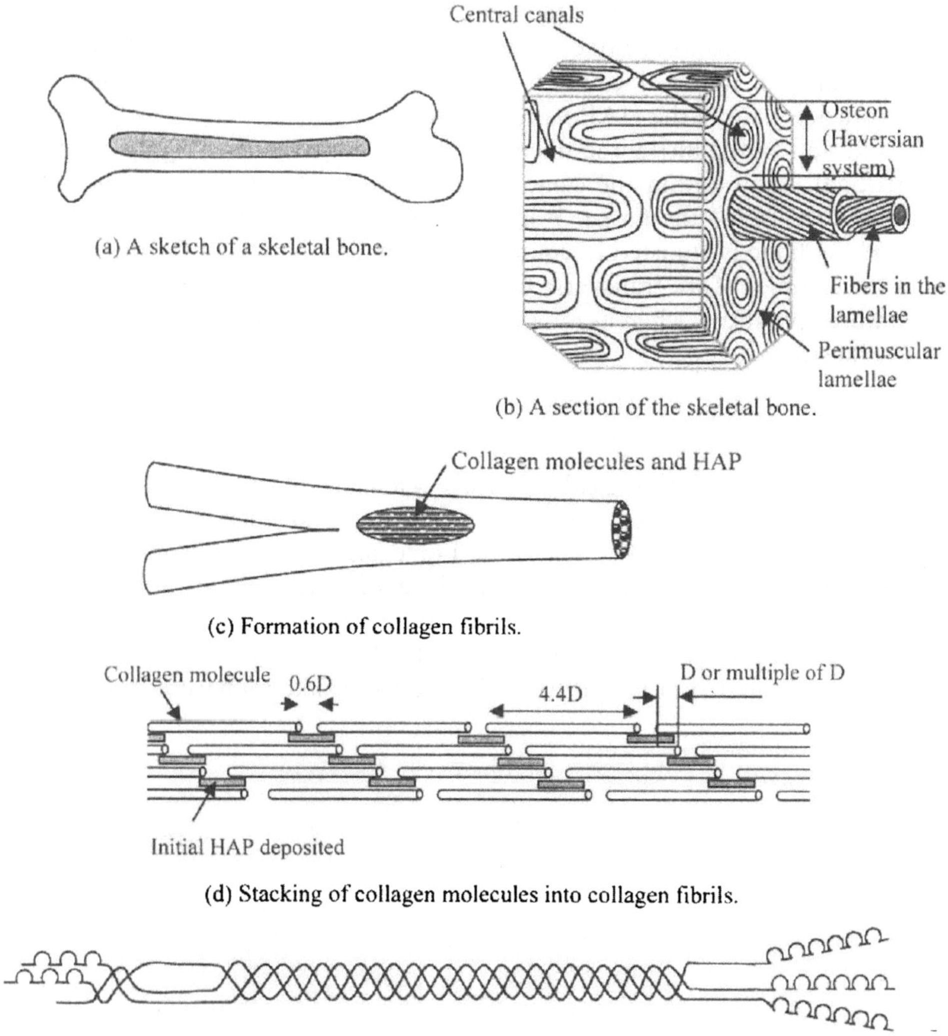

(a) A sketch of a skeletal bone.

(b) A section of the skeletal bone.

(c) Formation of collagen fibrils.

(d) Stacking of collagen molecules into collagen fibrils.

(e) Triple helix structure of a collagen molecule.

FIGURE 3.2. Hierarchical structure in bone tissue: (a) Skeletal bone. (b) A section of skeletal bone; close packed concentric lamellae structures around vascular channels. (c) Fibrils made of collagen fibers and HAP minerals. (d) Stacking of collagen molecules and HAP crystals. (e) Triple helix of peptides in a collagen molecules. This figure is produced after Refs. 5–8.

The highly organized structure of bone tissues is optimized for both compressive and tensile strengths, and for strength and toughness. The minerals provide stiffness, but the HAP material by itself is not an impressive structural material. Therefore, both the organic and inorganic phases contribute to the mechanical properties. The mechanical strength of bone tissues can be attributed to several unique aspects. First, the perivascular lamellar structure, with parallel fibrils within each layer, is adapted for maximum strength under compression and tension. The remarkable structure has yet to be matched by any man-made materials. Second, the 90-degree arrangement of the fibrils in adjacent layers is also desirable for laminated materials, a fact that has been realized by mankind only in modern times. Furthermore, bone tissue can tolerate considerable deformation without catastrophic failure due to deformation occurring on the nanometer scale, including cracking along the HAP nanoparticles.

In addition to the intriguing microstructures, bone tissue is highly adaptive and functional. Even within the same bone, the microstructure is optimized to achieve the best use of the tissue. For example, the orientation of the collagen fibrils in a long bone is adjusted to comply with the distribution of the mechanical stress in the sample. More transverse collagen fibrils are produced in the regions under compressive stress, whereas more longitudinal fibrils are produced in regions under tensile stress.[9]

3.2.3. Shell as a Composite Material

Among the natural biological materials, sea shells have attained a celebrity status because of their fascinating microstructure and relative simplicity as compared to other biological systems.[10,11] For example, the gastropod Haliotis rufescens (the red abalone) has two distinct layers in the shell: an outer layer of column-like calcite, and an inter layer of laminated aragonite. The nacreous layer,[12] also called the mother of pearl, are made of polygonal blocks of 0.5 μm in thickness, and about 15 μm in width, stacked together like miniature brick walls.

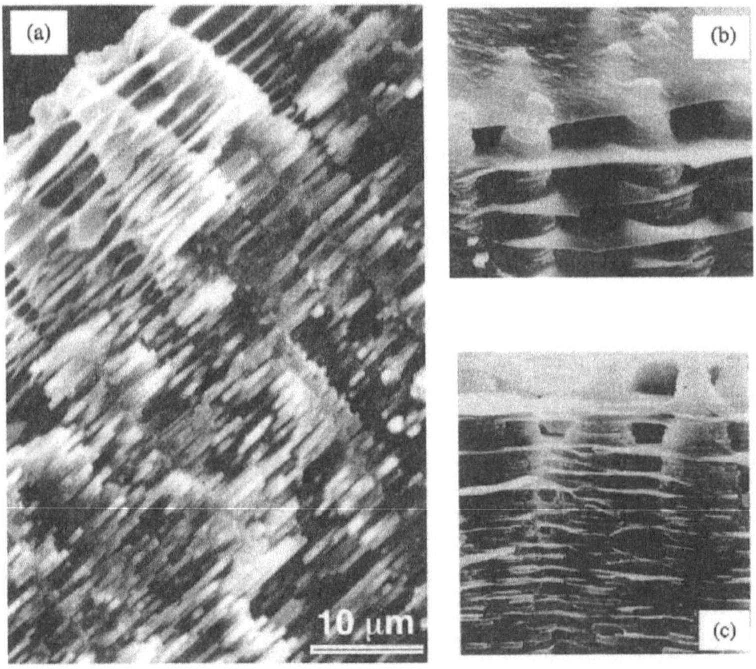

FIGURE 3.3. SEM micrographs of the nacreous layer in a red abalone shell: (a) Regions showing fully grown aragonite plates and partial grown plates near the growth tip. (b) and (c) Cross-sectional view, and tilted cross-sectional view showing the column growth patterns of the aragonite phase, and the inorganic membranes that separate the inorganic plates.

All the aragonite plates are oriented along the *c*-axis (001). Each plate is surrounded by an organic matrix of 40 nm or so. The polygonal plates are also closely packed within each plane. Altogether, the composite materials contain less than 5% of organic materials. The calcium carbonate and the organic matrix together form composite materials with amazing mechanical properties. The Young modulus is about 60 GPa, similar to typical ceramic materials.[13] However, the fracture strength is several thousand times greater than that of the aragonite mineral by itself.[14,15]

The microstructure of nacre is shown in Fig. 3.3. Near the growth tip, the aragonite plates are not completely grown, thus the organic matrix is clearly revealed. The column-like structures suggest the growth is propagated from one layer to another, growing from the center outwards. The organic membranes are probably formed first, as indicated by Fig. 3.3b and c. They act as envelops for the inorganic materials to fill.

3.3. MULTIFUNCTION OF THE ORGANIC PHASE IN BIOLOGICAL NANOSCALE MATERIALS

3.3.1. The Contribution to the Mechanical Properties

Biological materials, including bones and seashells, attain their unusual mechanical properties through many different mechanisms. They are not only strong, but also have greater tolerance for deformation and destruction without catastrophic failure. First, these materials are extremely uniform and precisely organized not only on the nanometer scale, but also on all other length scales, from angstrom and up. Second, the microstructure is optimized for the particular environment, such as the cross-perimuscular, and concentric lamellar structures in skeletal bones. In nacre, before the whole material fails, microcracks have to propagate through the nanolaminate in a zigzag fashion, which adsorbs considerable energy. Third, the organic matrix considerably improves the toughness. For cortical bones, the collagen fibrils are themselves part of the structural material. The elastic modulus of the composite is 18 GPa, between the elastic modulus of the inorganic mineral (165 GPa) and that of the collagen fiber (1.2 GPa). The HAP mineral provides the stiffness and the organic matrix provides the flexibility for toughness. In nacre, the organic phase makes for less than 5% of the total composite material. The vast improvement in the toughness (3000 times) cannot be explained by the rule of mixture. Synthetic materials containing close packed laminated ceramics with a few weight percent organic adhesive material do not exhibit a similar toughness as nacre.[16]

One possibility is that the organic matrix, even though in very small amounts, act as a glue that pulls the aragonite plates together during cracking. It is not difficult to conceive that the deformation of nacre goes through several steps. The initial phase is the elastic deformation. Beyond the elastic region, cracks are developed. These cracks are preferably nucleated between the aragonite plates, rather than in the aragonite crystals. The cracks will then propagate along the organic matrix. Due to the nanoscale arrangement of aragonite plates, many cracks can be developed and they can travel a long distance in the material without causing the whole structure to fall apart. This nanocracking, by itself, will absorb a lot of energy. As the deformation proceeds, sliding and pullout of the plates may occur. In order for the sliding and the pullout to occur, the plates must overcome the friction from the neighboring plates, as well as shearing of the organic matrix. Finally, in order for the crystals to separate from one another, they have to overcome the adhesion forces imparted by the organic matrix. Numerous studies have revealed elongated filaments of the organic matrix caused by viscous deformation around the crack region. These adhesion filaments act as bridges between the inorganic laminates and continue to hold the structure together before complete failure.

The microstructural arrangement also contributes to the unique interactions and mechanical strength. Recently, Wang *et al.*[17,18] studied the mechanical properties of nacre and found that nacre can be strained (inelastic deformation) up to 8% in shearing parallel to the plates, and more than 1% in tension. They suggested that the deformation in shear is through inter-lamellae sliding of the aragonite crystals, while tensile deformation is found through the formation of dilatation bands accompanied by the inter-lamellae sliding. When nacre deforms, a large interaction zone was formed around the notch (or crack) region. The formation of the large deformation zone suggests that nacre is able to redistribute the stress and release the stress concentration effect. On a microscopic scale, parallel bands of cracks (dilatation bands) were formed between the neighboring plates in the deformation zone (Fig. 3.4). The formation of these multitudes of cracks necessitated sliding of the plates relative to one another, which is not an easy task considering the rough interfaces between different layers and the interlocking positioning of the plates. The sliding and the deformation along the rough interfacial regions greatly increased the toughness of the material.

The organic matrix may have contributed to the toughness of the composites in a way previously not realized. Recently, AFM techniques were used to measure the adhesive force of the organic material on aragonite of nacre with a flat surface and a cantilever configuration.[19] Although these experiments may be different from single molecule measurement the resultant sawtooth force pattern of the force as a function of distance is very interesting. This sawtooth pattern, which is very desirable if the material is required to sustain very large deformation without causing complete failure, implies a multistep deformation process.

Schematically, several deformation processes have been suggested, as shown in Fig. 3.5.[17] If two aragonite plates are glued together by a stiff, short molecule that is covalently or ionicly bonded to the surface, the force will increase rapidly when the tow plates are pulled apart. When the force exceeds the strength to break such strong bonds, the material will fail. This composite material will have high mechanical strength, but low fracture toughness, which is related to the area under the force–distance curve. Such a material has very little tolerance for deformation, and will break at a very low strain (extension). If the two aragonite plates are glued together by a hypothetical inelastic polymer, which requires a higher energy to deform as the molecule is elongated, the force will increase gradually over a wide extension range until the breaking point is reached.

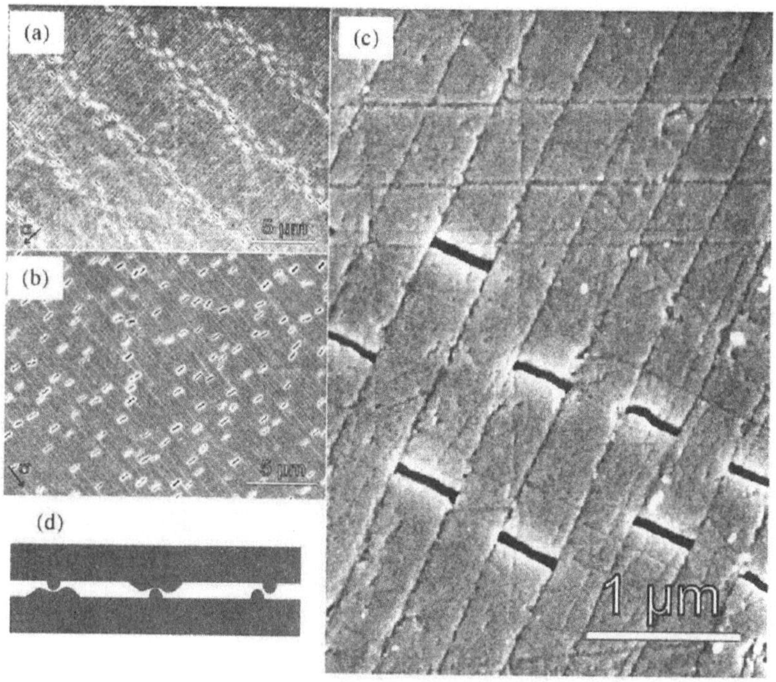

FIGURE 3.4. Toughening mechanisms in nacre: (a) Formation of dilatation bands in abalone. (b) Formation of dilatation bands in pearl oyster. (c) High magnification micrograph of abalone nacre showing the dilation bands and the rough interfacial region between the layers. (d) Resistance to interlamellae sliding due to the asperities of the rough interfaces. The photographs in this figure were provided by Prof. R. Z. Wang from University of British Columbia in Canada, with the permission of the Materials Research Society.

The force curves observed in nacre seem to be an optimization of the two extreme cases, with multiple breaking peaks. The area under the force–extension curve, representing the energy required to break the entire material, is larger than either that of the short stiff molecule or the inelastic polymer. From the illustration (Fig. 3.5), it is clear that the sawtooth curve is the most desirable. When the material is deformed to a certain limit, part of the material breaks (or relaxes), but the overall structure has not failed. Further deformation triggers the breakage (or the relaxation) of the next structural unit, and so forth. These deformation processes seem to be able to repeat through many circles.

On a molecular level, the sawtooth deformation pattern may be related to unfolding of protein domains. Similar sawtooth patterns have been observed for titin, a protein in striated muscle responsible for retaining the elasticity of the tissue, both experimentally with AFM techniques,[20] and by computer simulation.[21] Titin contains many repeating units of immunoglobulin (Ig) domains distributed along the molecule chain, as shown in Fig. 3.5c. When this protein is extended, the Ig domains unravel one by one in a sequential fashion. When the protein is stretched, one Ig domain begins to deform and the force required for the domain deformation gradually increases, until a holding force of a few hundred

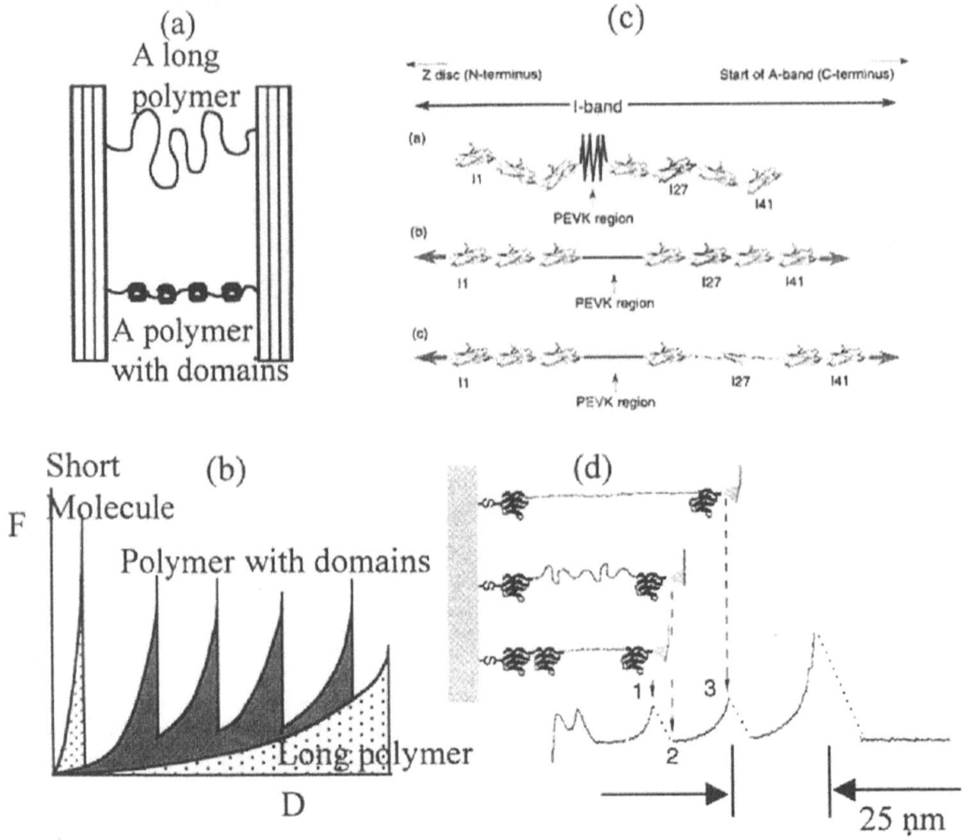

FIGURE 3.5. A possible molecular mechanism for the organic matrix to contribute to the mechanical toughness of nacre: (a) Two aragonite plates glued together by two hypothetical polymers, a long-chain polymer and a polymer with coiled domains. (b) Typical force–extension curves expected from different "glue" polymers. Only polymers with domain structures give both reasonable strength and toughness. (c) Domain structures in titin. (d) Force–extension curve observed from titin. Each peak corresponds to the unraveling of a single domain. The domains and the peaks are spaced 25 nm apart. This figure is reproduced and adapted from Refs. 17 and 18 with permission from *Science*.

pN is reached. Beyond this holding force, the Ig domain unfolds abruptly and the holding force is reduced rapidly. Continued extension triggers the deformation of another Ig domain and a new circle begins. In these experiments, the number of peaks corresponds well with the number of domains, and the peaks are spaced 25 nm apart, corresponding to the distance between the domains.

3.3.2. Molecularly Directed Self-Assembly of Inorganic Minerals

Although the detailed mechanisms differ, the nucleation and growth of organized inorganic minerals in natural composites involve three essential steps.[22] (1) *Nucleation*. In the first phase, there must be an assembly of material above certain critical mass concentration, which is similar to the final mineral to be deposited in orientation the crystalline structure. The nucleation may occur simply by aggregation of the ionic species that will make up the mineral, or may be stimulated by some molecules or aggregate of molecules (substrates) that favor the formation of a particular phase and orientation through heterogeneous nucleation. (2) *Mineral growth or expansion.* In this phase, the mineral expands along selected orientations. Organic molecules or proteins specifically adsorbed on a certain plane, may inhibit the growth on this plane. This leads to the formation of minerals with well-controlled morphology and orientation. (3) *Termination.* Finally, either due to the adsorbed molecules on the surface or through physical confinement, old crystals stop growing and new crystals nucleate in the growing front.

The organic phase, which controls the assembly of the inorganic mineral, is very complicated. This phase contains insoluble matrix, water-soluble proteins, and extracellular vesicles (or lipids). A very superficial review of the literature results suggest that the organic phase may play many roles in the formation of the organized composite materials. These roles include: (1) selectively initiating the nucleation of the desired inorganic phase, (2) controlling the orientation and growth rate of the crystals, (3) controlling the size, morphology, and ordering of the inorganic minerals, (4) terminating (inhibitating) the crystal growth of the mineral. The exact nature of the interfacial interactions between the inorganic and organic phases, and the role of the different components in the organic phase, have been the subject of extensive study and are still controversial. For example, the same organic constituent, such as certain proteins, may act as the initiator in the early stage, and as inhibitator in the later stage, but it is not known how the proteins perform the different roles. It is also known that both the molecular structure (primary structure) and the higher level ordering of the molecules are important. Early studies even suggested that the matching of the molecular ordering in the matrix or substrate with the atomic structure of the crystal is required for controlling the nucleation and growth. Although more recent studies have shown that this assumption is not necessarily true, there is no doubt that high level ordering of the organic molecules is still important.

In the nacreous layer of red abalone, at least four classes of soluble and insoluble proteins have been identified.[23] In the middle there is an insoluble, cross-linked, chitin-like protein matrix that forms organized compartments confining the inorganic minerals. The insoluble matrix is sandwiched between two layers of nucleation protein sheets with a β conformation. The insoluble matrix is organized as laminated compartments and helps maintain the long-range regularity and ordering of the material. Nucleation sheets control the nucleation of the first layer of the inorganic phase. In addition, there is a family of polyanionic proteins that plays an important role in controlling the nucleation and growth of the aragonite crystals.

Recently, both *in vivo*[24,25] and *in vitro*[26–28] studies of the growth of the red abalone have provided very interesting insights into the growth process of the minerals. These studies showed that at first, a sheet of nucleating proteins was deposited on the insoluble matrix. Initially a prime layer of calcite crystals, 10–20 μm in thickness, was grown on the nucleating proteins along the (104) orientation. The existence of these nucleating proteins is beneficial for the nucleation process. Although calcite would also be deposited on a substrate without nucleating proteins, the nucleation density was lower, and the crystals were less oriented. After the prime calcite layer was formed, a sudden phase transition was trigged. The new crystalline phase was grown on the initial calcite crystals as overgrowth and depended on the crystal-specific proteins in the solution. There were two types of water soluble polyanionic proteins, those purified from the nacreous (aragonite) region of the shell, and those from the calcite region of the shell. The aragonite proteins and calcite proteins have different molecular weights, and different charge distributions on the molecules. The presence of the aragonite proteins would favor the formation of aragonite, and the presence of the calcite proteins (or the absence of any soluble proteins) would favor the formation of calcite. In the presence of a mixture of aragonite and calcite proteins, aragonite would be the main product. If the aragonite crystals were allowed to grow after the soluble proteins were consumed, calcite overgrowth would occur on the aragonite crystals.

The *in vitro* study suggests that the soluble proteins play an even more important role than previously suggested.[29] These proteins not only control the specific mineral phase, but they may also regulate the high order morphology. The precise shape and arrangement of the aragonite plates in nacre might have been facilitated by both the pre-organized organic matrix and the soluble proteins in the solution.

In bone tissues, the Type I collagen makes up about 95% of the organic matrix. The rest includes numerous noncollagen proteins. Some studies have suggested that the HAP minerals were nucleated in the space between the collagen molecules. The platelets of HAP are in the nanometer range, and are parallel to one another. As the crystals grow, the width and length of the crystals increase, but the thickness (depth) remains unchanged. As a result, the HAP platelets fuse together to form large sheets interconnected with a perimuscular lamellar fashion. With the intercalation of more minerals and the intergrowth of the minerals, both the HAP minerals and the organic matrix become more disordered. However, the detailed mechanism of bone mineralization is much less understood as compared with sea shells. Some of the critical issues debated now include the initial sites of nucleation and growth, and the role of collagen and noncollagen proteins.

In the literature, evidence has been reported that the nucleation is associated with the extracellular matrix vesicles (phospholipids) and with the collagen fibers.[30] The initial nucleation process may not be unique, and may occur through different pathways. In addition to collagen proteins, noncollagen proteins can also have an effect on the nucleation, growth, and the inhibition process.[31] The role of noncollagen proteins my be similar to the role of soluble proteins in sea shells, as discussed earlier.

3.3.3. Sensing, Motion, and Response

3.3.3.1. Protein Conformation Change and Its Implications for Responsive Materials

As mentioned earlier, the conformational change (folding and unfolding) of protein molecules is a very important phenomenon. Most of the biological functions are in some way related to protein conformation change. Recently, there has been intensive interest in how protein conformational change is transformed into mechanical energy on the molecular level.[32] The action of multidomained protein molecules, as rotary motors[33] or linear motors,[34] has been studied. In some cases, nearly 100% efficiency of energy conversion is reported.[35]

In the last decade or so, direct conformational change in protein molecules, namely elastic proteins (elastins), has been extensively studied as a powerful mechanism leading to useful responsive polymeric materials.[36,37] These elastic protein polymers are found in lungs, skins, arteries, and ligaments.

Typically, a protein molecule contains both hydrophilic (polar) and hydrophobic (apolar) segments. For example, –COOH group, in the protonated from, is hydrophobic, while –COO^-, in the ionized form, is hydrophilic. One of the important driving forces for the protein to transform from the disordered, unfolded to the folded ordered states is the entropy change associated with the hydrophilic and hydrophobic interactions. Below a critical temperature T, the hydrophobic groups do not interact strongly with the water molecules that penetrate the protein chain. The water molecules near the hydrophobic groups associate with one another to form structured water like ice. These molecules are arranged to form hollow balls made of pentagons. The structured water layer near the hydrophobic group is quite stable and keeps the protein from folding, because protein folding requires energy to break the pentagonal water structure. As the temperature is increased beyond the critical temperature, the water structure around the protein molecules becomes disordered. The protein chains can easily fold without breaking the pentagonal water structure that would otherwise exist at a lower temperature.

Therefore, by increasing the temperature, the water molecules becomes more disordered, but the protein chains become more ordered. The net entropy change is still positive, not violating the thermodynamic law. The transition from an unfolded state to a folded state upon an increase in temperature is called the ΔT transition. On the macroscopic scale, the change from an unfolded state to a folded state represents a large change in molecular dimension. This change in molecular dimension can be used to perform mechanical work. When the temperature changes, biomolecules can lift up to 1000 times their own dry weight.

The transition temperature T_t in protein polymers ranges from 0°C to 80°C. The exact transition temperature depends on the balance of the hydrophilic and hydrophobic groups. Hydrophilic groups on the polymer tend to exert a strong pull on the water molecules and, therefore, reduce the tendency for the water molecules to aggregate into pentagonal structures around the hydrophobic groups. Consequently, the transition temperature will be increased when more hydrophilic groups are added to the polymer because the water molecules are already disrupted and the temperature effect on the structured water around the hydrophobic groups is reduced. At the same time, adding hydrophilic groups also reduces the amount of energy needed to induce folding because the number of structured water molecules is reduced through the pulling of the hydrophilic groups. On the other hand, adding hydrophobic groups to the polymer reduces the transition temperature and increases the amount of energy to induce folding. This is due

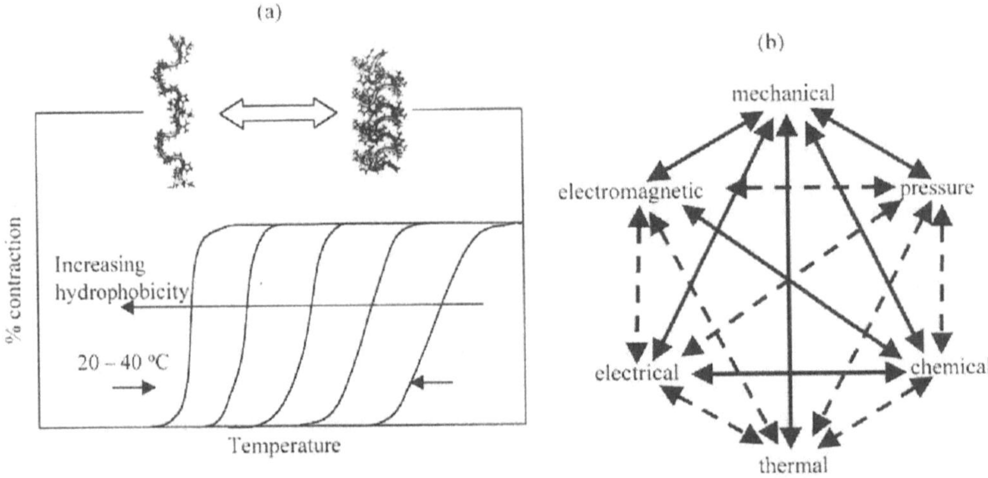

FIGURE 3.6. Generalized mechanism of protein polymer contraction: (a) Protein conformation change and the corresponding contraction curves as a function of temperature. The protein changes from unfolded state to folded state when the temperature is increased over a certain range. Increasing the hydrophobicity decreases the transition temperature. (b) The ΔT transition can be induced by a wide range of external stimuli. This phenomenon can be used to convert one energy form to another, such as thermal energy to mechanical energy (temperature change), and chemical energy to mechanical energy (pH change), and electrical energy to mechanical energy (applying external electrical field) (adapted from Ref. 37 with permission from *J. Phys. Chem.*).

to an increase of the number of structured water molecules around the hydrophobic groups.

Beside temperature, many other factors can induce the order–disorder transition. For example, when pH changes, the organic acid group can be either ionized or deionized. At a high pH, charged carboxylate groups are hydrophilic. At a low pH, an uncharged carboxyl group is hydrophobic. Therefore, a polymer containing this group can change from unfolded state at a high pH to a folded state at a low pH. Similarly, by adding other functional groups that are sensitive to an electrical field, pressure, etc., the biopolymers can be made sensitive to a wide range of external stimuli. Using this mechanism, many different energy forms can be converted to mechanical energy, and vise versa, as shown in Fig. 3.6.

3.3.3.2. Motor Proteins, Filaments, and Microtubules: Mechanics of Motion in Biology

Motor proteins, filaments, and microtubules are special proteins in biological systems responsible for many critical functions. On a molecular level, these proteins are critical for chemical–mechanical energy conversion, signal transfer, cell repair, and division. On a higher level, they are responsible for many critical biological and physiological functions such as body movements. On a more fundamental level, these proteins play an important role in maintaining the stability of biological systems. Thermodynamically, biological systems are not in the most stable states. The ability of molecular machines to harness the chemical energy and distribute and organize materials is key in maintaining the functional integrity of the system.

Actin filaments[38] (Fig. 3.7a) are linear protein polymers existing in a two stranded, right hand double helix structure. The individual strand (protofilament) contains globular protein domains about 5.5 nm in size. These globular proteins are packed through van der Waals forces, hydrophobic and ionic interactions along the actin filament. The two strands wrap around each other with a periodicity of 72 nm ($2 \times 36\,nm^2$). One full period contains 26 subunits (actin protein monomers), and makes 12 complete turns. The diameter is about 8 nm wide and the length can reach a few hundred micrometers. These actin filaments are polarized with plus and minus ends.

Microtubules[36] are also long protein polymers. The basic building blocks for microtubules are dimmers of α- and β-tubulin protein molecules that are bound together through non-covalent bonding. The tubulin subunits then stack together

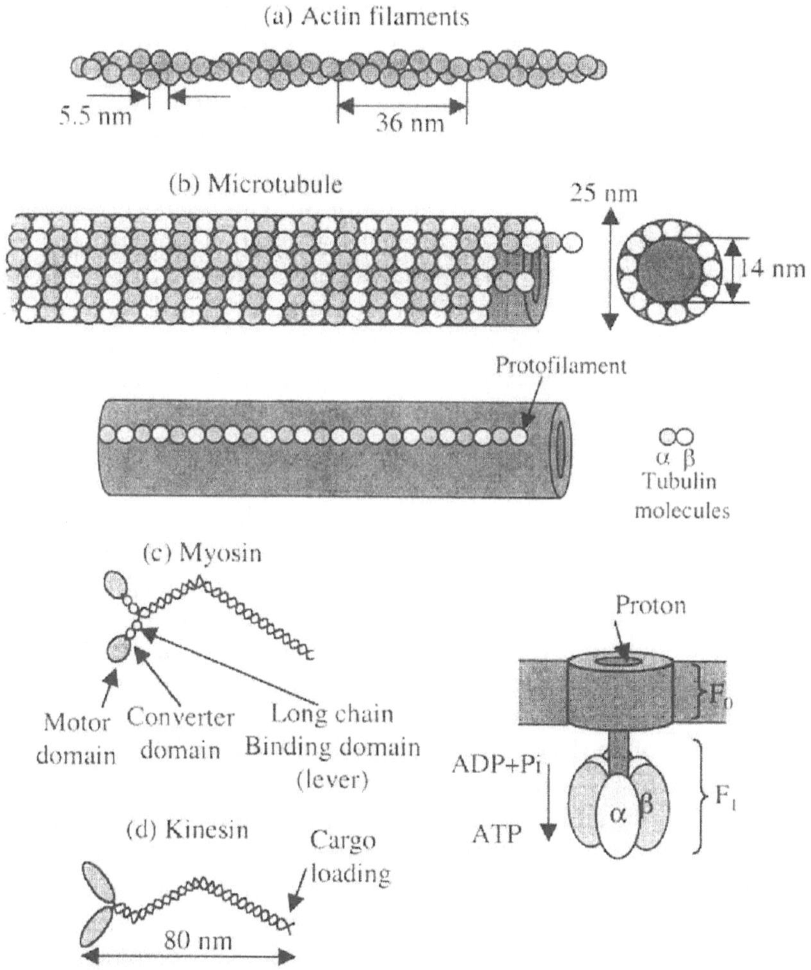

FIGURE 3.7. Schematic structures of motor proteins and microtubules: (a) Actin filaments. (b) Microtubules. (c) Myosin motor protein. (d) Kinesin motor protein. (e) F1-ATPase rotary protein.

through non-covalent bonding to form the hollow cylindrical microtubules. The microtubule structure can be regarded as 13 parallel protofilaments wrapped around a tube, each protofilament made of alternative α and β tubulins. The protofilament always start with an α-tubulin and ends with a β-tubulin, with the β-tubulin end growing faster than the α-tubulin end. Because of the asymmetric arrangement of the dimmers, the whole microtubule structure is highly polarized with the β-tubulin end as the plus end. The microtubule can also be regarded as a three stranded helix structure, with a light offset between the neighboring protofilaments. Microtubules normally have an external diameter of 25 nm, and an internal diameter of 14 nm.

Motor proteins are protein molecules that can direct movement though conformational change utilizing adenosine triphosphate (ATP) hydrolysis. An ATP molecule contains three negatively charged phosphate groups as the side chain. These highly charged side groups hydrolyze to produce adenosine diphosphate (ADP) and an inorganic phosphate (Pi). The energy produced during the hydrolysis is utilized by motor proteins to produce motion. In the last ten years, the family of motor proteins discovered has expanded exponentially, but the functions of these proteins can be represented by three different types of motor proteins:[39] the kinesin motor proteins that can transport over a long distance, the myosin proteins that produce short distance sliding, and the rotary motors such as F_1-ATPase that produce rotation.

The detailed molecular structure of the motor proteins are fairly complicated, but studies in the last few years have begun to shed light on the relationship between the molecular structures and the motility generated by these proteins. In particular, comparison of the structures between myosin and kinesin has led to better understanding of the function of the proteins. For example, myosin proteins contain two motor domains. The head area of the myosin protein contains several functional subdomains (Fig. 3.7c):[40] the motor domain, the converter domain, and the light-chain binding domain (lever). Each domain may consist of several α-helix strands and β-sheets. The motor domain portion contains an actin binding site and a nucleotide binding pocket. The motility of myosin may have originated from two different states in the head area: the nucleotide bound state and the nucleotide free state. In the nucleotide bound state ADP–Pi is bound to the nucleotide binding pocket, and the light-chain binding domain is tilted towards the minus end of the actin filament. In the nucleotide free state, the dissociate of Pi causes a 70° rotation of the converter domain. The rotation is amplified by the swing of the light-chain binding domain to up to 10 nm displacement.

Kinesin motor protein (Fig. 3.7d),[38] which are about half the size of myosin, is made of two identical kinesin heavy chains that entangle into a long-chain molecule 80 nm in length, and two light chains with carboxylic terminal groups that function as the cargo binding sites. The force generating device is made of two globular protein regions ($4 \times 7\,nm^2$ each). Although the amino sequence of the proteins is quite different from that of myosin, the molecular structure of the motor domain is very similar to that of myosin. The head area contains essentially similar packing of α-helix strands and β-sheets with a nucleotide binding pocket. This suggests that the origin of motility might be the same as myosin. However, the long light-chain binding domain is absent. The absence of a lever mechanism implies kinesin needs to employ additional means to generate long distance movement, which is accomplished by using two motor domains at the same time, not just one motor domain as in the case of myosin.

Motility is realized with both motor proteins and protein filaments such as microtubules and actin filaments. Essentially, the motor proteins can be regarded as molecular trains that move along molecular rail tracks of filaments and microtubes. Figure 3.8 illustrates how the motor motility is achieved.

Although the molecular structure and the origin of motility of myosin and kinesin are similar, the physical movement generated on the nanometer scale is very different. Myosin and actin are the most important protein constituents in skeletal muscle, and are responsible for the extension and the contraction of muscle. Only one of the two heads is involved in the action. The myosin sliding over actin filament involves the following sequence (Fig. 3.8a):[41,42] (i) In the initial ADP–Pi bound state, the myosin head approaches the binding site on the actin filament. (ii) When the myosin head docks on the binding site, the binding triggers the release of the phosphate. (iii) The release of the phosphate causes the rotation of the converter domain of the head region, which in turn is translated into up to 10 nm lateral movement of the actin fiber. (iv) After completing the movement, ADP dissociates and ATP binds to the head. At the same time, the head group binding to actin is weakened. Finally, hydrolysis of ATP to ADT–Pi returns the head group to the original state. Unlike myosin, kinesin utilizes both heads to complete the motion because it does not have the lever mechanism as in

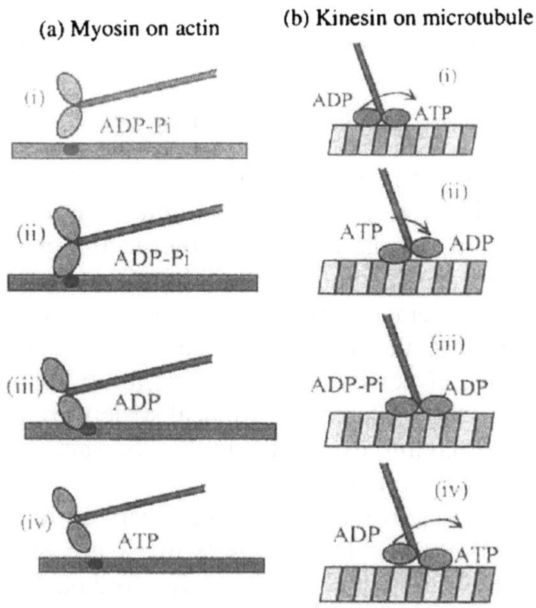

FIGURE 3.8. Motor protein motility: (a) Myosin on actin filament. (i) Initial state. (ii) Binding to the binding sites on actin. (iii) Release of the nucleotide (Pi) and rotation of one motor domain to generate a sliding of the actin filament. (iv) ATP binding and detachment from the actin filament. (b) Kinesin on microtubule. Redrawn after Ref. 41.

the myosin. The two heads "walk" in a coordinated and processional manner along the track, binding alternatively to the tubulin molecules. As shown in Fig. 3.8b, the docking of the forward ATP bound head causes the ADP bound partner head in the back to be thrown forward by 16 nm in the forward direction. After random diffusion and searching, the new leading head docks to the binding site on the tubulin. Therefore, in each cycle, the kinesin moves 8 nm on the track. The binding of the kinesin head also causes the ADP to release from the new leading group. At the same time, the ATP molecule on the other head group hydrolyzes into ADP–Pi. Finally, after ADP release from the leading head, ATP binds to the same group. The phosphate is released from the trailing group, which is ready to be thrown forward again.

Microtubules are also critical components of cells as cytoskeletons. These semi-rigid polymers, and the interactions with motor proteins play many roles in biology:[43]

1. They are responsible for the structural integrity of the cells. They maintain the structural stability of cell when subject to external stress and attacks. They influence the shapes of the cell and its transformation.
2. They are responsible for inter- and intra-cellular transport. The arrangement of matter within the cell is not determined by random diffusion. They are arranged through the use of motor protein that moves along the microtracks as determined by the functional needs of the cell.
3. The microtubules, filaments, and motor proteins play an important role in cell growth and cell division.
4. They are responsible for the directed cell movement.
5. They are critical in the signal transfer in neural systems.

3.3.3.3. Nanoscale Structure and Response in Muscle

Closely associated with cortical bones is skeletal muscle.[44,45] Skeletal muscle is also called striated muscle, voluntary muscle because of the characteristic striations in the microstructure, and its ability to subject themselves to conscious command of the body. In general, muscle performs four basic functions: producing movement, maintaining posture, stabilizing joints, and generates heat. Contrary to the bone hard tissues, muscle is a special tissue that exhibits very high degree of flexibility and contractibility. Similar to bone tissues, skeletal muscles have highly ordered hierarchical structures, but the ability of the muscle to respond (internal and external stimuli), contract, extend, and recover is different from that of bone tissue.

Human skeletal muscle is made of muscle fibers up to 30 cm in length, and with a diameter of 10–100 μm. These muscle fibers are bundled together. The bundles are wrapped by a layer of connective tissue and packed into a hierarchical arrangement. Each muscle fiber contains parallel fine fibrous structures called myofibrils. The myofibrils are about 0.1–2 μm in diameter. When viewed under high magnification electron microscopes, the myofibrils contain two kinds of filaments: thick filaments made of bundles of myosin proteins (M, about 30 nm in length, and 2.5 nm in diameter), and thin filaments made of actin (about 160 nm in length, and 3.2 nm in diameter) and myosin. The actin and myosin filaments are organized in parallel patterns as shown in Fig. 3.9a, in which the thick myosin filaments form the A-band. The space between the myosin filaments is called the I-band, and the gaps between the actin filaments is called the H-band. Under an electron microscope, a dark line can be seen in the middle of the I band. This dark line is called the Z-line (Z-disc). In the Z-plane, a coin-shaped giant protein molecule (not shown) anchored to actin connects all the fibrils throughout the width of the muscle cells. The Z-line, with a spacing of about 2 μm, defines the basic muscle unit (the sarcomere). Not shown in Fig. 3.9 is another protein called titin (connectin). Titin is an elastic filament that connects to both actin and myosin, and is thought to be responsible for the muscle's ability to spring back after the stretching force is removed.

The muscle response and contraction originate from the molecular chemical reaction and nanometer scale conformation change through a sliding mechanism. The actin is negatively charged and the myosin is positively charged. Actin can bind to myosin through cross-bridge binding. The binding sites on actin can assume three states: (1) blocked state (unable to bind to cross-bridge, namely

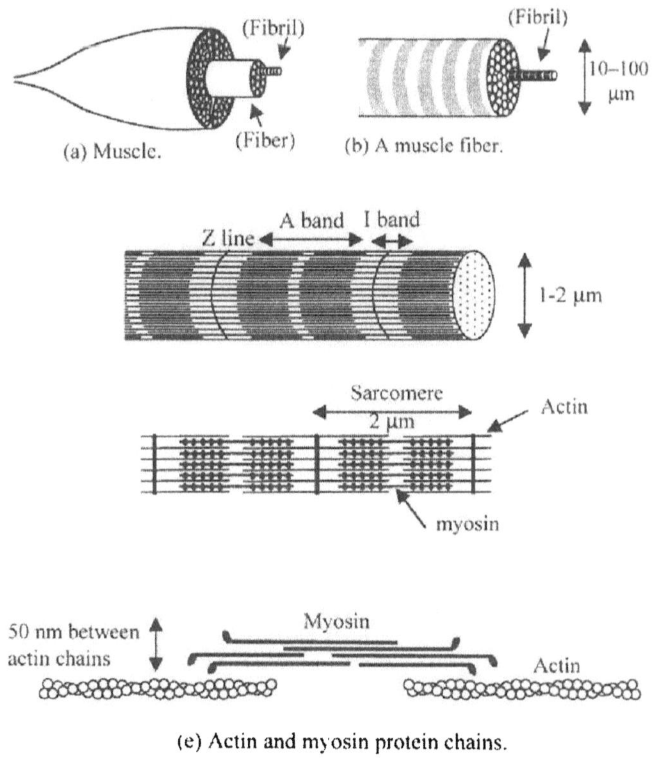

FIGURE 3.9. Hierarchically ordered structures in skeletal muscles. (a) Muscle organ. (b) A single muscle fiber. (c) Myofibrils in a fiber. (d) Arrangements of actin and myosin in myofibrils. (e) Molecular chains of myosin and actin (only one head of the myosin motor domains is shown. Redrawn after Refs. 44 and 45.

the heads of myosin), (2) closed state (able to weakly bind to cross-bridge) and (3) open state (able to strongly bind to cross-bridge). In the blocked state, the binding sites on the actin to the cross-bridge of myosin are blocked by a protein called tropomyosin. The energy source comes from the free energy of hydrolysis of ATP molecules. ATP contains three negatively charged phosphate groups as the side chain. These highly charged side groups are not stable and easily react with water (hydrolyze) to produce ADP, and an inorganic phosphate (Pi). This chemical reaction produces the energy required for the muscle contraction. In general, muscle contraction involves several chemical and physical changes.[46] (1) When the nerve impulse reaches the muscle cells, the calcium channels in the sarcomere membranes open, allowing calcium ions to enter from

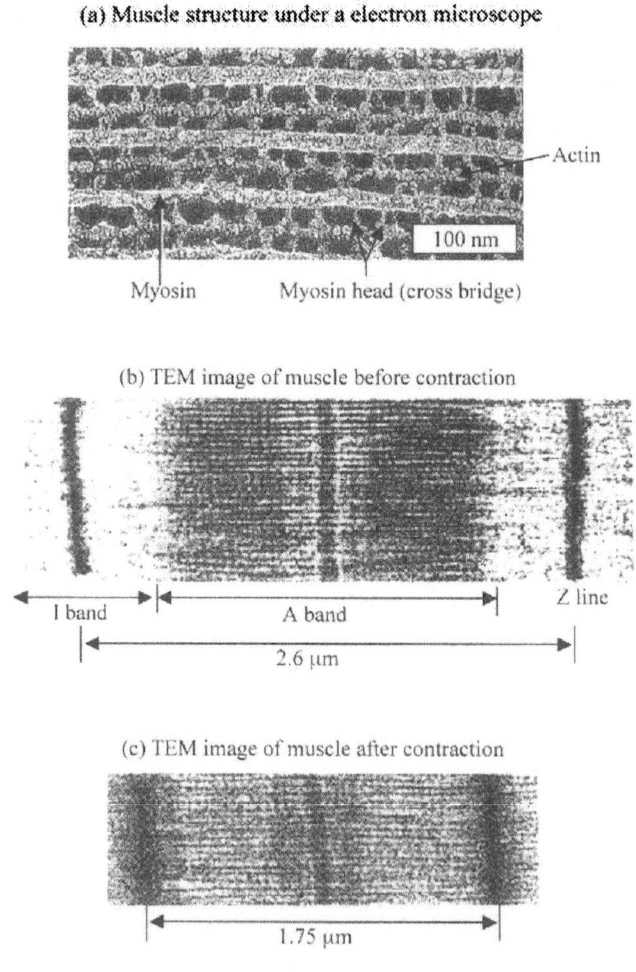

FIGURE 3.10. Electron micrographs of muscle structures: (a) An electron microscopy image showing the actin and the myosin filaments, as well as the cross-bridge heads. (b) A TEM image showing the filament and band structure before contraction. (c) The filament and the band structure after contraction (TEM image). Reproduced from Ref. 47 with permission from Prof. John Heuser and Dr. H. E. Huxley.

the extracellular fluids. This causes an increase of calcium ion concentration in the muscle cytoplasm from less than 10^{-7} to about 10^{-5} M. (2) The calcium ions bind to a protein molecule on actin named troponin, and displaces the tropomyosin that blocks the binding sites on actin, therefore opening the binding sites. At this stage, binding sites on actin are open. (3) M–ADP–Pi (ADP is the hydrolysis product of ATP, Pi is the inorganic phosphate produced from the same hydrolysis reaction) formed during the precious contraction cycle binds to the binding sites through the cross-bridge (the heads) on myosin. (4) The myosin heads bend to pull the thin filament and perform the work, and at the same time release ADP and Pi. (5) A small number of strongly attached cross-bridges activate additional myosin binding and isomerization to form more strongly bound cross-bridges. (6) New ATP attaches to the myosin head and the cross-bridge detaches. (7) ATP hydrolyzes to ADP and Pi, providing the energy needed to return the myosin heads to the upright high energy position, and readying the muscle for the next contraction cycles. The end result of the cross-bridge sliding is the contraction of muscles as shown in Fig. 3.10.[47]

REFERENCES

1. B. Alberts, D. Bray, A. Johnson, J. Lewis, M. Raff, K. Roberts, and P. Walter, Protein structures and functions, in: *Essential Cell Biology* (Garland Publishing Inc., New York and London, 1998), Chapter 5, pp. 133–182.
2. H. Lodish, D. Baltimore, A. Berk, S. L. Zipursky, P. Matsudaria, and J. Darnell, *Molecular Cell Biology* (Scientific American Books, New York, 1995).
3. E. N. Marrieb, Chemistry comes alive, in: *Human Anatomy and Physiology* (Benjamin/Cummings Science Publishing, Menlo Part, 1998), Chapter 2, pp. 47–50.
4. E. N. Marrieb, Cells: The living units, in: *Human Anatomy and Physiology* (Benjamin/Cummings Science Publishing, Menlo Part, 1998), Chapter 3, p. 66.
5. S. C. Marks Jr. and D. C. Hermey, The structure and development of bone, in: *Principles of Bone Biology*, edited by J. P. Bilezikian, L. G. Raisz, and G. A. Rodan (Academic Press, San Diego, 1996), Chapter 1, pp. 3–14.
6. J. Rossert and B. de Crombrugghe, Type I collagen: structure, synthesis, and regulation, in: *Principles of Bone Biology*, edited by J. P. Bilezikian, L. G. Raisz, and G. A. Rodan (Academic Press, San Diego, 1996), Chapter 10, pp. 127–142.
7. W. J. Landis, K. J. Hodgens, J. Arena, M. J. Song, and B. F. McEwen, Structure relationship between collagen and mineral in bone as determined by high voltage electron microscopic tomography, *Microsc. Res. Tech.* **33**, 192–202 (1996).
8. J. D. Currey, Biocomposites: Micromechanics of biological hard tissues, *Curr. Opin. Solid State Mater. Sci.* **1**(3), 440–445 (1996).
9. P. Bianco, Structure and mineralization of bone, in: *Calcification in Biological Systems*, edited by E. Bonucci (CRC Press, Boca Taton, 1992), Chapter 11, pp. 243–268.
10. K. M. Wilbur and G. Owen, Growth, in: *Physiology of Mollusca*, edited by K. M. Wilbur and C. M. Yonge (Academic Press, New York, 1964), Chapter 8, pp. 243–282.
11. K. M. Wilbur, Shell formation and regeneration, in: *Physiology of Mollusca*, edited by K. M. Wilbur and C. M. Yonge (Academic Press, New York, 1964), Chapter 7, pp. 211–242.
12. G. Grégoire, Ultrastructure of the nautilus shell, in *Nautilus, The Biology and the Paleobiology of a Living Fossil*, edited by W. B. Saunders, and N. H. Landman (Plenum Press, New York, 1987), pp. 463–486.
13. A. P. Kackson, J. F. V. Vincent, and R. M. Turner, The mechanical design of nacre, *Proc. R. Soc. London, Ser. B* **234**, 415–440 (1998).

14. J. D. Currey, Mechanical properties of mollusc shell, in: *The Mechanical Properties of Biological Materials*, Vol. XXIV, edited by J. F. V. Vincent and J. D. Currey (Cambridge University Press, Cambridge, 1980), pp. 75–97.
15. A. G. Evans and D. A. Marshall, The mechanical behavior of the ceramic matrix composites, *Acta. Met.* **37**, 2567–2583 (1989).
16. N. Almqvist, N. H. Thomson, B. L. Smith, G. D. Stucky, D. E. Morse, and P. K. Hansma, Methods for fabricating and characterizing a new generation of biomimetic materials, *Mater. Sci. Eng.* C **7**(1), 37–34 (1999).
17. R. Z. Wang, Z. Suo, A. G. Evans, N. Yao, and I. A. Aksay, Deformation mechanisms in nacre, *J. Mater. Res.* **16**, 2485–2493 (2001).
18. A. G. Evans, Z. Suo, R. Z. Wang, I. A. Aksay, M. Y. He, and J. W. Hutchinson, A model for the robust mechanical behavior of nacre, *J. Mater. Res.* **16**, 2475–2484 (2001).
19. B. L. Smith, T. E. Schaffer, M. Viani, J. B. Thompson, N. A. Frederick, J. Kindt, A. Belcher, G. D. Stucky, D. E. Morse, and P. K. Hansma, Molecular mechanistic origin of the toughness of natural adhesives, fibers and composites, *Nature* **399**, 761–763 (1999).
20. M. Rief, M. Cautel, F. Oesterhelt, J. M. Fernandez, and H. E. Guab, Reversible unfolding of individual titin immunoglobulin domain by AFM, *Science* **276**(5315), 1295–1297 (1997).
21. H. Lu, B. Isralewitz, A. Krammer, V. Vogel, and K. Schulten, Unfolding of titin immunoglobulin domains by steered molecular dynamics simulation, *Biophys. J.* **75**(2), 667–671 (1998).
22. A. L. Boskey, Biomineralization: Conflicts, challenges, and opportunities, *J. Cell. Biochem.* **30/31**, 83–91 (1998).
23. X. Shen, A. M. belcher, P. K. Hansma, G. Stucky, and D. E. Morse, Molecular cloning and characterization of lustrin A, a matrix protein from shell and pearl nacre of haliotis rufescens, *J. Biochem.* **272**(51), 32472–32481 (1997).
24. M. Fritz, Am. M. Belcher, M. Radmacher, D. A Walters, P. K. Hansma, G. D. Stucky, D. E. Morse, and S. Mann, Flat pearls from biofabrication of organized composites on inorganic substrates, *Nature* **371**, 49–51 (1994).
25. C. M. Zaremba, A. M. Blcher, M. Fritz, Y. L. Li, S. Mann, P. K. Hasma, D. E. Morse, J. S. Speck, and G. D. Stucky, Critical transitions in the biofabrication of abalone shells and flat pearls, *Chem. Mater.* **8**, 679–690 (1996).
26. A. M. Belcher, P. K. Hansma, G. D. Stucky, and D. E. Morse, First steps in harnessing the potential of biomineralization as a route to new high-performance composite materials, *Acta Mater.* **46**(3), 733–336 (1998).
27. M. Fritz and D. E. Morse, The formation of highly organized polymer/ceramic composite materials: The high-performance microaluminate of the molluscan nacre, *Curr. Opin. Colloid. Interface Sci.* **3**(1), 55–62 (1998).
28. A. M. Belcher, X. H. Wu, R. J. Christensen, P. K. Hansma, G. D. Stucky, and D. E. Morse, Control of crystal phase switching and orientation by soluble mollusc-shell proteins, *Nature* **381**(2), 56–58 (1996).
29. G. Falini, S. Albeck, S. Weiner, and L. Addadi, Control of aragonite or calcite polymorphism by mollusk shell macromolecules, *Science* **271**(5245), 67 (1996).
30. M. Golber and A. L. Boskey, Lipids and biomineralizations, *Prog. Histochem. Cytochem.* **31**, 1–187 (1997).
31. A. L. Boskey, Will biomimetics provide new answers to old problems of calcified tissues, *Calcified Tissue Int.* **63**, 179–182 (1998).
32. A. D. Metha, M. Rief, J. A. Spudich, D. A. Smith, and R. M. Simmons, Single-molecular biomechanics with optical methods, *Science* **283**(5408), 1689 (2000).
33. H. Noji, R. Yasuda, M. Yoshida, and K. Kinosita Jr., Direct observation of the rotation of F-1-ATPase, *Nature* **386**, 299–302 (1997).
34. R. Simmons, Molecular motors, single-molecule mechanics, *Curr. Biol.* **6**, 392–394 (1996).
35. R. Yasuda, H. Noji, K. Kinsita Jr., and M. Yashida, F-1-ATPase is a highly efficient molecular motor that rotates with discrete 120 degrees steps, *Cell* **93**, 1117 (1998).
36. D. W. Urry, Elastic biomolecular machines, *Sci. Am.* 64–69 (1995).
37. D. W. Urry, Physical chemistry of biological free energy transduction as demonstrated by elastic protein-based polymers, *J. Phys. Chem. B.* **101**, 11007–11028 (1997).

38. J. Howard, Structure of motor proteins, in: *Mechanics of Motor Proteins and Cytoskeleton*, Chapter 12 (Sinauer Associates Inc., Sunderland, Massachusetts. 2001), pp. 197–212.
39. A. D. Metha, M. Rief, J. A. Spudich, D. A. Smith, and R. M. Simmons, Single-molecule biomechanics with optical methods, *Science* **283**, 1689–1695 (1999).
40. J. Howard, Structures of cytoskeleton filaments, in: *Mechanics of Motor Proteins and Cytoskeleton*, Chapter 7 (Sinauer Associates Inc., Sunderland, Massachusetts, 2001), pp. 121–134.
41. R. D. Vale and R. A. Milligan, The way things move: Looking under the hood of molecular motor proteins, *Science* **288**, 88–95 (2000).
42. J. Howard, Molecular motors: Structural adaptations to cellular functions, *Nature* **389**, 561–567 (1997).
43. S. Leibler, Collective phenomena in motosis: A physicists' perspective, in: *Physics of Biomaterials: Fluctuation, Selfassembly, and Evolution*, edited by T. Riste and D. Scherrington (Kluwer Academic Publishers, Dordecht, Netherlands, 1996), pp. 135–151.
44. W. Hughes, The behavior of striated muscles, in: *Aspects of Biophysics* (John Wiley 298.
45. A. G. Lowe, Energetics of muscle contraction in: *Biochemical Thermodynamics*, edited by M. N. Jones (Elsevier Scientific Publication Co. 1979), Chapter 10, pp. 308–332.
46. A. M. Gordon, E. Homsher, and M. Regnier, Regulation of contraction in striated muscle, *Physiol. Rev.* **80**(2), 853–924 (2000).
47. E. N. Marrieb, Chemistry comes alive, in: *Human Anatomy and Physiology* (Benjamin/Cummings Science Publishing, Menlo Part, 1998), Chapter 9, pp. 261–301.

4

Nanocrystal Self-Assembly

4.1. NANOCRYSTALS

A nanocrystal is a crystalline entity that in some cases has specific shape and specific number of atoms. The unique chemical and physical properties of nanocrystals are determined not only by the large portion of surface atoms, but also by the crystallographic structure of the particle surface. The former is determined by the size of the particles, and the latter relies on particle shape. One typical example is that the melting temperature of nanocrystals strongly depends on the crystal size and is substantially lower than the bulk melting temperature. The melting point of Au particles of core sizes 2.5 nm, for example, is ~40% lower than that of bulk gold.[1] Similar behavior has been observed for sodium clusters[2] and CdS nanocrystals.[3] The other interesting property is the quantum confinement effect in small size metallic and semiconductor quantum dots (QDs).[4,5] The shift of electron energy levels as a function of particle size gives rise to emission of photons with unique wavelengths.

In nanomaterials, the effect from the surface is comparable in some cases to the chemical composition in influencing the chemical, electronic, magnetic, and optical behaviors. A first-order approximation of a particle shape is spherical. The ratio between the numbers of surface atoms (N_s) to the volume atoms (N_v) is

$$\frac{N_s}{N_v} = \frac{3\rho_s}{r\rho_v} \tag{4.1}$$

where ρ_s and ρ_v are the densities of surface atoms and volume atoms, and r is the particle radius. Figure 4.1 shows a plot of the N_s/N_v as a function of particle diameter calculated for Au. More than 30% of the atoms are on the surface when the particle size is smaller than 3 nm.

4.2. SHAPES OF POLYHEDRAL NANOCRYSTALS

Surface energies associated with different crystallographic planes are usually different, and a general sequence may hold, $\gamma\{111\} < \gamma\{100\} < \gamma\{110\}$, for face

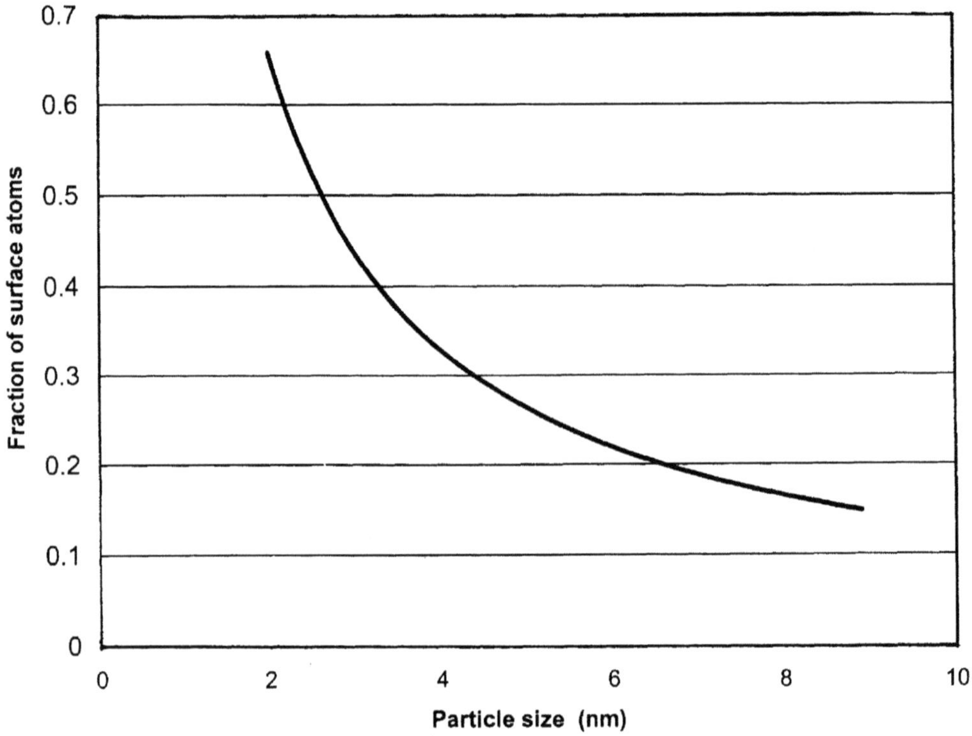

FIGURE 4.1. Calculated surface-to-volume atom ratio for a spherical gold nanoparticle as a function of particle size. The calculation was made based on an assumption that 50% of the particle surface is covered by {100} and the another 50% is by {111}, thus the density of the surface atom is $\rho_s = (1 + 2/3^{1/2})\, a^2$, where a is the lattice constant, the result is $N_s/N_v = 1.616\, a/r$.

centered cubic (fcc) structured metallic particles. For a spherical single-crystalline particle, its surface must contain high index crystallography planes, which possibly result in a higher surface energy. Facets tend to form on the particle surface to increase the portion of the low index planes. Therefore, for particles smaller than 10–20 nm, the surface is a polyhedron. Figure 4.2a shows a group of cubo-octahedral shapes as a function of the ratio, R, of the growth rate in the $<100>$ to that of the $<111>$. The longest direction in a cube is the $<111>$ diagonal, the longest direction in the octahedron is the $<100>$ diagonal, and the longest direction in the cubo-octahedron ($R = 0.87$) is the $<110>$ direction. The particles with $0.87 < R < 1.73$ have the {100} and {111} facets, which are named the truncated octahedral (TO). The other group of particles has a fixed (111) base with exposed {111} and {100} facets (Fig. 4.2b). An increase in the area ratio of {111} to {100} results in the evolution of the particle shape from a triangle-based pyramid to a tetrahedron.

Nanocrystals deposited on the surface of a substrate may have random orientation. If the particle is oriented along a low index zone axis, the distribution of atoms on the surface can be imaged in profile by high-resolution transmission electron microscopy (HRTEM)[6], which will be introduced in detail in Chapter 6. This is a powerful technique for direct imaging of the projected shapes

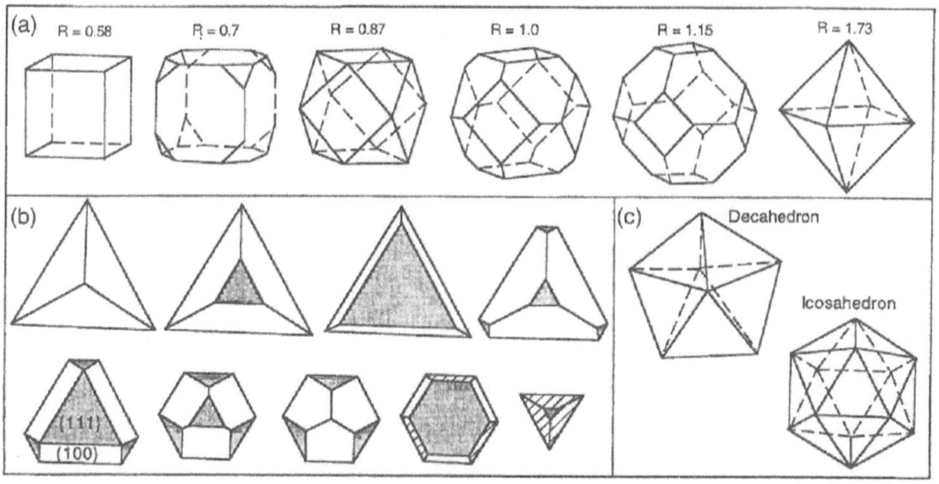

FIGURE 4.2. (a) Geometrical shapes of cubo-octahedral nanocrystals as a function of the ratio, R, of the growth rate along the $<100>$ to that of the $<111>$. (b) Evolution in shapes of a series of (111) based nanoparticles as the ratio of {111} to {100} increases. The beginning particle is bounded by three {100} facets and a (111) base, while the final one is a {111} bounded tetrahedron. (c) Geometrical shapes of multiply twinned decahedral and icosahedral particles.

of nanoparticles particularly when the particle size is small. We now use the results provided by this technique to illustrate a few typical particle shapes.

4.2.1. Cubic-Like Nanocrystals

Cubic shape nanocrystals are formed for some of the metallic nanocrystals under certain experimental conditions.[7] A TEM image given in Fig. 4.3 shows the cubic-like shape of a Pt particle oriented along [100] and [110].[8] With the resolution power of HRTEM, we are able to directly image the projection of atom rows in each individual particle if it orients with a low index zone axes parallel to the incident beam. The arrangement of the atoms on the surfaces parallel to the beam is imaged edge-on. The missing or distortion of surface atoms can be seen from the variation of image contrast provided the influence from the amorphous carbon support could be suppressed.

The image shown in Fig. 4.3a indicates that the particle is bounded by {100} facets and there is no defect in the bulk of the particle. The distances between the adjacent lattice fringes is the interplanar distance of Pt {200}, which is 0.196 nm, and the bulk structure is fcc. Surface relaxation, if any, is probably restricted to the first one or two atomic layers. The surface of the particle may have some steps and ledges particularly at the regions near the corners of the cube. To precisely image the defects and facets on the cubic particles, a group of particles oriented along [110] are selected (Fig. 4.4). This is the optimum orientation for imaging cubic structured materials. All these particles are single crystals without twinning or defects. Atom-high surface steps are seen on {100} facets, and the {110} facets are rather rough. {111} Facets are seen as well as some high index surfaces are

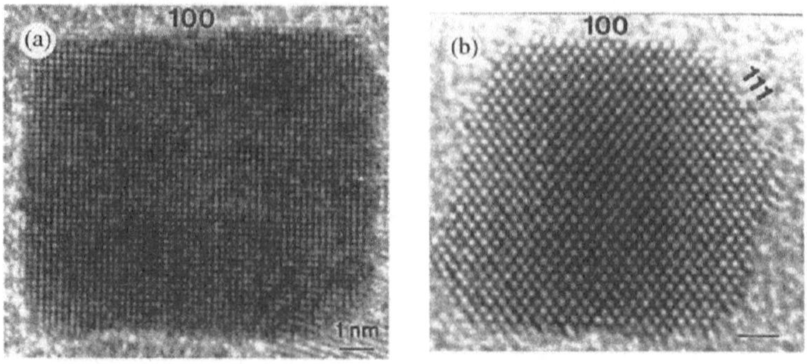

FIGURE 4.3. HRTEM images of cubic Pt nanocrystals oriented along (a) [001] and (b) [110], showing surface steps/ledges and the thermodynamically inequilibrium shapes.

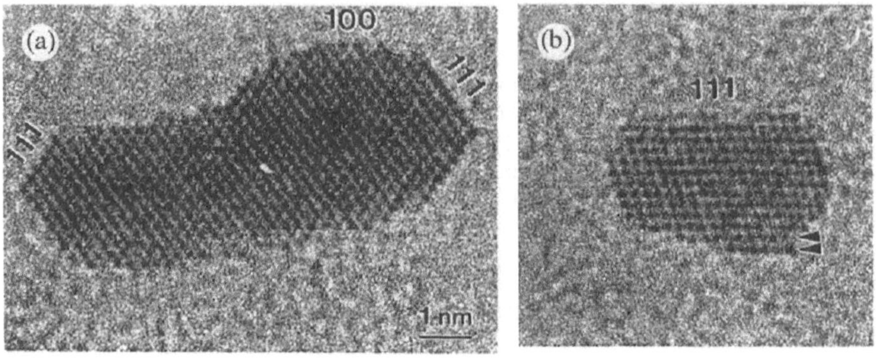

FIGURE 4.4. HRTEM images of Pt nanocrystals oriented along [110], showing the reconstructed {100} surfaces and surface relaxation.

also observed. The shapes of these particles are not for particles at thermal equilibrium since the specimen was prepared at room temperature. Thus, the particles have many surface atoms with different degrees of unsaturated valences and exhibiting different efficiencies and selectivity.

4.2.2. *Tetrahedral Nanocrystals*

Tetrahedral is a commonly observed particle shape for fcc structured metallic particles, because it is enclosed by the densely packed {111} facets and is energetically favorable. Here, we use Pt nanoparticles to illustrate the characteristics of tetrahedral shape. Figure 4.5 shows HRTEM images of tetrahedral Pt particles oriented along [110]. The particle is a truncated-tetrahedron showing a {100} (top) and two {111} facets (bottom). The atomic scale roughness at the two truncated {111} facets are apparent.

The general structural characteristics of a truncated tetrahedral particle is illustrated in Fig. 4.6, where the edges and corners of the tetrahedron are cutoff, resulting in the formation of {111} and {100} facets. If the area ratio of {100} to

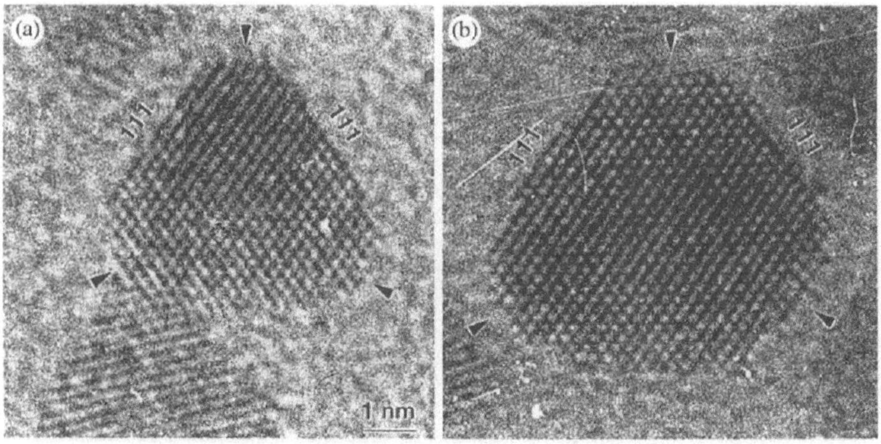

FIGURE 4.5. HRTEM images of truncated tetrahedral Pt nanocrystals oriented along [110]. The surface steps and ledges at the truncated corners are clearly resolved.

{111} can be changed, particles with a variety of shapes can be generated (see Fig. 4.2).

4.2.3. Octahedral and Truncated Octahedral Nanocrystals

Octahedral and TO shapes are likely to be the most popular metallic nanocrystals. An octahedron has eight {111} facets (Fig. 4.7b), four {111} facets are edge-on if viewed along [110] (Fig. 4.7b). If the particle is a truncated octahedron, six {100} facets are created by cutting the corners of the octahedron, two of which are edge-on while viewing along [110]. These expected results are well

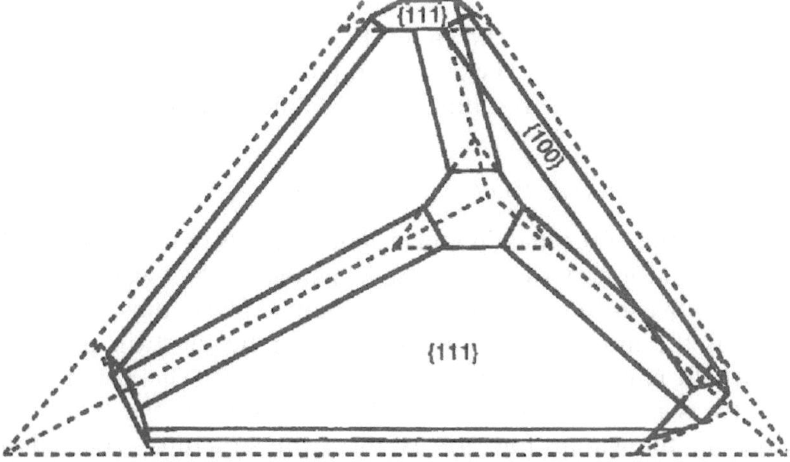

FIGURE 4.6. A schematic diagram showing the shape of a truncated tetrahedral nanoparticle.

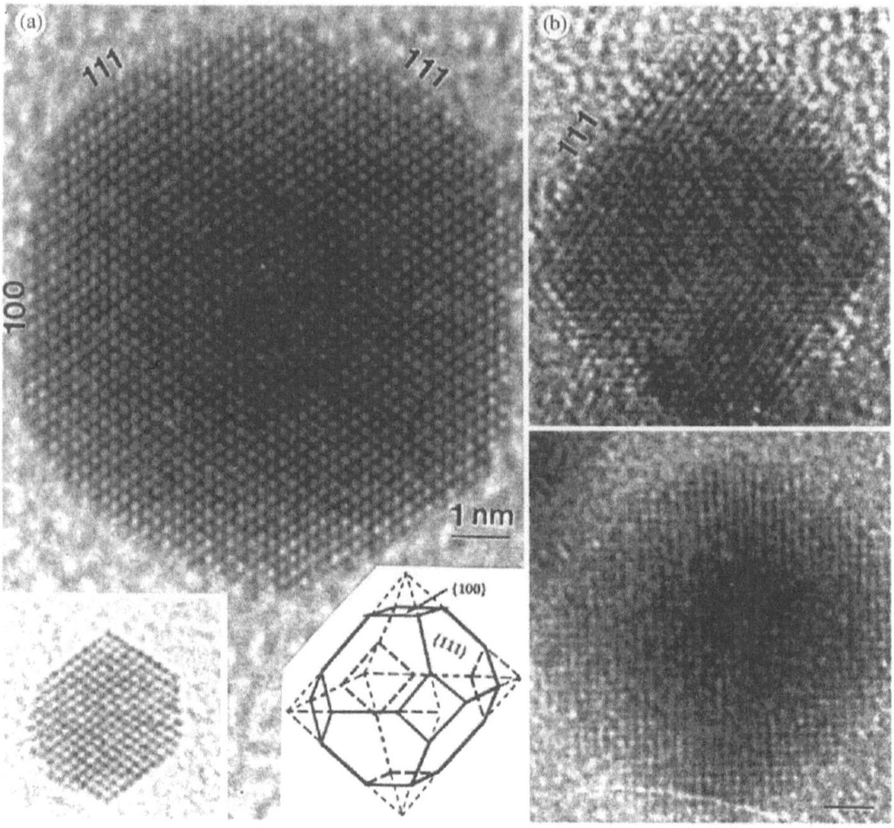

FIGURE 4.7. HRTEM images of Pt nanocrystals (a) with a truncated octahedral shape and oriented along [110], and (b) with an octahedral shape and oriented along [110] and [001]. The inset in (a) is a model of the particle shape.

reproduced experimentally, as shown in Fig. 4.7a, in which three truncated octahedral particles are shown. The {111} and {100} facets are seen. The area ratio of {100} to {111} can change, resulting in slight differences in particle shapes.

4.2.4. *Twinning and Stacking Faults*

Twinning is one of the most popular planar defects in nanocrystals, and it is frequently observed for fcc structured metallic nanocrystals. Twinning is the result of two subgrains sharing a common crystallographic plane, thus, the structure of one subgrain is the mirror reflection of the other by the twin plane. The fcc structured metallic nanocrystals usually have {111} twins. The easiest direction to observe the twins is along the $<110>$ direction so that the twin planes are imaged edge-on. Shown in Fig. 4.8 is the twin structure observed in FePt nanocrystals. The (111) twin plane is a common crystallographic plane of the two twinned crystals, and it is also called "mirror plane," although some displacement

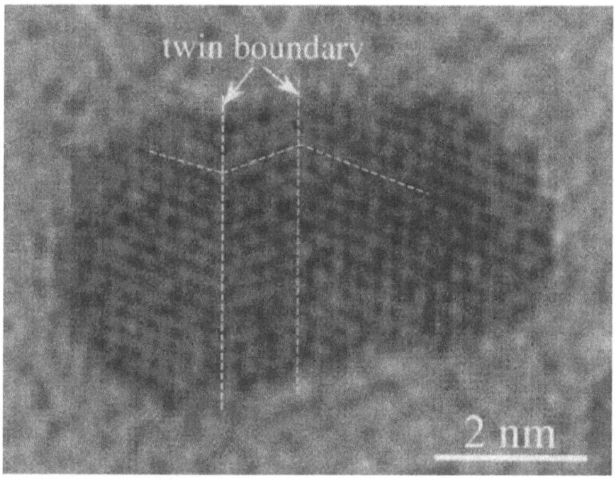

FIGURE 4.8. Twins observed in FePt nanocrystals.

at the twin plane is possible. If the orientation of one crystal is known and the twin plane is defined, a rotation of 180° around the twin axis gives the orientation of the other mirror crystal. In the diffraction pattern, the twin axis is parallel to the **g** vector of the twin plane; a rotation of 180° about the twin axis gives the diffraction pattern of the mirror crystal. The superposition of the two patterns gives the experimental diffraction pattern.

Stacking faults are typical planar defects present in functional materials. Stacking faults are produced by a distortion on the stacking sequence of atom planes. The (111) plane stacking sequence of a fcc structure follows A–B–C–A–B–C–A–B–C–. If the stacking sequence is changed to A–B–C–A–B–A–B–C–, a stacking fault is created. The crystal lattices on the both sides of the stacking fault is shifted by $\mathbf{R} = 1/3[1\bar{1}0]$, which is defined as the displacement vector of the stacking fault.

Figure 4.9a shows a Pt nanocrystals that has a single twin. The twin plane is (111). Figure 4.9b is a particle that has a twin (T) and a stacking fault (S). Twining is probably the most popular defect observed in nanocrystals, because nanocrystal is too small to have dislocation in the volume.

4.2.5. Multiply Twinned Icosahedral and Decahedral Particles

The two typical examples of multiply-twinned particles (MTP) are decahedron and icosahedron.[9,10] Theoretical calculation shows that the particle energy is minimized by forming miltiple twins when their sizes are smaller than a critical value[9], assuming bulk values for the cohesive, surface, twin boundary and elastic strain energies and for the adhesive energy to the surface. Mainly due to the balance between energy gained by having low-surface-energy, external facets and energy spent in creating twin boundaries and also due to the increasing influence of elastic strain energy with size, Ino found that below a certain critical size the icosahedral is the most stable, while the decahedral particle is quasi-stable. In reality, both decahedral and icosahedral particles have been observed.

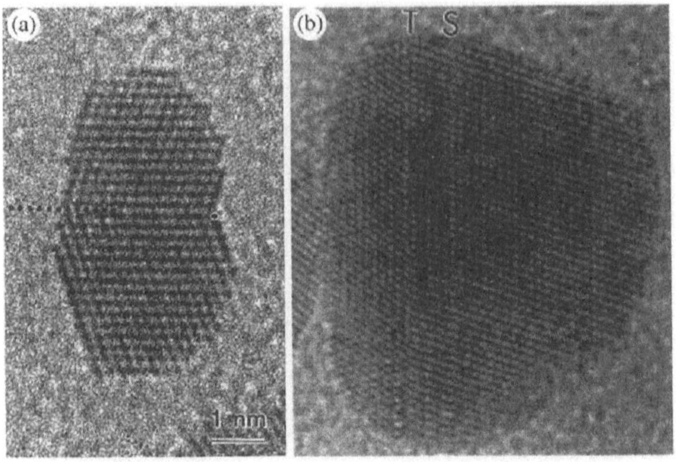

FIGURE 4.9. HRTEM image recorded from Pt nanocrystals showing twin and stacking fault planar defects.

Starting from an fcc structured tetrahedron, a decahedron is assembled from five tetrahedral sharing an edge (Fig. 4.10a). If the observation direction is along the fivefold axis and in an ideal situation, each tetrahedron shares an angle of 70.5°, five of them can only occupy a total of 352.6°, leaving a 7.4° gap. Therefore, strain must be induced in the particle to fill the gap.[11] Therefore, a decahedral particle has high strain and there is lattice distortion. An icosahedron

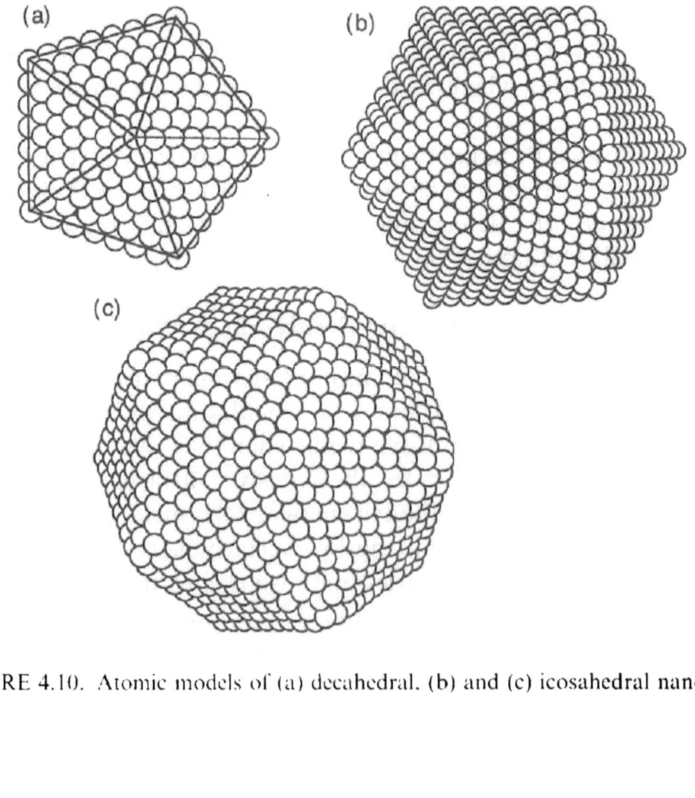

FIGURE 4.10. Atomic models of (a) decahedral, (b) and (c) icosahedral nanocrystals.

is assembled using 20 tetrahedra via sharing apexes (Fig. 4.10b). The icosahedral and decahedral particles are the most extensively studied twinned nanocrystals.[12,13] The easiest orientation for identifying the MTPs is along the fivefold symmetry axis.

Figure 4.11 shows two MTP Au particles, where the fivefold axis is clearly seen. The particle given in Fig. 4.11a is a single decahedral particle, while the one given in Fig. 4.11b is a particle constituted of two decahedra. Twinning is the mechanism for forming these particles. The multiply twinning is the favorable structural configurations when the particles are small, possibly because of the smaller surface and volume energies.

The decahedral and icosahedral particles are "quasicrystals," which means that they have the fivefold symmetric axis and cannot exist in bulk crystals. These structures are most commonly observed for small size particles. Gold clusters of sizes ~2 nm are dominated by decahedral.[14]

4.2.6. *Faceted Shape of Rod-Like Nanocrystals*

Gold nanorods have been synthesized using an electrochemical technique.[15] A gold metal plate is used as the anode and a platinum plate is used as the cathode. Both electrodes were immersed in an electrolytic solution consisting of a cationic surfactant, hexadecyltrimethylammonium bromide ($C_{16}TAB$), and a rod-inducing surfactant. The $C_{16}TAB$ serves not only as the supporting electrolyte, but also as the stabilizer for nanoparticles to prevent their further growth. An appropriate amount of acetone was added into the electrolytic solution, which may facilitate the incorporation of cylindrical-shape-inducing surfactant into the $C_{16}TAB$ micelle framework and inducing the cylindrical growth to form the Au–$C_{16}TAB$–TC_8AB system.

An Au rod is bounded by facets, and it is a thermodynamically unstable crystal growth. Almost all of the Au nanorods are single crystalline and contain no twins or dislocations.[16] When dispersing in liquid drop onto a flat carbon substrate, the nanorods are preferentially aligned and oriented along [110]

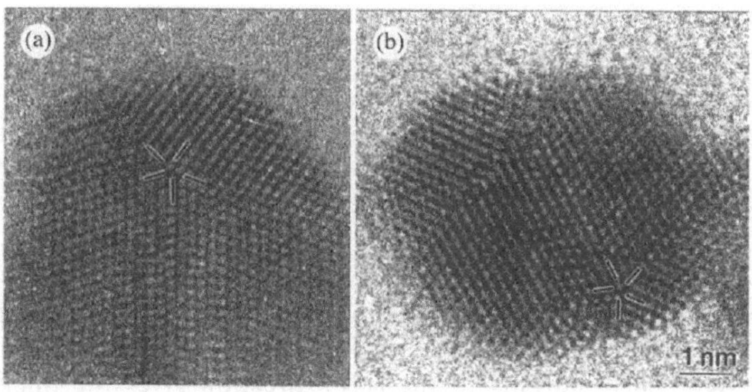

FIGURE 4.11. TEM images of decahedral Au nanocrystals when the incident electron beam is parallel or nearly parallel to the fivefold symmetry axis.

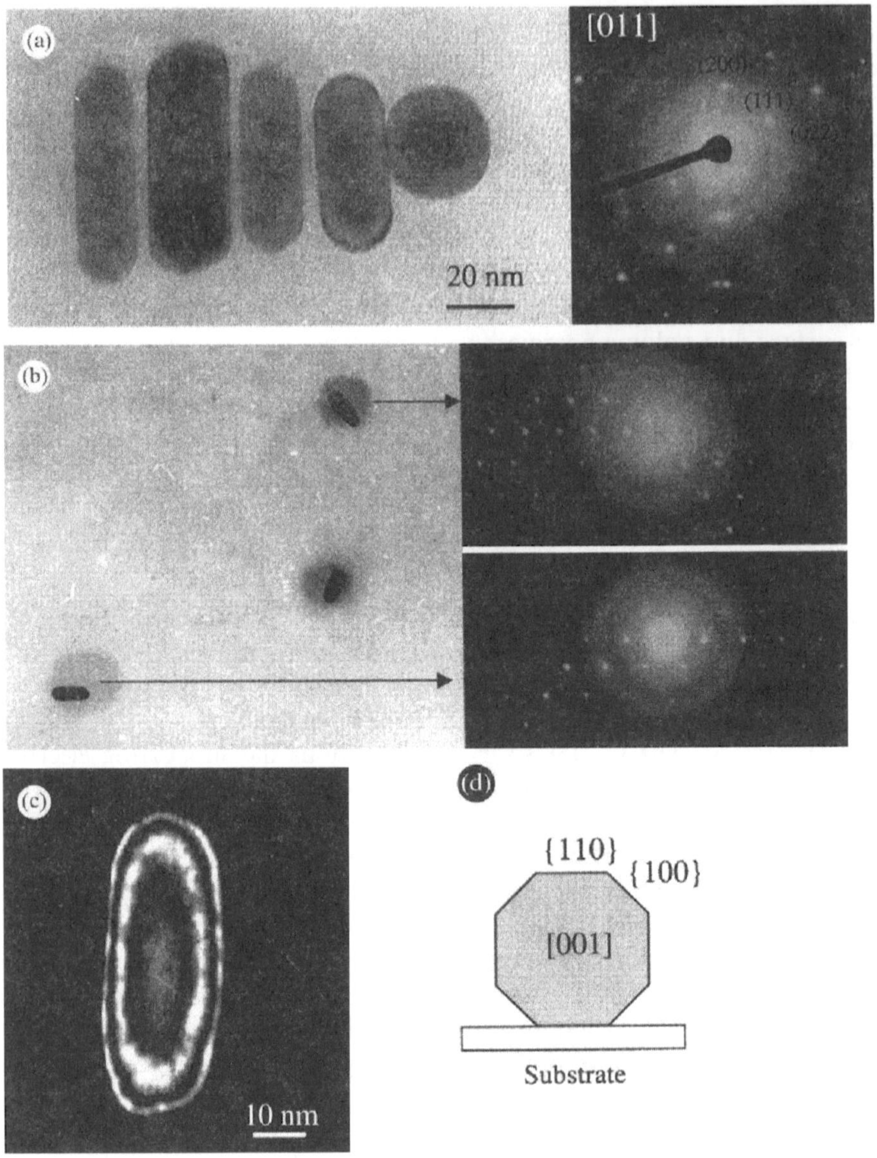

FIGURE 4.12. (A) Bright-field TEM image of Au nanorods and the corresponding electron diffraction pattern recorded from the first three rods aligned on the left, proving the unified [110] orientation of the rods. (b) Bright-field TEM image of individually dispersed au nanorods, showing the [110] orientation perpendicular to the substrate. (c) Dark-field TEM image of an Au rod oriented nearly parallel to [110]. These images unambiguously suggest the formation of {110} facets on the nanorods.

perpendicular to the substrate (Fig. 4.12a). The single Au nanorods also prefer to orient along [110] (Fig. 4.12b). This is possible only if the nanorods have {110} facets. A dark-field TEM image recorded using a {111} reflected beam when the rod is oriented nearly parallel to the [110] direction gives the thickness fringes (Fig. 4.12c) owing to thickness variation across the specimen. The intervals between the fringes would have an equal distance if the nanorods were bound only

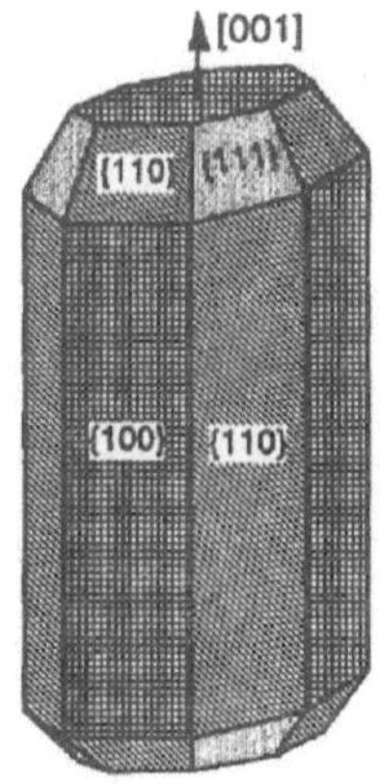

FIGURE 4.13. Geometrical and crystallographic structure of an Au nanorod.

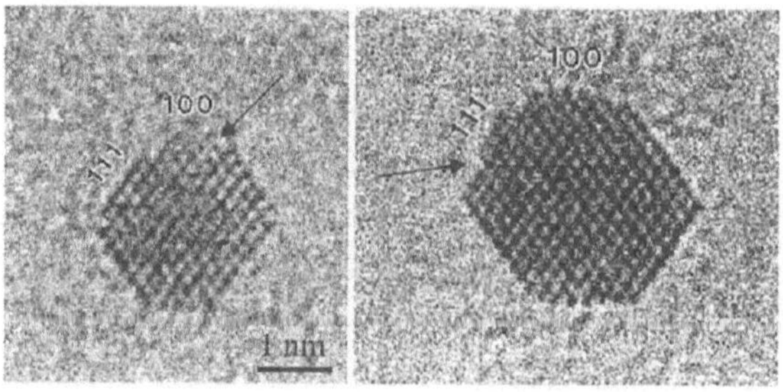

FIGURE 4.14. HRTEM image recorded in profile from Pt nanocrystals, showing surface defects, as indicated by arrowheads.

by four {100} facets. The finite width of the band at the center and its uniformity in intensity indicate that the thickness at the center region is nearly constant with a width of ~5 nm, strongly supporting the presence of {110} facets. A model for the geometrical and crystallographic structure of an Au nanorod is given in Fig. 4.13.

4.2.7. Surface Defects

Nanocrystal surfaces usually have point defects, such as vacancies and adsorbed atoms, and the rounded corners, edges, and apexes. Surfaces are most active for atom diffusion, mass transport, and chemical reaction. Defects on surfaces also include surface steps, ledges, and kinks, which are introduced by the growth process of the surface. Observation of surface defects is carried out using TEM. Scanning Tunneling microscope (STM) can also be used to image nanoparticles, but achieving atomic resolution for a single nanoparticle is very challenging. Figure 4.14a shows HRTEM images recorded from a Pt nanocrystals oriented along [110]. The surface atoms are clearly resolved, but some rows of surface atoms are missing, indicating the presence of point defects and missing-row of atoms on the surface. This type of defects is often observed for nanocrystals, especially the as-synthesized nanocrystals.

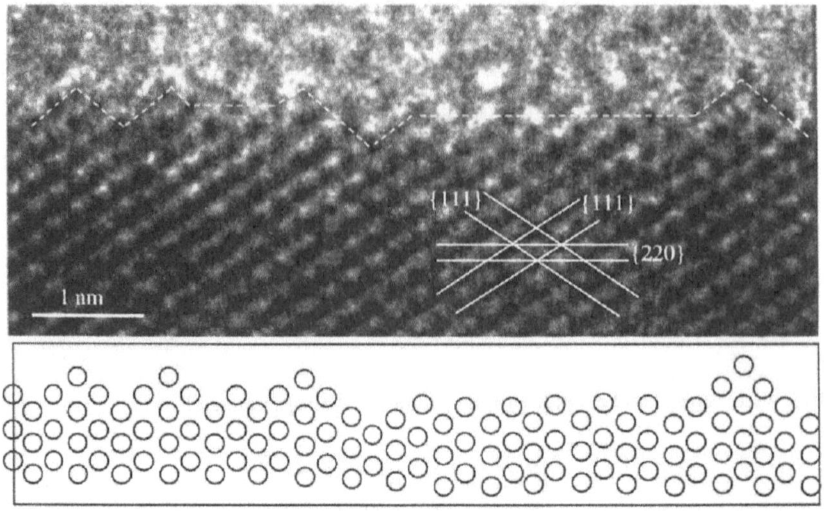

FIGURE 4.15. (a) HRTEM images recorded from an Au nanorod oriented along [110] and the corresponding positions of the projected atom rows, showing the rearrangement of the surface atoms. (b) Atomic models of (a) an ideal (110) surface and (b) the reconstructed (110) surface with missing-rows.

4.2.8. Surface Reconstruction of Nanocrystals

Simply speaking, surface atoms have less number of bonds in comparison to the atoms in the bulk because of the loss in nearest neighbors. Surface atoms tend to find new equilibrium positions to balance the forces, resulting in surface reconstruction. Surface reconstruction has been extensively studied for large area surfaces in the literature, but the reconstruction of nanocrystal surfaces remain to be investigated. The {110} surface of an Au nanorod, for example, has been found to reconstruct so that the surface is composed of zigzag stripes of {111} facets, for example, the Au atoms are missing every other row parallel to the [110] direction, as shown in Fig. 4.15a, where the teeth type structure is the result of surface reconstruction.[17] Figure 4.15b shows a schematic model for the perfect (110) surface built from the fcc lattice. If a row of atoms is missing along [110], the (110) surface transforms into strips of {111} facets. It is known that the {111} surface is the most densely packed surface for fcc structure and is thus the most stable surface.

4.2.9. Ultra-Small Particles and the Magic Number

Gold nanocrystals are likely to be the most extensively studied structures. Systematic theoretical calculation and experimental studies show that Au nanocrystals of truncated octahedral shapes with numbers of atoms 116, 140, 225, 314, and 459 are energetically favorable and are stable in practice.[18] Some of the Au clusters with 79, 38, or even 28 atoms have been observed.[19,20] The number of atoms in each nanocrystal is fixed, and the geometrical shape of the nanocrystal

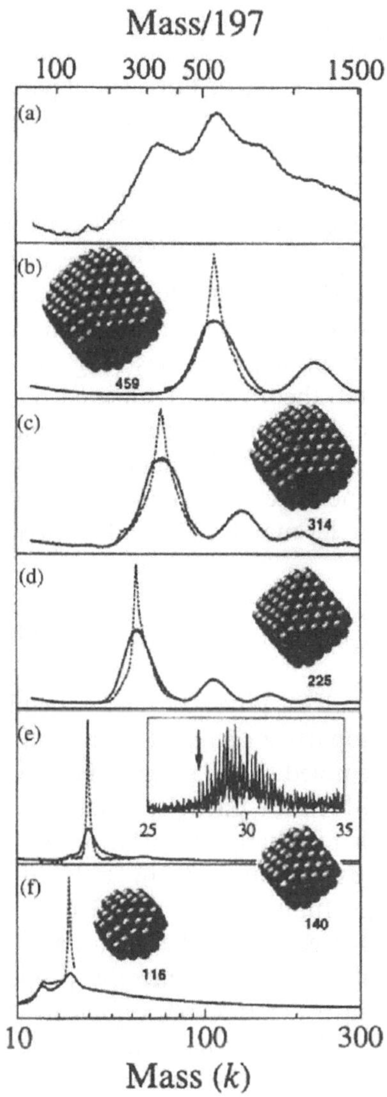

FIGURE 4.16. Mass spectra (abundance vs. mass, in $k = 10^3$ amu, and in mass-equivalent number N of gold atoms, $m_{Au} = 197$ amu) for mass separated Au nanocrystals. The inset structures are predicted optimal core structures containing magic number of atoms (Courtesy Whetten *et al.*[18]).

is specific. These numbers are frequently observed in mass spectra and are thus called *magic numbers*. The most powerful technique for verifying the magic numbers is through mass spectrometry. Figure 4.16 shows a series of mass spectra from the dominant separated Au nanocrystals, corresponding to the mass of the specific shapes of nanocrystals predicted theoretically. The second peak arises from aggregation occurring in the mass spectrometer. In the conventional lower-resolution spectra (solid lines), the widths of the peaks arise from an undetermined influence of instrumental factors and mass dispersion resulting, for example, from a variable desorption of the surfactant monolayer. Enhanced-resolution spectra (dotted lines) exhibit a significant narrowing of the main peaks. The nanocrystals have specific shape and size and they are a molecular matter and can be used as building blocks for self-assembly.

The contrast of a small particle in TEM is greatly influenced by the substrate used to support the particle. Therefore, a question is how small a particle can be before HRTEM becomes ineffective in identifying its shape? There seems not to be a definite answer for this question depending on the performance of the instrument. Here, we take an example from a JEOL 4000EX (400 kV, point-to-point resolution 0.17 nm) to illustrate this question. We have tried to image the Pt atom clusters that were nucleated at the beginning of the growth. A lower magnification image of the particles dispersed on a carbon substrate shows the co-existence of large size as well as small size particles (Fig. 4.17a), where the smaller particles are dominated by tetrahedra and truncated octahedra, while the large particles are dominated by cubic. This shape transformation from tetrahedral-to-octahedral-to-cubic is the result of the kinetically controlled growth.[21] Small size clusters of truncated octahedra can be imaged (Fig. 4.17b). The number of atoms presented in these particles may be 225 and 79 provided the particle shapes are perfect.

4.3. SELF-ASSEMBLY OF NANOCRYSTALS

The physical and chemical functional specificity and selectivity of nanoparticles suggest them as ideal building blocks for 2D and 3D cluster self-assembled superlattice structures (in forms of powder, thin film, and solid bulk), in which the particles behave as well-defined molecular matter and are arranged with long-range translation and even orientation order.[21] Well-defined ordered solids prepared from tailored nanocrystalline building blocks provide new opportunities for optimizing, tuning, and/or enhancing the properties and performance of the materials. Tuning particle size and interparticle distance could effectively tune the properties of the nanocrystal system. This is a new initiative of research on *cluster engineered materials*.

Research has successfully fabricated self-assembly passivated nanocrystal superlattices (NCSs) or nanocrystal arrays (NCA) of metal, semiconductor, and oxide clusters, which are a new form of materials with fundamental interests and technological importance. There are four key steps in developing these materials: preparation of size and shape selected and controlled nanoparticles, controlling the self-assembling process to produce large size well-ordered NCSs, structural characterization, and modeling the dynamic behavior of the system. Indeed, understanding and optimizing the structures of the NCSs is an important step towards a systematic exploration of the nature of patterned superstructures assembled from such building units, and formulation and implementation of methods for designing and controlling such novel assembled materials and their operational characteristics.

Self-assembled arrays involve self-organization into monolayers, thin films, and superlattices of size-selected nanoclusters encapsulated in protective compact organic coatings. A key step in this process is the fabrication of size and shape controlled NCAs that have the potential to grow into large sizes for technological

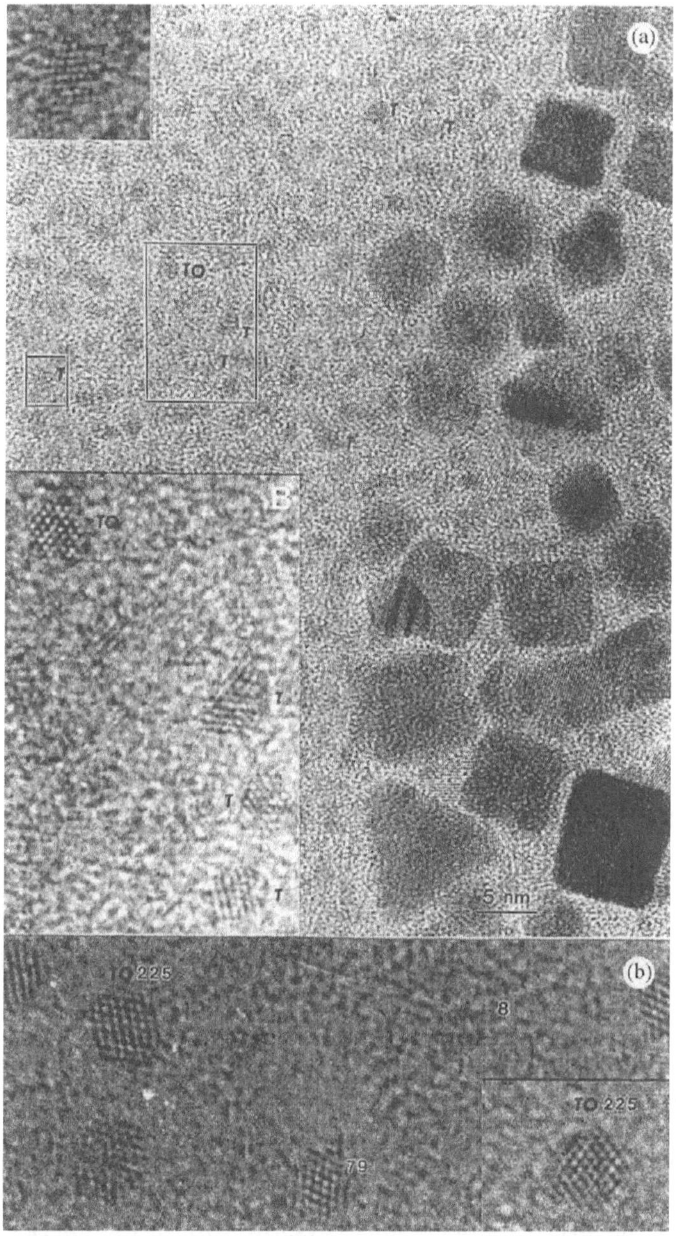

FIGURE 4.17. HRTEM image of Pt nanocrystals whose shapes were dominated by tetrahedra when they are small, but a transformation to cubic shape occurred as they grow larger. (b) The possible "smallest" nanocrystals whose shape still can be identified by HRTEM. The numbers represents the possible of numbers of atoms in the particles if they are perfect truncated octahedra without point defects.

applications. Colloidal chemistry or wet chemistry has played a powerful role in this process. Self-organization of nanoparticles is a new route for synthesis of superlattice materials—solid, periodic arrays built using nanocrystals as building blocks,[18,23,24] achieved by preparing the size and shape selected nanocrystals using

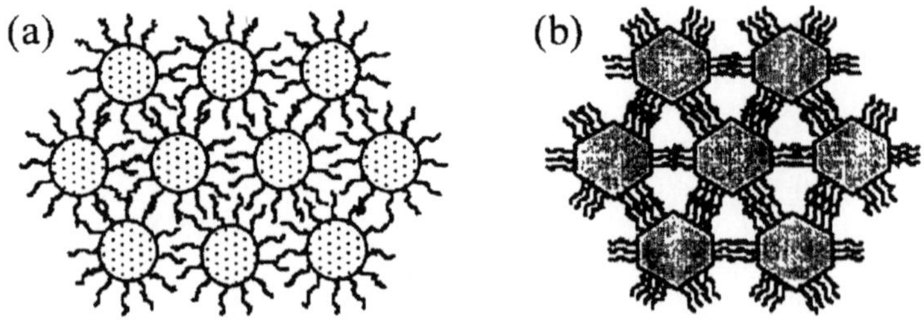

FIGURE 4.18. Schematic illustration of the self-assembled nanocrystals. To form ordered self-assembly, several conditions need to be met: the nanocrystals should be monodispersive; the surfactant molecules should be strong enough to separate the individual nanocrystal; and the drying rate should be so slow that the nanocrystals can move to the suitable positions. The above plot is 2D assembly of faceted nanocrystals and the bottom one is assembly of non-faceted nanocrystals.

colloidal chemistry and then using self-assembly as a way of joining them together[25] (Fig. 4.18). The macroscopic properties of the NCS is determined not only by the properties of each individual particle, but also by the coupling/interaction between nanocrystals interconnected and isolated by a monolayer of thin organic molecules. By changing the length of the molecular chains, quantum transitions and insulator to conductor transition could be introduced.

NCSs are characterized by unprecedented size-uniformity and translation and even orientation order, created through a multistage processing strategy involving self-assembly, annealing, etching of defects, reversible passivation by self-assembled monolayers, and macroscopic separation by size of the resulting assemblies. Particles that can be self-assembled usually have sizes smaller than 10 nm, it is in this size range that many exciting and unusual physical and/or chemical properties are enhanced. The nanocrystal thin films reported here are formed primarily on a solid substrate via a self-assembling process of nanocrystals. An alternative technique for forming monolayer nanocrystal thin films is at the liquid surface using the Langmuir–Blodgett films, which has shown great potential in synthesis and assembling nanocrystals.

Self-assembled nanostructure is a new form of materials with fundamental interest and potential technological applications.[26–32] In nanocrystal self-assembled structures, each individual nanocrytal passivated with surfactants is the fundamental building unit, and it serves as an "artificial atom" for constructing the ordered structure. The analogy of a nanocrystal with an atom is obvious. If the size of nanocrystals is below 10 nm, the energy levels of each individual nanocrystal are discrete, as in the case of atoms. The energy level spacing and other "atomic" properties can be adjusted by changing the size of the nanocrystals, in contrast to those of the atoms. Size and shape controlled nanocrystals can be viewed as molecular matter with specific shape and electronic structure, and a self-assembly of the nanocrystals can form "nanocrystal solids" with translation and even orientation order. Based on this approach, several types of nanocrystal self-assembled structures have been synthesized, including CdSe,[33,34] InP,[35] CdS,[36] Au,[18,37,38] Ag,[39,40] Pt,[41] Co,[42–44] FePt,[45] Ni,[46,47] TiO_2,[48] CoO,[49,50] Fe_2O_3[5] and Ag_2S[52,53] etc.

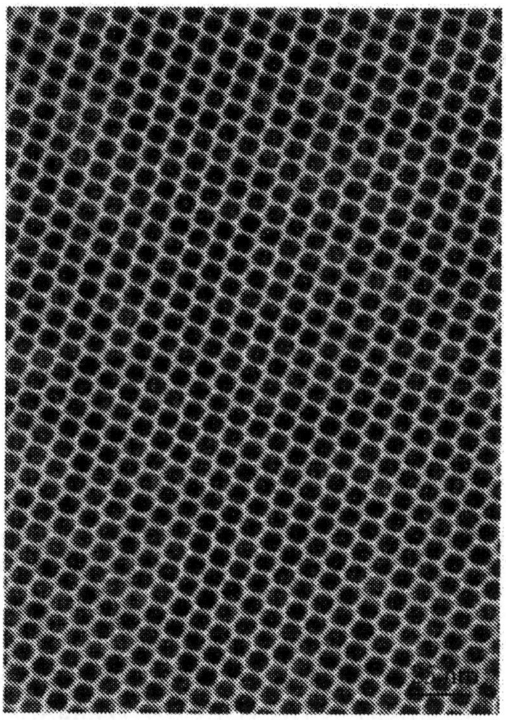

FIGURE 4.19. Ordered array of Co nanocrystals.

4.3.1. *Surface Passivation*

At nanoscale, both the physical and chemical properties of the materials have been profoundly changed. The relative large surface area of each bare nanocrystal suggests that nanocrystals are very reactive, thus, surfactant molecules are needed to cap and stabilize the surfaces of nanocrystals. In the evaporation processing method, for instance, if no surfactant is applied, the freshly made gold nanocrystals coalescence with one another to form heavily twinned larger nanocrystals, but if the surfactant molecules are applied into the reaction chamber, gold nanocrystals can keep the size and shape. After an inorganic nanocrystal is coated with a densely packed monolayer of surfactant molecules, the surface of nanocrystals becomes hydrophobic, and this kind of nanocrystal–surfactant combination is soluble in non-polar solvents and suspend in the solution, forming stable colloids. After the solvent is evaporated or removed, the passivated nanocrystals rearrange themselves to form assemblies instead of fusing together, because they are separated by a thin layer of molecules. If the size of nanocrystals is monodispersive, ordered self-assembly of nanocrystals will be formed. Figure 4.19 shows an ordered array of Co nanocrystals, which are dominated by an anisotropic shape as determined elsewhere.[54]

Among all of the systems, gold particles are probably the most advanced molecular crystal system that has been widely studied. Gold nanocrystals are usually passivated by the alkanelthiolate surfactant (SR, where $R = n\text{-}C_nH_{2n+1}$,

$n = 4, 6, 8, 12, \ldots$). The surfactant specifies a compact and ordered monolayer passivation of the thiolate over the nanocrystal surface. The thiolate serves not only as the protection layer for the particles to avoid direct contact between the particles with a consequence of collapsing, but the interparticle bonding. The strength of this bond characterizes the structural stability and the maximum operation temperature up to which the materials can sustain. The core size of the particles can be controlled by changing the gold-to-thiolate ratio.[55] Recent studies have shown that the size of the Au core and the length of the thiolate are the two parameters, which determine the 3D crystallography of the nanocrystal assembling (i.e., phase diagram). The studies of Andres *et al.*[37] have demonstrated that the alkanelthiolate can be replaced by arenedithiol or arenediisonitrile molecules for forming covalently bonded NCAs to enhance the structural stability and mechanical strength.

The choice of passivation surfactant is a key in forming the NCS as well as controlling their properties. For magnetic nanocrystals, such as cobalt, three major steps were introduced.[42] Dioctylether solution of superhydride ($LiBEt_3H$) is prepared by mixing the tetrahydrofuren (THF) superhydride solution with dioctylether and evaporating THF under vacuum. The particle synthesis began with the injection of 2 M dioctylether superhydride solution into a hot (200°C) $CoCl_2$ dioctylether solution in the presence of oleic acid (octadec-9-ene-1-carboxylic acid, $CH_3\,(CH_2)_7CH{=}CH(CH_2)_7COOH$) and trialkylphosphine (PR_3, $R = n\text{-}C_4H_9$, or $n\text{-}C_8H_{17}$). Continued heating at 200°C allowed steady growth of these clusters into nanometer sized, single crystals of cobalt. Average particle size was coarsely controlled by selecting the type of alkylphosphine used in combination with oleic acid during the growth. Bulky $P(C_8H_{17})_3$ limited growth to produce particles (2–6 nm) and less bulky $P(C_4H_9)_3$ led to larger (7–11 nm) particles. Fine tuning of the particle size was done by fractionating the coarsely adjusted samples and selecting the desired size fraction.

The strength of the chain molecule is in the order of 0.1 eV, comparable to the kinetic energy of atom thermal vibration. A key question here is about the stability and the phase transformation behavior of the weakly bonded NCA since the melting point of nanoparticles is much lower than that of the bulk. With consideration of the even lower melting point of the passivating organic molecules, the stability of the superlattices (not only the structure of nanocrystals, but also the "crystallinity" of the superlattice) above ambient temperature is a serious concern because the NCS is likely to be used in the areas such as microelectronics and data storage. The *in situ* behavior of monolayer self-assembled NCAs of cobalt oxide nanoparticles has been observed using TEM[56]. The results proved the high stability of a monolayer assembling to temperatures as high as 600°C, whereas the multilayer assembly coalesced at 250°C. The substrate exhibited strong adhesive effect on the stability of the nanocrystals.

4.3.2. Interparticle Bonds

Due to the capping monolayer of molecules, nanocrystals are held together mainly by van der Waals forces. Three requirements need to be met to make ordered self-assemblies of nanocrystals: (1) building blocks (monodispersive size and shape controlled nanocrystals); (2) passivation layers (suitable surfactants and solutions); and (3) a controllable and slow drying process to allow diffusion of the passivated nanocrystals in solution to find their equilibrium position. If the drying rate is too high, the disordered arrangement of nanocrystal self-assembly will form, just like a rapid quenching can lead to amorphous phase during solidification.

Of these, the interparticle interactions appear to be sensitive to the chemical properties of the particle protecting monolayers. For instance, for particles that are stabilized by neutral alkanethiolate monolayers, it is found by TEM measurements that the interparticle (edge-to-edge) distance is only slightly longer than the length of a single protecting molecule, which is due to ligand intercalation.[57] However, in the case that charge groups are introduced into the particle protecting monolayers, the closest particle gap is equal to two molecular lengths of the protecting ligands.[58] This is attributed to the electrostatic interactions between the peripheral charged groups. It is, therefore, anticipated that the former will exhibit stronger mechanical stability than the latter, despite the fact that in both cases ordered self-assembled structures can be obtained.

In addition, the dynamics of self-assembling will also play an important role in governing the particle structures, where slow evaporation of the solvent tends to favor the formation of long-range ordered assemblies of particle molecules. Thus, one can see that the resulting particle assemblies can be manipulated by using solvents of varied volatility. Alternatively, allowing the evaporation process in an atmosphere of the identical solvent provides another route in controlling the solvent evaporation rate.

4.4. SOLUTION-PHASE SELF-ASSEMBLY OF PARTICLES

The controlled fabrication of very small structures at scales beyond the current limits of lithographic techniques is a technological goal of great practical and fundamental interest. Important progress has been made over the past few years in the preparation of ordered ensembles of metal and semiconductor nanocrystals. Many methods have been developed to process different kinds of nanocrystals, single phase or compounds. Here, only the methods based on solution chemistry that can produce or have the potential to produce monodispersive nanocrystals or quasi-monodispersive nanocrystals, which have been used for making self-assembled structures, will be discussed.

4.4.1. Metallic Nanocrystals

Synthesis of metallic nanocrystals, such as gold, can be tracked as early as Faraday's days. Gold nanocrystals have been taken as a model system for theoretical and experimental investigations of the properties of nanocrystal materials, such as size effect on color, size induced quantum effect, light absorption and many more. Experimentally, gold is an ideal choice because of no impurity, no oxidation, and possibly no by-reaction. Theoretically, gold is easier to be described using quantum theory because all of its inner shells are filled orbitals and its most outer shell only has one electron; thus, the hydrogen-like model may be adopted for theoretical calculation. Therefore, gold nanocrystals are likely to be most extensively studied and well understood systems, although there are still many exciting research activities being carried out.

The reduction of metal ions in solution is the most popular and economic technique for preparing metallic particles, though evaporation of metals at high temperature can also yield metal nanoclusters. Faraday [59] first used the two-phase method to prepare stable colloidal metal nanocrystals, in which he reduced an aqueous gold salt with phosphorus in carbon disulfide and obtained a ruby colored aqueous solution of dispersed gold nanoparticles. In a typical process, the gold nanoparticles growing from metal ions $AuCl_4^-$ are reduced at the oil–water interface in the presence of an alkanethiolate surfactant (SR, where R = n-C_nH_{2n+1}, $n = 4, 6, 8, 12, \ldots$) and a reducing agent such as sodium borohydride.[60,61] $AuCl_4^-$ was transferred from aqueous solution to toluene using tetra-n-octylammonium bromide as the phase-transfer reagent and reduced with aqueous sodium borohydride in the presence of dodecanethiol. On addition of the reducing agent, the organic phase changes color from orange to deep brown within a few seconds. The overall reaction is summarized as

$$AuCl_4^-(aq) + N(C_8H_{17})_4^+(C_6H_5Me) \longrightarrow N(C_8H_{17})_4^+AuCl_4^-(C_6H_5Me)$$

$$mAuCl_4^-(C_6H_5Me) + nC_{12}H_{25}SH(C_6H_5Me) + 3me^-$$

$$\longrightarrow 4mCl^-(aq) + (Au_m)(C_{12}H_{25}SH)_n(C_6H_5Me)$$

where the source of electrons is BH_4^-. The strategy used here is to finish processing of gold nanocrystals and attaching surfactant molecules in one step. By an extended exposure to excess reducing agent defective structures initially formed can be etched away.

Platinum nanocrystals can also be processed by reduction of platinum ions with the presence of capping materials.[7] For example, the shapes and sizes of platinum nanoparticles are controlled by changes in the ratio of the concentration of the capping polymer material (sodium polyacrylate) to the concentration of the platinum ions (from K_2PtCl_4) used in the reductive synthesis of colloidal particles in solution at room temperature. In this method, hydrogen gas is bubbled at a high flow rate through the solution. Tetrahedral, cubic, irregular-prismatic, icosahedral, and cubo-octahedral particle shapes were observed,[8] whose distribution was dependent on the concentration ratio of the capping polymer material to the platinum cation.

Pileni's group has extensively carried out studies of metallic nanocrystals and they have focused on the collective properties of self-assembly of magnetic nanocrystals. More details can be found in their recent papers.[62,63] Ag particles coated with silica shell can also be self-assembled.[64]

The aerosol method can also be used to process monodispersive metal nanocrystals.[37,40] Gold atoms are evaporated first from a carbon crucible in a resistively heated carbon tube, which are entrained in He and induced to condense

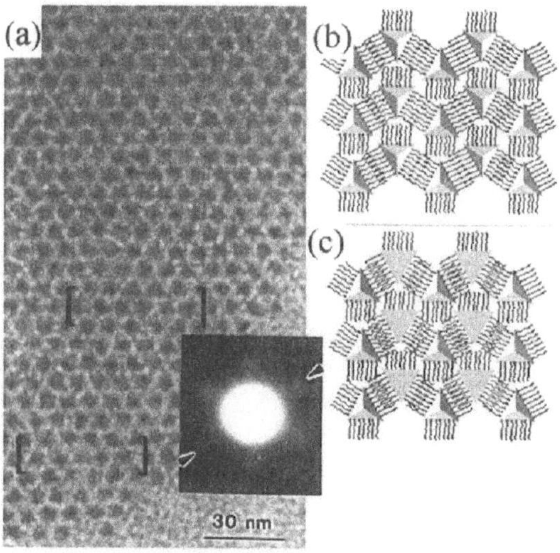

FIGURE 4.20. (a) Ordered self-assemblies of silver nanocrystals. The inset is an electron diffraction pattern, showing the ordered self-assembly. (b,c) Structural models of the self-assembly.

into nanoclusters by mixing the hot flow from the oven with a room temperature stream of He. Controlling conditions in the oven and the flow downstream from the oven controls the mean cluster size. The clusters are molten and recrystallized while still in the gas phase. They are scrubbed from the gas phase by contact with a mist of organic solvent containing 1-dodecanethiol and collected as a stable colloidal suspension. In order to assure that all of the clusters are single crystalline, a dilute aerosol stream of clusters suspended in inert gas is passed through 1-m long tube in which the clusters are first heated above their melting temperature and then cooled to room temperature. Figure 4.20a is NCS of Ag nanocrystals produced be aerosol, which have a rather uniform size distribution after size selection. The Ag nanocrystals are dominated by tetrahedral and their assembly is affected by the shape. Figure 4.20b and c represents two possible structural models.[65]

4.4.2. Semiconductor Nanocrystals

The band structure of semiconductor nanocrystals is quite different from that of the bulk material (Fig. 4.21). Nanocrystals have discrete excited electronic states and an increased band gap in comparison to the bulk semiconductor materials. The smaller the size, the larger is the difference. The difference can be easily differentiated by optical absorption spectroscopy. But for indirect band gap semiconductors like silicon and rocksalt CdSe, the optical spectra are continuous, though the individual valence and conduction band eigenstates are discrete. The bandgap is tunable via controlling nanocrystal size, providing an effective way in adjusting the electronic structure in addition to controlling of particle chemistry.

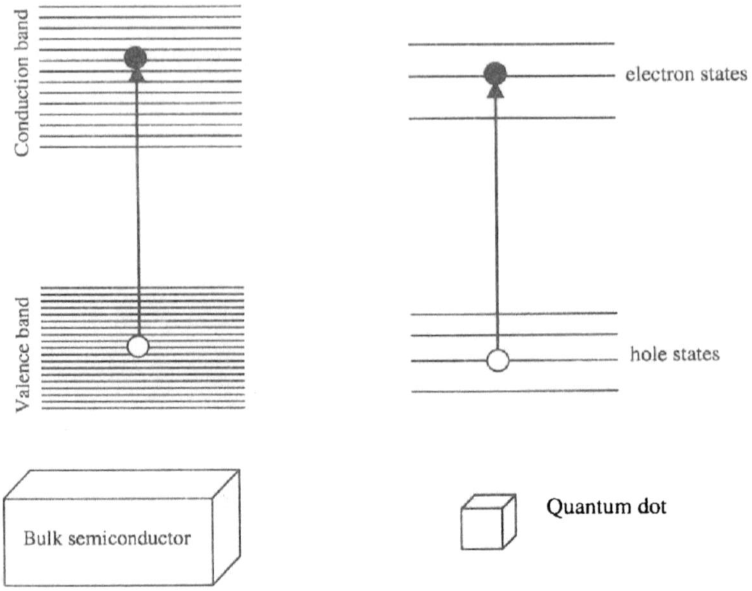

FIGURE 4.21. Schematic comparison of the band structure and electronic states of semiconductor nanocrystals and bulk semiconductor.

The most typical example is silicon nanocrystals. The dynamics and spectroscopy of silicon nanocrystals that emit at visible wavelengths were analyzed.[66] Size-selective precipitation and size-exclusion chromatography clearly separate the silicon nanocrystals from larger crystallites and aggregates and provide direct evidence for quantum confinement in luminescence. Measured quantum yields are as high as 50% at low temperature, principally as a result of efficient oxide passivation. Despite a 0.9 eV shift of the band gap to higher energy, the nanocrystals behave fundamentally as indirect band gap materials with low oscillator strength.

Due to the oxidation on the surface of silicon nanocrystals, intensive research focuses on compound (VI–II and V–III type) semiconductor nanocrystals. In 1988, Brus' group reported a process for synthesis of pure and stable organic capped CdSe nanocrystals using an inverse micelle method.[67] And up to now, CdSe is still the most intensively studied semiconductor nanocrystals.[68–70] CdSe nanocrystals are synthesized by co-dissolving dimethylcadmium and selenium powder in a tri-alkyle phosphine (-butyl or -octyl), and the solution is injected into hot (340–360°C) trioctyl phosphine oxide (TOPO); nucleation occurs rapidly, followed by growth (280–300°C)[71,72] By separating the nucleation stage from the growth stage and the following precipitation process, the CdSe nanocrystals are nearly monodispersive. A control over the kinetics results in the formation of CdSe nanorods.[73]

CdS nanocrystals are prepared by injection of sodium polyphosphate as the stabilizing agent and $Cd(ClO_4)_2$ into deionized water. Introduction of H_2S and an adjustment of the pH by NaOH results in the formation of CdS nanocrystals.[74]

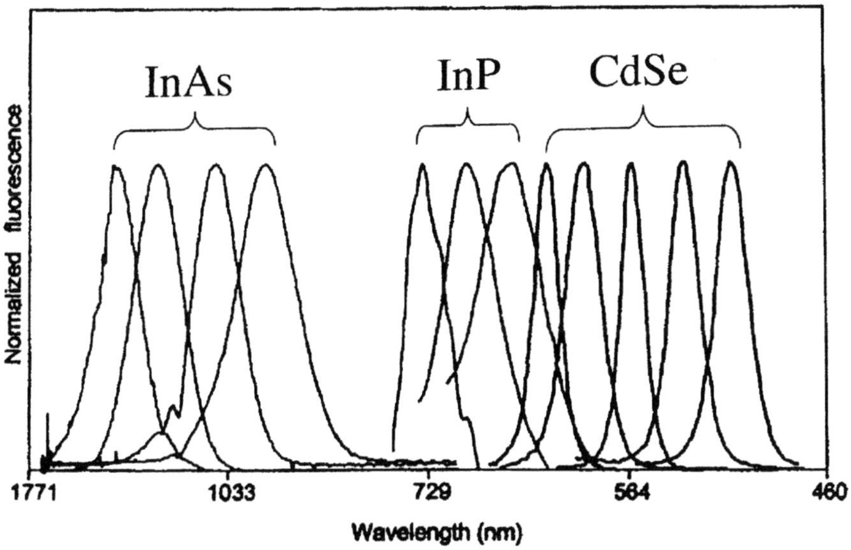

FIGURE 4.22. The emission spectra of several surfactant-coated monodispersive semiconductor nanocrystals. The blue series represents different sizes of CdSe nanocrystals with diameters of 2.1, 2.4, 3.1, 3.6, and 4.6 nm (from right to left). The green series is of InP nanocrystals with diameters of 3.0, 3.5, and 4.6 nm. The red series is of InAs nanocrystals with diameters of 2.8, 3.6, 4.6, and 6.0 nm.nanocrystal probes in aqueous buffer, all illuminated simultaneously with a handheld ultraviolet lamp (Courtsey of Dr. A. P. Alivisatos).[69]

Semiconductor nanocrystals have discrete excited electronic states and an increased band gap in comparison to bulk materials. The size-dependant emission peaks collected from several monodispersive semiconductor nanocrystals are shown in Fig. 4.22. For the indirect band-gap semiconductors such as silicon and rock-salt CdSe nanocrystals the optical absorption spectrum is continuous, even though the band gap increases and the conduction band eigen states are discrete. This is because many overlapping discrete transitions appear to be present with roughly equal intensity via electron–phonon interaction.

The self-organization of CdSe nanocrystallites into 3D semiconductor quantum dot superlattices (colloidal crystals) has been demonstrated.[75] The size and spacing of the dots within the superlattice are controlled with near atomic precision. This control is a result of synthetic advances that produces CdSe nanocrystallites exhibiting monodisperse within the limit of atomic roughness. The methodology is not limited to semiconductor QDs but provides general procedures for the preparation and characterization of ordered structures of nanocrystallites from a variety of materials.

4.4.3. Metallic Magnetic Nanocrystals

Patterned magnetic nanocrystals are of vital interests both scientifically and technologically because of their potential applications in information storage, color imaging, bioprocessing, magnetic refrigeration, and ferrofluids. In ultra-compact information storage, for example, the size of the domain determines the

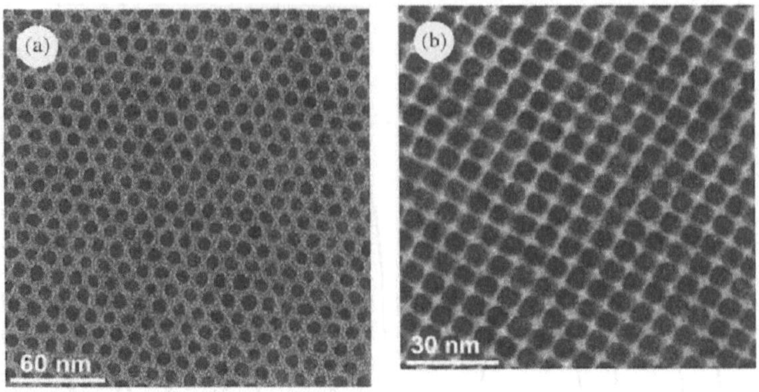

FIGURE 4.23. Self-assembly ordered array of FePt nanocrystals (Courtesy of Dr. S. Sun).

limit of storage density while the sharpness of the domain boundaries is closely related to the media noise. This issue is critically important in the 300 Gb/in.^{2} information storage predicted for the twenty-first century.[76] The noise reduction can be achieved by the segregation of a non-magnetic phase at the grain boundaries, thus, the media are composed of at least two materials. The self-assembly passivated nanocrystal superlattice is a potential candidate for solving this problem, in which the passivated surfactant serves not only as an isolation layer but also as a protection layer for the nanomagnets.

The magnetic nanocrystals like iron, cobalt, and nickel can be processed by the decomposition of metal carbonyl in organic solution. It was noticed that thermal decomposition of metal carbonyl in organic solution (with surfactant) often led to metal nanocrystals with a very narrow size distribution.[77,78] Thermal decomposition of cobalt carbonyl in different kinds of solution (toluene, xylene, etc.) with different kinds of long C–H chains and strong ionic group (sulfonate) surfactant have been systematically studied.

Ordered magnetic nanocrystal self-assembly was processed using the traditional reverse micelle technique.[42] Choosing the cationic surfactant, didodecryl ammonium bromide (DDAB) with toluene as a binary system, DDBA as the reducing agent, the Co^{2+} from $CoCl_2$ can be reduced to form cobalt nanocrystals with the assistance of PR_3 ($R{=}n\text{-}C_4H_9$, $n\text{-}C_8H_{17}$) as the capping material. Dioctylether solution of superhydride ($LiBEt_3H$) is prepared by mixing the THF superhydride solution with dioctylether and evaporating THF under vacuum. The particle synthesis began with the injection of dioctylether superhydride solution into a hot (200°C) $CoCl_2$ dioctyether solution in the presence of oleic acid (octadec-9-ene-1-carboxylic acid, $CH_3(CH_2)_7CH{=}CH(CH_2)_7COOH$) and PR_3. Figure 4.23 shows a self-assembled structure of FePt nanocrystals.[45] These particles have interesting collective magnetic properties and they can be used for magnetoresistance data storage. This self-assembled approach has been recommended by IBM as a potential approach for achieving ultrahigh density data storage. This is also a model system for understanding the interparticle magnetic interaction.

4.4.4. *Oxide Nanocrystals*

Oxide nanocrystals have many interesting applications in the area such as sensors, catalysis and surface coating. Among the self-assembled structures, synthesis of shape and size controlled nanocrystals is a challenging task. Among a few reported systems, self-assembling of α-Fe_2O_3 was first achieved by accident.[51] After the ferrofluid of iron nanocrystals was exposed to air for about one month (processed by thermal decomposition of iron carbonyl in the mixture of decalin with oleic acid as the surfactant), it was found that the iron nanocrystals have been transformed into iron oxide. After drying the colloidal solution, a hexagonal close packing of hematite (α-Fe_2O_3) nanoparticles (antiferromagnet) formed on the carbon substrate. The nanocrystals had a very narrow size distribution (in the case of mean particle size 6.9 nm, the standard deviation is 0.4 nm). Dispersive e-Fe_3N fine particles synthesized by a vapor–liquid chemical reaction between $Fe(CO)_5$ and ammonia have shown a narrow size distribution and they can form nicely locally ordered monolayer array.

A simple method to control the growth of TiO_2 nanocrystallites and the formation of nanostructured TiO_2-based materials has also been developed. The method used to form these materials is based on controlling the hydrolysis and polycondensation of titanium alcoxide using organic ligands in order to build and stabilize intermediate building units (slabs). Anatase structured TiO_2 particles with different sizes and shapes are obtained simply by changing the titanium/cation ratio. The small clusters agglomerate into condensed "snow-ball" structure, which in turn self-assemble into superlattices.

Cobalt oxide nanocrystals were processed by thermal decomposition of $Co_2(CO)_8$ in toluene with the presence of oxygen. Homogenous nucleation and growth were maintained to grow monodispersive nanocrystals. To avoid agglomeration of the freshly nucleated nanocrystals, sodium bis(2-ethylhexyl) sulfosuccinate (in short, Na(AOT)) was added as a surface active agent at the beginning of the growth, forming a monolayer passivation over the nanocrystal surface. Following the systematic study of Papiper *et al.*,[78] the average size of nanocrystals was optimized by controlling the temperature and concentration. Then, the mixed solution was diluted by adding three volume parts of toluene into one part of the solution. The as-prepared solution may contain some impurities and oxides. A size and phase selection was required to obtain Co nanocrystals with specific size and even shape.[49] Since cobalt oxides (CoO and Co_3O_4) are anti-ferromagnetic, a small magnetic field, generated by a horse-shoe permanent magnet, was applied in the vertical direction to make the phase selection of Co from its oxides. The cobalt nanoparticles floated to the top surface of the liquid under the driving force of the magnetic field, while the oxide particles were left in the solution. Though there was gravity force, the smaller size particles could still suspend in the liquid because of the Brownian motion, while the larger ones sank to the bottom. By selecting the particles suspended at the surface of the solution after 24 h in the magnetic field, the pure cobalt nanoparticles were obtained. Figure 4.24 shows a monolayer assembly of CoO nanocrystals. The CoO particles have a tetrahedral shape, as seen in the inserted image and display a triangle

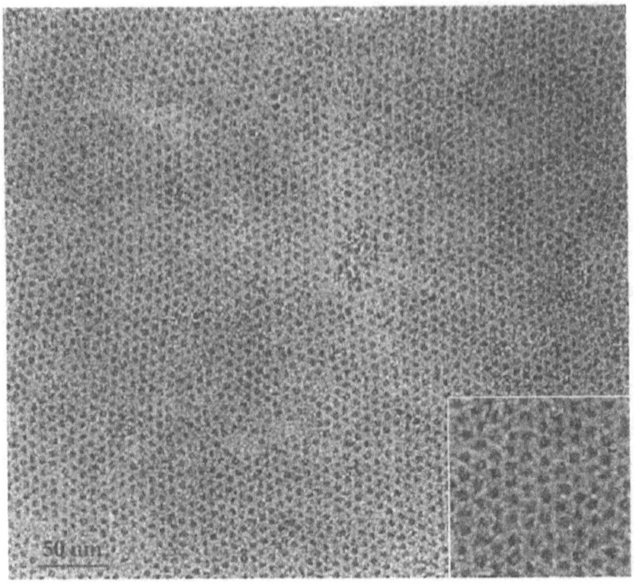

FIGURE 4.24. Monolayer self-assembly of CoO nanocrystals, which are dominated by tetrahedral shape.

contrast in the TEM image. We found that by size selection, it is possible to achieve shape selection in some cases, because certain shape of nanocrystals are stable only in specific size range.

Shape controlled $BaCrO_4$ nanocrystals have been synthesized by fusing reverse micelles and microemulsion droplets containing fixed concentrations of barium and chromate ions. Barium bis(2-ethylhexyl)sulphosuccinate ($Ba(AOT)_2$) reverse micelles were added to sodium chromate (Na_2CrO_4)-containing NaAOT microemulsiopn droplets.[77] The precipitation of the solution at a fixed ration of molar ratios for Ba and Cr results in the formation of rod-like nanocrystals. Self-assembled TiO_2 nanocrystals have been prepared by adding titanium isopropoxide to an aqueous solution of tetramethylammonium hydroxide, and the particles show hexagonal close-packed structure.[80] PbS nanocrystals have been synthesized by hydrolysis of $TiCl_4$ at low temperature, and the cubic-like shape of the nanocrystals can form self-assembly.[81]

4.5. TECHNICAL ASPECTS OF SELF-ASSEMBLING

4.5.1. *Size Selection of Monodispersive Nanocrystals*

Whatever processing methods to be used, the nanocrystal sizes usually have a broad distribution even if the processing conditions can be strictly controlled. Thus, size selection and possibly shape selection are critical for ordered self-assembly. Lyophobic colloidal nanocrystals attract each other via the van der Waals force. The attraction is strong because of the near linear addivity of forces between pairs of unit cells in different nanoparticles. The efficiency of the steric stabilization is strongly dependant on the interaction of the alkyl groups with the solvent. Gradual addition of the nonsolvent can produce size-dependant flocculation of the nanocrystal dispersion.

Several of solvent/nonsolvent systems can be used to make size selection, for example, hexane/ethanol, chloroform/methanol, pyridine/hexane, etc. The addition of the nonsolvent increases the average polarity of the solvent and reduces the energy barrier to flocculation. The large nanoparticles have a higher probability to overcome the reduced energy barrier and precipitate. With more nonsolvent added, the size distribution becomes narrower and narrower. After the addition of methanol, the light absorption peak becomes sharper and the position shifts towards lower energy. From the quantum theory, the smaller the nanocrystals, the wider the energy gap than that of the bulk. The shift of the peak means that more and more larger sized nanocrystals have precipitated from the colloidal solution. In addition to CdSe semiconductor nanocrystals, this method has also been successfully applied to gold nanocrystal separation. One point that needs to be addressed is that this method can only be applied to nanoparticles with relatively small polydispersity.

Photocorrosion is also a very useful method to narrow the size distribution of some nanocrystal systems. For example, polydispersive CdS nanocrystals can be selected to have an average diameter of 4.2 nm with a standard deviation of 1.9 nm by a sequential irradiation with a monochromatic light whose wavelength was changed step by step from 490 to 430 nm in air-saturated sodium hexametaphosphate solution. Analysis of the amount of sulfate ions produced by photocorrosion of quantum-CdS colloids revealed that the number of CdS particles in the colloid decreased with promotion of photocorrosion, suggesting that during the course of photocorrosion process photocorroded CdS particles were agglomerated to give larger particles which were further photocorroded. The molar absorption coefficient of CdS particles at the first exciton peak was found to be independent of the particle size.[82]

4.5.2. Assembling of Nanoparticles with Mixed Sizes/Phases

A key requirement in self-assembly is the size selection of the nanocrystals. If the solution contains two distinctly different sizes of nanocrystals, and in a general case, can the mixture of nanocrystals with different sizes form self-assembling packing? Several recent papers addressed this question.[38,83] First take the simplest situation: the mixture of two monodispersive nanocrystals (A and B) with particle size R_A and R_B. On micrometer scale, when $0.482 < R_A/R_B < 0.624$, the mixture can form stable structure, and when $R_A/R_B \sim 0.58$, stable AB_2 structure will be formed. For $0.458 < R_A/R_B < 0.482$, phase separation occurs. Can these theories be applied to self-assembling of nanocrystals? The self-assembly of nanoscale bimodal packing was observed (Fig. 4.25), which agrees well with the micrometric scale colloidal crystals. So if the polydispersive gold nanocrystals to examine the mechanism of self-assembling.[83] It was found that self-assembling is an entropy-driven crystallization process.

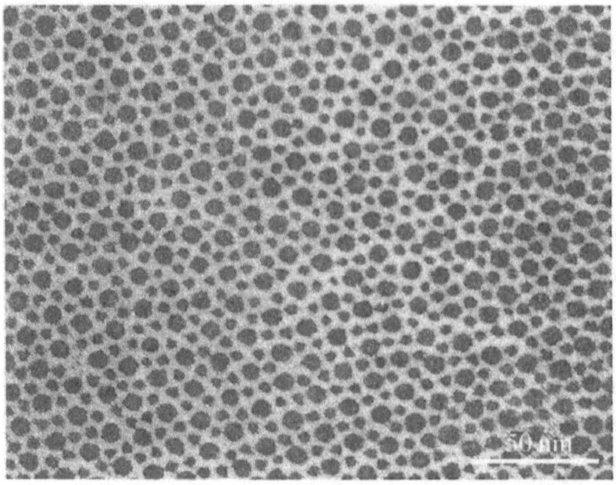

FIGURE 4.25. A monolayer ordered superlattice of thiol-stabilized gold nanocrystals with two distinctive sizes and the particle diameter ratio about 0.58 (Courtesey of Dr. D. J. Schiffrin).[38]

FIGURE 4.26. Self-assembly of Ag nanocrystals have two distinct sizes, showing the segregation of particles of the same size.

In a general case, particles of different sizes tend to segregate via size distribution during self-assembly. Shown in Fig. 4.26 is a TEM image illustrating the self-assembling of 5 and 3 nm Ag nanocrystals. It is apparent that the same size of nanocrystals tend to form self-assembly with them. This might be driven by energy minimization.

4.5.3. Growth Mechanism of Nanocrystal Self-Assembly

Although NCS has been grown for several different types of materials, its growth mechanism is still an open question. In most TEM observations, nanocrystals are uniformly adhered on the substrate surface and there is no "piling up" effect among pieces of NCSs, suggesting that the NCSs are formed on the liquid surface during the slow drying process. For TEM observation, a droplet of diluted solution is deposited on a carbon film, a slow drying process is critical for forming large size, ordered NCSs. Nanocrystals are believed to be uniformly suspended in the solution at first. If the diffusion speed of the nanoparticles in the liquid is slower than the evaporation speed of the liquid surface, the particle concentration is expected to increase *locally* right underneath the liquid surface, possibly resulting in the self-assembling of a 2D monolayer at the liquid surface, and the surface tension might be a force to hold on the assembling. As the evaporation continues, the second layer can be formed on the top of the first layer if there are enough particles in the solution. Finally, when the evaporation reaches the substrate, the 2D assembled NCS layer(s) may be broken due to the unmatched curvature between the liquid droplet and the flat substrate surface, but no overlap between the large size NCSs would be possible.

Figure 4.27 shows such a process for Ag NCS. From the structural point of view, most of the NCSs comprised spherical nanoparticles have been found to show the platelet structure and large size monolayer,[40] indicating the growth is a layer-by-layer "epitaxial" growth. This growth is possible if the nanocrystals are driven to diffuse towards the liquid surface. This is evident by examining the crystallinity of Ag NCSs grown by a slow, dilute drying process. Moreover, for faceted nanocrystals, the self-assembled NCSs frequently show platelet structure and more importantly, planar defects such as twins and stacking faults are usually observed, indicating that each layer is grown row-by-row and particle-by-particle, in supporting the diffusion model. From the experimental observation of Harfenist *et al.*,[41] the triangular platelet structure of Ag NCSs is bounded by three $\{110\}$ facets and the plate normal is $[111]_S$. This type of structure is likely to be grown from a single nucleation site followed by a row-to-row 2D growth along $<110>$, forming a triangle monolayer; a continuous layer-by-layer epitaxial growth on the top of this triangle layer, as the particles are provided may form the platelet structure. The smooth morphological structure of NCSs indicates the particle diffusion is much faster than the arriving rate of the particles, and the "hit-and-stick" model does not exist.[83] This growth mechanism is proposed based on the current limited experimental data. More observations are required to examine this model in detail.

4.6. PROPERTIES OF NANOCRYSTAL SELF-ASSEMBLY

Due to the ordered structure of self-assembly, they are expected to have many interesting properties different from either individual nanocrystals or the bulk materials. One of the major challenges in making full use of the self-assembled nanocrystals is making defect-free, larger-area and structurally controlled self-assemblies. But in reality, this self-assembly process depends on many conditions and it seems that defects are almost inevitable. Thus, a major

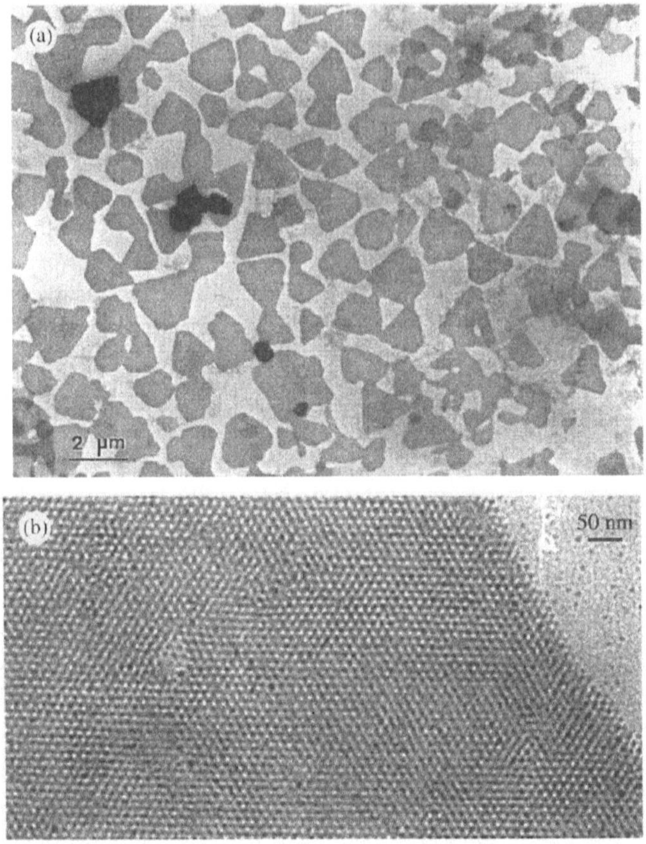

FIGURE 4.27. Platelet structure of uniform thickness formed by self-assembly of Ag nanocrystals. The triangular shape is the result of {111}$_S$ facets. (b) A magnified TEM image of the self-assembly, showing particle rows in the structure and the (111)$_S$ faceted edge.

question is if we can work with the self-assembled structures with defects. A recent report sheds some light in solving this problem.

Teramac, a massively parallel experimental computer, was built at Hewlett-Packard Laboratories to investigate a wide range of different computational architectures.[85] This machine contains about 2,20,000 hardware defects, any one of which could prove fatal to a conventional computer, and yet it operated 100 times faster than a high-end single-processor workstation for some of its configurations. The defect-tolerant architecture of Teramac, which incorporates a high communication bandwidth that enables it to easily route around defects, has significant implications for any future nanometer-scale computational paradigm. It may be feasible to chemically synthesize individual electronic components with less than a 100% yield, assemble them into systems with appreciable uncertainty in their connectivity, and still create a powerful and reliable data communications network. Future nanoscale computers may consist of extremely large-configuration memories that are programmed for specific tasks by a tutor that locates and tags the defects in the system.

Heterostructured diode at nanoscale is also an important application. Metal/self-assembled monolayer heterostructured diodes connected by conjugated molecular wire had been made.[86] Electronic transport measurements showed a distinctive rectifying behavior from the asymmetry of molecular heterostructure. If these conductive conjugated molecular wires can be used to link metal nanocrystals, the dimension of one diode can be decreased below 10 nm. To make such devices successful, it is crucial to understand the bonding between metal nanocrystals and the surfactant molecules, the charging transfer process in and between metal nanocrystals and the relative distance between the nanocrystals.

Close-packed planar arrays of nanometer-diameter metal clusters that are covalently linked to each other by rigid, double-ended organic molecules have been successfully self-assembled.[36] Gold nanocrystals, each encapsulated by a monolayer of alkyl thiol molecules, were cast from a colloidal solution onto a flat substrate to form a close-packed cluster monolayer. Organic interconnects (aryl dithiols or aryl di-isonitriles) displaced the alkyl thiol molecules and covalently linked adjacent clusters in the monolayer to form a 2D superlattice of metal QDs coupled by uniform tunnel junctions. Electrical conductance through such a superlattice of 3.7-nm-diameter gold clusters, deposited on a SiO_2 substrate in the gap between two gold contacts and linked by an aryl di-isonitrile [1,4-di(4-isocyanophenylethynyl)-2-ethylbenzene], exhibited nonlinear Coulomb charging behavior. The electrical conductance through gold clusters interconnected by aryl dithiol and aryl di-isonitrile molecules has been measured by a STM, which was used to measure the current–voltage characteristics of a bare gold cluster deposited on a dithiol SAM grown on a flat Au(111) surface. The results are in good agreement with semiclassical predictions. Clusters with diameters <2 nm exhibited "Coulomb staircase" behavior in conductance at room temperature.

Another significant progress was made in the property measurement by light absorption and impedance spectroscopy in the ordered self-assembled silver monolayer thin film.[30,87] *In situ* measurements were conducted on both linear and nonlinear optical properties of organically functionalized silver nanocrystal Langmuir monolayers as a continuous function of interparticle separation distance. The results are shown in Fig. 4.28. As the monolayer was compressed from an average separation between the surfaces of the metal cores of 12 (± 2) to ~5 (± 2) Å, the linear and nonlinear optical properties reveal evidence of both classical and quantum interparticle coupling phenomena. Below ~5 Å, evidence for a sharp insulator-to-metal transition is observed in both optical signals. The nonlinear optical response abruptly decreases to a nearly constant value, and the linear reflectance drops precipitously until it matches that of a continuous metallic film. This transition is reversible: The particles can be re-dissolved back into a colloid, or, if the trough barriers are opened, the film is again characterized by the optical properties of near-isolated silver nanocrystals. The transition between the metal-like and insulator-like can be summarized using a new parameter $D/2r$, in which D is the interparticle separation and r is the radius of monodispersive nanocrystal. For $D/2r > 1.3$, the optical response abides by the classical coupling model; for $1.2 < D/2r < 1.3$, quantum coupling dominates the trends in the linear and non-linear optical responses; for $D/2r < 1.2$, a sharp metal/insulator transition was

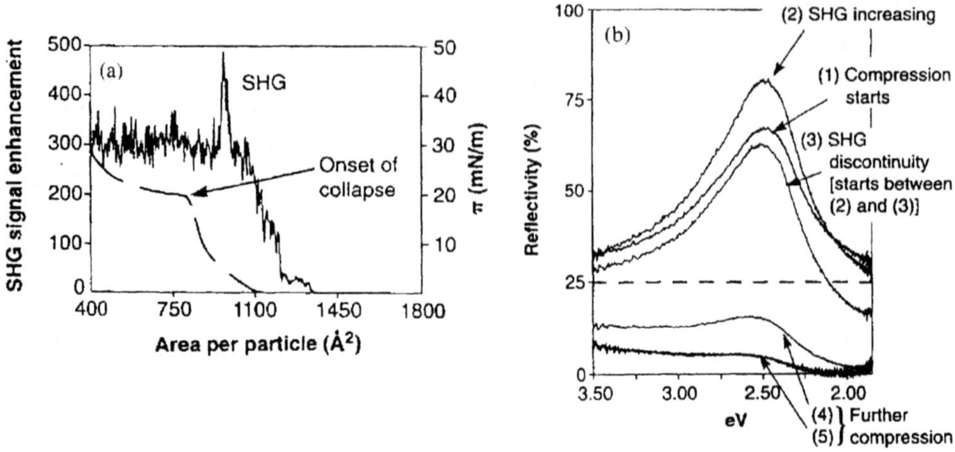

FIGURE 4.28. (a) Surface pressure of the Langmuir film (π) and SHG signal enhancement as a function of area per particle for hexanethiol-capped particles. The change from insulator connection between thiol capped silver nanocrystals (diameter 3.5 nm) to metallic connection was demonstrated. (b) UV absorption spectra collected *in situ* as film was compressed. Correlation with the nonlinear optical response is indicated. After excessive compression, the strong plasma peak vanished.[38]

observed. More work is required to verify if this general conclusion applies to other metal nanocrystal systems. Practically, this method also gives a good method to handle the self-assembled thin films. First large area self-assembled nanocrystal films can be formed on the surface of a Langmuir trough, and then transfer to a suitable substrate so that different kinds of properties can be measured.

A strategy for the synthesis of "nanocrystal molecules", in which discrete numbers of gold nanocrystals are organized into spatially defined structures based on Watson–Crick base-pairing interactions, were proposed by Alivisatos *et al.*[88] Single-stranded DNA oligonucleotides of defined length and sequence were attached to individual nanocrystals, and these assemble into dimers and trimers on addition of a complementary single-stranded DNA template. This approach should allow the construction of more complex 2D and 3D assemblies between organic and inorganic materials.

The charging effect of metal nanocrystals in the self-assembly is very important to understand the basic properties of self-assembly. A transition from metal-like double-layer capacitive charging to redox-like charging was observed in electrochemical ensemble Coulomb staircase experiments from solutions of gold nanoparticles of varied core sizes.[89] The monodisperse gold nanoparticles are stabilized by short-chain alkanethiolate monolayers and have 8–38 kDa core mass (1.1–1.9 nm in diameter). Larger cores display Coulomb staircase responses consistent with double–layer charging of metal-electrolyte interfaces, whereas smaller core nanoparticles exhibit redox chemical character, including a large central gap. The change in behavior is consistent with new near-infrared spectroscopic data showing an emerging gap between the highest occupied and lowest unoccupied orbitals of 0.4–0.9 eV. In addition, because the staircase behavior is closely related to MPC core electronic energy structure, it may aid understanding of other nanophase properties, such as the metal–insulator transition of silver nanoparticles upon compression. Finally, although differences in fundamental properties reside in the metal-like and molecule-like charging behaviors, we anticipate that their electrochemical, thermodynamic, and kinetic properties will, upon further study, prove to fit within a common formal representation.

If the building block in the self-assembly is magnetic nanocrystals, some extra magnetic effects may be brought in. Preliminary results showed that the blocking temperature of self-assembled magnetic nanocrystals have some differences from the bulk materials. It was reported[29] that the blocking temperature of 2D monolayer of cobalt nanocrystal increased from 58 K (for isolated cobalt nanocrystals with an average size of ~5.8 nm) to 63 K. It was predicted that the difference may be even higher in the case of 3D nanocrystal self-assembly. Larger area 2D magnetic nanocrystal self-assembly and the corresponding magnetic properties have also been reported.[42] The magnetic measurement showed that broad transition from superparamagnetic to ferromagnetic at 165 K is due to the magneto-static particle interactions in the close-packed nanocrystal network. But for the diluted solution with dispersed magnetic nanocrystals, the transition temperature is at 105 K. All these results demonstrate that the interactions between the nanocrystals in the self-assembling influence the magnetic properties.

Quantum-confined Stark effect in quantum wells is very useful for optical modulations. In the case of semiconductors, the Stark effect was found very strong at nanoscale because the energy levels can be fine tuned by an externally applied electric field. Depending on the size of the particles, semiconductor nanocrystals can give different colors of emission due to quantum effect. This characteristic can be applied to labeling molecules or cells, so that their diffusion and movement trace in biological tissues, for example, can be tracked.[69] This is a typical example of how nanocrystals can be used for biomedical research.

4.7. TEMPLATE ASSISTED SELF-ASSEMBLY

Interaction between nanoparticles is likely dominated by van der Waals force. This weak work force can only induce a limited constrain over the ordering of self-assembly, and in many cases the ordering is determined by the size of the particles, the thickness of the surface passivation layer and the shape of the particles. A key challenge in self-assembly of nanostructures is to create large size, long-range ordered and defect free self-stacked structure. Solution chemistry can achieve to a certain degree of perfectness in self-assembly, but large size self-assembly is possible with the utilization of templates. The template serves as the host for the formation of ordered structure, allowing creation of large size and structurally controlled self-assembly. This section describes a few commonly used template assisted self-assembly techniques.

4.7.1. NanoChannel Array Guided Self-Assembly

The most typical template for self-assembly of wire-like or rod-like structures is the alumina nanoholes created by electrochemical etching. Anodic porous alumina, which is prepared by the anodic oxidation of aluminum in an acidic electrolyte, is one of the typical self-organized fine structure with a nanohole arrays.[90,91] Anodic porous alumina has a packed array of columnar hexagonal cells with central, cylindrical, and uniform sized holes. These holes are nanocells for fabrication of a variety of nanostructures, such as magnetic nanorod arrays for longitudinal magnetic data storage,[92,93] optical devices,[94] functional electrodes, and electroluminescence display devices. Using these as templates, aligned carbon nanotubes with Y shape[95] and semiconductor nanowires[96] have been synthesized.

Long, nanometer-size metallic wires can be synthesized by injection of the conducting metal into nanochannel insulating plates. Large-area arrays of parallel wires 200 nm in diameter and 50 μm long with a packing density of 5×10^8 cm^{-2} have been fabricated in this way.[97]

Ultra-small size channels can be created using the molecular structure. The encapsulation of graphite-type carbon wires in the regular, 3-nm-wide hexagonal channels of the mesoporous host MCM-41 has been reported.[98] Acrylonitrile monomers are introduced through vapor or solution transfer and polymerized in the channels with external radical initiators. Pyrolysis of the intrachannel polyacrylonitrile results in filaments whose microwave conductivity is about 10 times that of bulk carbonized polyacrylonitrile.

4.7.2. Natural Structure Hosted Self-Assembly

Many biological structures are the result of molecular level or cell level self-assembly. Bacterial S-layers represent an almost universal feature of archaebacterial cell envelopes. S-layers are 2D crystalline structures of single protein or glycoprotein monomers and exhibit either oblique, square or hexagonal lattice symmetry with spacings between the morphological units in the range of 3–30 nm.[99] Most of the S-layers are 5–15 nm in thickness, possess pores of identical size and morphology in 2–6 nm range. The S-layers have been demonstrated to serve as a template for the *in situ* nucleation of ordered 2D arrays of CdS nanocrystals of 5 nm in size.[100]

Nanostructured titanium oxide mask can be derived from a protein crystal template.[101] Pattern transfer from the biological crystal to the metal oxide film and finally to the graphite substrate are accomplished entirely by parallel processing. A periodic array of 10-nm-diameter holes is fabricated by fast-atom beam milling of a smooth graphite surface in which a 3.5-nm-thick titanium oxide screen was used as a mask. This mask technology has been applied to synthesis of silicon QDs on silicon surfaces using controlled chemical etching.[102]

Light-directed nanocrystal self-assembly is also an effective method to control the self-assembly of nanocrystals.[103] Using the light sensitive substrate and suitable mask, different kinds of self-assembling patterns can be formed on a substrate. For example, because the surface-bound amino group can be derivatized to give a thiol terminus, the light directed assembly of CdS, CdSe, and other semiconductors or metal nanocrystals is possible.

4.7.3. Catalysis Guided Growth of Carbon Nanotube Arrays

Growth of carbon nanotubes via chemical vapor deposition relies on metal catalysts, such as Fe and Ni. The metal particles serve as the catalytic active sites for producing the carbon atoms. The metal particles can locate the roots of the carbon nanotubes or at their growth front depending on the strength of bonding between the nanoparticles and the substrate. One nanoparticle is usually responsible for the nucleation and growth of one carbon nanotube. If the distribution of catalytic particles can be patterned on the surface of the substrate, aligned carbon nanotube arrays with a lateral pattern determined by the prepatterned catalysis would be formed.[104-107] Figure 4.29 shows aligned growth carbon nanotubes grown by a CVD process. The growth of the aligned carbon nanotubes is a self-assembly process. This type of structure has unique applications in electron field emission display and environmental purification.

4.7.4. Stress Constrained Self-Assembly of Quantum Dots

Growth of heterostructures is largely affected by the mismatch between the grown film and the substrate. If the material to be grown has a much different atom size from the substrate atoms, islands will be grown on the surface instead of forming a uniform film.[108] The lattice mismatch between the grown material and the substrate results in a 2D network of coherently matched regions and largely

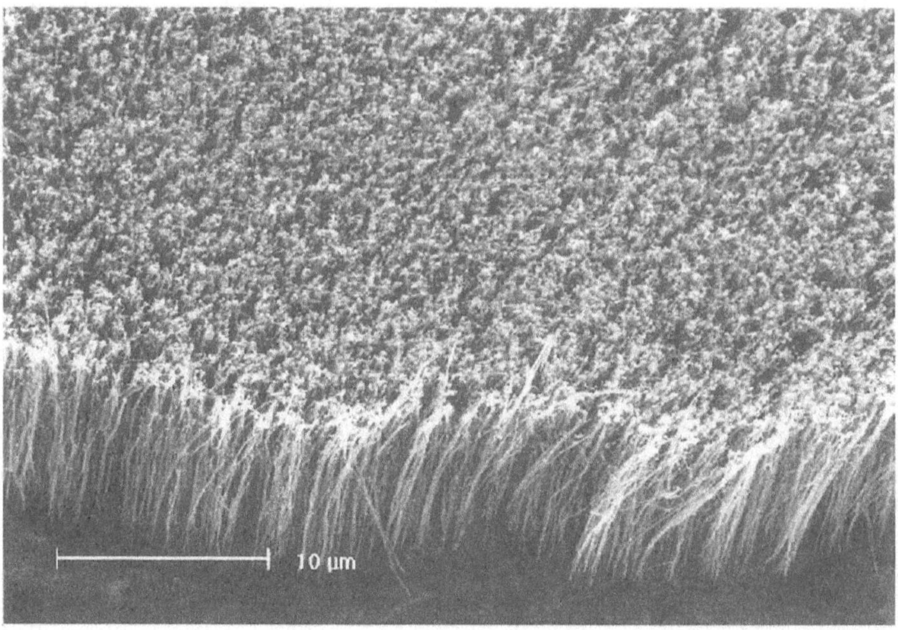

FIGURE 4.29. Aligned carbon nanotubes grown by a CVD process (Courtsey of DR. L. M. Dai).

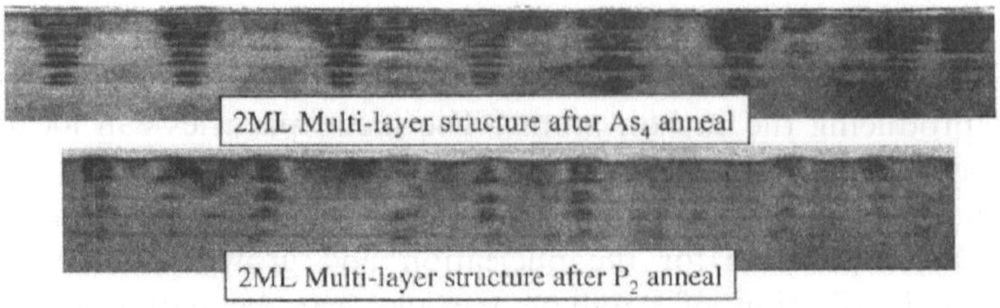

FIGURE 4.30. Self-assembled InAs quantum dots grown in GaAs layers after different post-growth treatments (Courtesy of Drs. A. Brown and Y. Q. Wang).

mismatched regions, leading to the growth of islands at the net points of coherent matched regions. The islands usually have a shape and their edges are the areas with maximum strain. Growth of InAs on GaAs is a typical example of QDs.[109,110] The growth of InAs on GaAs for the first few monolayers would form nanoislands on the flat substrate. The stress on a crystal surface provides a natural driving force for forming an ordered self-assembly of QDs. Figure 4.30 shows the InAs QDs embedded in GaAs layers, where a vertical alignment is apparent. The image clearly shows the lateral distribution of the QDs, but their shape may not be easily seen directly from the image because of strain contrast. The electronic structures of the QDs are determined not only by the size and shape of the QDs, but also by the strain.

QDs combine the advantages offered by bulk semiconductor and single atoms. Interaction of carriers in QDs is defined by quantum mechanics, and the main information transfer between QDs is single electron tunneling and Coulomb interaction of carriers in neighboring dots. This gives the possibility of building novel devices based on QDs, such as single electronics and QDs cellular automata computing. QDs can be used for infrared detectors, light-emitting diodes, lasers, and solar cells.[111,112]

4.8. SUMMARY

Self-assembling process is fundamental in biological systems in nature. Synthesis of new materials using self-assembly is an effective approach that has the potential of producing high quality, large quantity, and chemically and structurally controlled new materials. This chapter reviewed the current status of the ordered self-assembled nanocrystals, ordered mesoporous nanostructured materials, and hierarchically ordered materials. Most of the results are rather exciting, but substantial research is needed to improve and control the synthesis process to attain chemically and structurally well-characterized materials of high quality at a large yield. Growth of large-size single-crystalline self-assembled nanocrystal structures is crucial for their applications. Template assisted synthesis techniques are possible solutions.

REFERENCES

1. Ph. Buffat and J. P. Borel, Size effect on the melting temperature of gold particles, *Phys. Rev. A* **13**, 2287–2298 (1976).
2. M. Schmidt, R. Kusche, B. von Issendorff, and H. Haberland, Irregular variations in the melting point of size-selected atomic clusters, *Nature* **393**, 238–240 (1998).
3. A. N. Goldstein, C. M. Echer, and A. P. Alivisatos, Melting of semiconductor nanocrystals, *Science* **256**, 1425–1427 (1992).
4. A. P. Alivisatos, Semiconductor clusters, nanocrystals, and quantum dots, *Science* **271**, 933–937 (1996).
5. C. B. Murray, D. J. Norris, and M. G. Bawendi, Synthesis and characterization of nearly monodisperse Cde (E = S, Se, Te) semiconductor nanocrystallites, *J. Am. Chem. Soc.* **115**, 8706–8715 (1993).
6. P. Buseck, L. Eyring, and J. Cowley (eds) *High Resolution Transmission Electron Microscopy, Theory and Applications* (Oxford University Press, London, 1989).
7. T. S. Ahmadi, Z. L. Wang, T. C. Green, A. Henglein, and M. A. El-Sayed, Shape-controlled synthesis of colloidal platinum nanoparticles, *Science* **28**, 1924–1926 (1996).
8. Z. L. Wang, T. S. Ahmadi, and M. A. El-Sayed, Steps, Ledges and kinks on the surfaces of platinum nanoparticles of different shapes, *Surf. Sci.* **380**, 302–310 (1997).
9. S. Ino, Epitaxial growth of metals on rocksalt faces cleaved in vacuum. II. Orientation and structure of gold particles formed in ultrahigh vacuum, *J. Phys. Soc. Jpn.* **21**, 346–362 (1966).
10. J. G. Allpress and J. V. Sanders, The structure and orientation of crystals in deposits of metals on mica, *Surf. Sci.* **7**, 1–25 (1967).
11. L. D. Marks, Experimental studies of small-particle structures, *Rep. Prog. Phys.* **57**, 603–649 (1994), and the references therein.
12. C. Y. Yang, M. J. Yacaman, and K. Heinemann, Crystallography of decahedral and icosahedral particles, *J. Cryst. Growth* **47**, 283–290 (1979).
13. P.-A. Buffat, M. Flüeli, R. Spycher, P. Stadelmann, and J.-P. Borel, Crystallographic structure of small gold particles studied by high-resolution electron-microscopy, *Faraday Discuss.* **92**, 173–187 (1991).
14. C. L. Cleveland, U. Landman, T. G. Schaaff, M. N. Shafigullin, P. W. Stephens, and R. L. Whetten, Structural evolution of smaller gold nanocrystals: The truncated decahedral motif, *Phys. Rev. Lett.* **79**, 1873–1876 (1997).
15. Y. Y. Yu, S. S. Chang, C. L. Lee, and C. R. Wang, Gold nanorods: Electrochemical synthesis and optical properties, *J. Phys. Chem. B* **101**(34), 6661–6664 (1997).
16. Z. L. Wang, M. Mohamed, S. Link, and M. A. El-Sayed, Crystallographic facets and shapes of gold nanorods of different aspect ratios, *Surf. Sci.* **440**, L809–L814 (1999).
17. Z. L. Wang, R. P. Gao, B. Nikoobakht, and M. A. El-Sayed, Surface reconstruction of the ubstable {110} surface in Gold Nanorods, *J. Phys. Chem. B* **104**, 5417–5420 (2000).
18. R. L. Whetten, J. T. Khoury, M. M. Alvarez, S. Murthy, I. Vezmar, Z. L. Wang, P. W. Stephens, C. L. Cleveland, W. D. Luedtke, and U. Landman, Nanocrystal gold molecules, *Adv. Mater.* **8**, 428–433 (1996).
19. T. G. Schaaff, G. Knight, M. N. Shafigullin, R. F. Borkman, and R. L. Whetten, Isolation and selected properties of a 10.4 kDa Gold: Glutathione cluster compound, *J. Phys. Chem. B* **102**, 10643–10646 (1998).
20. R. L. Whetten, M. N. Shafigullin, J. T. Khoury, T. G. Schaaff, I. Vezmar, M. M. Alvarez, and A. Wilkinson, Crystal structures of molecular gold nanocrystal arrays, *Acc. Chem. Res.* **32**, 397–406 (1999).
21. J. M. Petroski, Z. L. Wang, T. C. Green, and M. A. El-Sayed, Kinetically controlled growth and shape formation mechanism of platinum nanoparticles, *J. Phys. Chem. B* **102**, 3316–3320 (1998).
22. Z. L. Wang, Structural analysis of self-assembling nanocrystal superlattices, *Adv. Mater.* **10**, 13–30 (1998).
23. P. V. Braun, P. Osenar, and S. I. Stupp, Semiconducting superlattices templated by molecular assemblies, *Nature* **380**, 325–328 (1996).

24. H. Weller, Self-organized superlattices of nanoparticles, *Angew. Chem.* **35**, 1079–1081 (1996).
25. R. G. Nuzzo and D. L. Allara, Adsorption of bifunctional organic disulfides on gold surfaces, *J. Am. Chem. Soc.* **105**, 4481–4483 (1983).
26. L. Brus, Quantum crystallites and nonlinear optics, *Appl. Phys. A* **53**, 465–474 (1991).
27. J. H. Fendler and F. C. Meldrum, The colloid-chemical approach to nanostructured materials, *Adv. Mater.* **7**, 607–632 (1995).
28. A. P. Alivisatos, Perspectives on the physical chemistry of semiconductor nanocrystals, *J. Phys. Chem. B* **100**, 13226–13239 (1996).
29. M. P. Pileni, Nanosized particles made in colloidal assemblies, *Langmuir* **13**, 3266–3276 (1997).
30. C. P. Collier, R. J. Saykally, J. J. Shiang, S. E. Henrichs, and J. R. Heath, Reversible tuning of silver quantum dot monolayers through the metal–insulator transition, *Science* **277**, 1978–1981 (1997).
31. C. J. Brinker, Y. Lu, A. Sellinger, and H. Fan, Evaporation-induced self-assembly: Nanostructures made easy, *Adv. Mater.* **11**, 579 (1999).
32. J. Y. Ying, C. P. Mehnert, and M. S. Wong, Synthesis and applications of supramolecular-templated mesoporous materials, *Angew. Chem. Int. Ed. Eng.* **38**, 56–77 (1999).
33. C. B. Murray, C. R. Kagan, and M. G. Bawendi, Self-organization of CdSe nanocrystallites into three-dimensional quantum dot superlattices, *Science* **270**, 1335–1338 (1995).
34. W. Shenton, D. Pum, U. B. Sleytr, and S. Mann, Synthesis of cadmium sulphide superlattices using self-assembled bacterial S-layers, *Nature* **389**, 585–587 (1997).
35. A. A. Guzelian, J. E. B. Katari, and A. V. Kadavanich, Synthesis of size-selected, surface-passivated InP nanocrystals, *J. Phys. Chem. B* **100**, 7212–7219 (1996).
36. H. Hu, M. Brust, and A. J. Bard, Characterization and surface charge measurement of self-assembled CdS nanoparticle films, *Chem. Mater.* **10**, 1160–1165 (1998).
37. R. P. Andres, J. D. Bielefeld, J. I. Henderson, D. B. Janes, V. R. Kolagunta, C. P. Kubiak, W. J. Mahoney, and R. G. Osifchin, Self-assembly of a two-dimensional superlattice of molecularly linked metal clusters, *Science* **273**, 1690–1963 (1996).
38. C. J. Kiely, J. Fink, M. Brust, D. Bethell, and D. J. Schiffrin, Spontaneous ordering of bimodal ensembles of nanoscopic gold clusters, *Nature* **396**, 444–446 (1998).
39. S. A. Harfenist, Z. L. Wang, M. M. Alvarez, I. Vezmar, and R. L. Whetten, Highly oriented molecular Ag-nanocrystal arrays, *J. Phys. Chem. B* **100**, 13904–13910 (1996).
40. S. A. Harfenist, Z. L. Wang, M. M. Alvarez, I. Vezmar, and R. L. Whetten, Hexagonal close packed thin films of molecular Ag-nanocrystal arrays, *Adv. Mater.* **9**, 817–822 (1997).
41. P. N. Provencio, J. E. Martin, J. G. Odinek, and J. P. Wilcoxon, Studies of hexagonal Pt and Au nanocluster superlattices, *Micros. Microanal.* **4**(Suppl. 2), 734–735 (1998).
42. S. Sun and C. B. Murray, Synthesis of monodisperse cobalt nanocrystals and their assembly into magnetic superlattices, *J. Appl. Phys.* **85**, 4325–4330 (1999).
43. J. S. Yin and Z. L. Wang, Preparation of self-assembled cobalt nanocrystal arrays, *Nanostruct. Mater.* **11**, 845–852 (1999).
44. C. Petit, A. Taleb, and M. P. Pileni, Self-organization of magnetic nanosized cobalt particles, *Adv. Mater.* **10**, 259 (1998).
45. S. H. Sun, C. B. Murray, D. Weller, L. Folks, and A. Moser, Monodisperse FePt nanoparticles and ferromagnetic FePt nanocrystal superlattices, *Science* **287**, 1989–1992 (2000).
46. H. B. Sun, S. Matsuo, and H. Misawa, *Appl. Phys. Lett.* **74**, 786 (1999).
47. H. B. Sun, Y. Xu, S. Matsuo, and H. Misawa, Microfabrication and characteristics of two-dimensional photonic crystal structures in vitreous silica, *Opt. Rev.* **6**, 396–398 (1999).
48. T. Moritz, J. Reiss, D. Diesner, D. Su, and A. Chemseddine, Nanostructured crystalline TiO_2 through growth control and stabilization of intermediate structural building units, *J. Phys. Chem. B* **101**, 8052–8053 (1997).
49. J. S. Yin and Z. L. Wang, Ordered self-assembling of tetrahedral oxide nanocrystals, *Phys. Rev. Lett.* **79**, 2570–2573 (1997).
50. J. S. Yin and Z. L. Wang, Synthesis and structure of self-assembled cobalt oxide nanocrystal materials, *J. Mater. Res.* **14**, 503–508 (1999).
51. M. D. Bentzon, J. Van Wonterghem, S. Mrup, A. Thölen, and C. J. W. Koch, Ordered aggregates of ultrafine iron-oxide particles – super crystals, *Phil. Mag. B* **60**, 169–178 (1989).

52. L. Motte, F. Billoudet, E. Lacaze, and M.-P. Pileni, Self-organization of size-selected nanoparticles into three-dimensional superlattices, *Adv. Mater.* **8**, 1018–1020 (1996).
53. L. Motte, F. Billoudet, E. Lacaze, J. Douin, and M. P. Lipeni, Self-organization into 2D and 3D superlattices of nanosized particles differing by their size, *J. Phys. Chem. B* **101**, 138–144 (1997).
54. Z. L. Wang, Z. R. Dai, and S. Sun, Polyhedral shapes of cobalt nanocrystals and their effect on ordered nanocrystal assembly, *Adv. Mater.* **12**, 1944–1946 (2000).
55. M. M. Alvarez, J. T. Khoury, T. G. Schaaff, M. Shafigullin, I. Vezmar, and R. L. Whetten, *Chem. Phys. Lett.* **91**, 266 (1997).
56. J. S. Yin and Z. L. Wang, In situ structural evolution of self-assembled oxide nanocrystals, *J. Phys. Chem. B* **101**, 8979–8983 (1997).
57. M. J. Hostetler and R. W. Murray, Colloids and self-assembled monolayers, *Curr. Opin. Colloid. Interface Sci.* **2**, 42–50 (1997).
58. D. E. Cliffel, F. P. Zamborini, S. M. Gross, and R. W. Murray, Mercaptoammonium-monolayer-protected, water-soluble gold, silver, and palladium clusters, *Langmuir* **16**, 9699–9702 (2000).
59. M. Faraday, Experimental relations of gold (and other Metals) to light, *Philos. Trans. R. Soc. London* **147**, 145–181 (1857).
60. M. Brust, M. Walker, D. Bethell, D. J. Schiffrin, and R. Whyman, Synthesis of thio-derivatized gold nanoparticles in a two-phase liquid–liquid system, *J. Chem. Soc. Chem. Commun.* 801–802 (1994).
61. M. Brust, J. Fink, D. Bethell, D. J. Schiffrin, and C. Kiely, Synthesis and reactions of functionalised gold nanoparticles, *J. Chem. Soc. Chem. Commun.* 1655–1656 (1995).
62. M. P. Pileni, Nanocrystal self-assemblies: Fabrication and collective properties, *J. Phys. Chem. B* **105**, 3358–3371 (2001).
63. M. P. Pileni, Self-assemblies of nanocrystals: Fabrication and collective properties, *Appl. Surf. Sci.* **171**, 1–14 (2001).
64. M. Giersig, T. Ung, L. M. Liz-Marzan, and P. Mulvaney, Direct observation of chemical reactions in silica-coated gold and silver nanoparticles, *Adv. Mater.* **9**, 570–575 (1997).
65. Z. L. Wang, S. A. Harfenist, I. Vezmar, R. L. Whetten, J. Bentley, N. D. Evans, and K. B. Alexander, Superlattices of self-assembled tetrahedral Ag nanocrystals, *Adv. Mater.* **10**, 808–812 (1998).
66. W. L. Wilson, P. F. Szajowski, and L. E. Brus, Quantum confinement in size-selected, surface-oxidized silicon nanocrystals, *Science* **262**, 1242–1244 (1993).
67. M. L. Steigerwald, A. P. Alivisatos, J. M. Gibson, T. D. Harris, R. Kortan, A. J. Muller, A. M. Thayer, T. M. Duncan, D. C. Douglass, and L. E. Brus, Surface derivatization and isolation of semiconductor cluster molecules, *J. Am. Chem. Soc.* **110**, 3046–3050 (1988).
68. M. Nirmal, B. O. Dabbousi, M. G. Bawendi, J. J. Macklin, J. K. Trautman, T. D. Harris, and L. E. Brus, Fluorescence intermittency in single cadmium selenide nanocrystals, *Nature* **383**, 802–804 (1996).
69. M. Bruchez, M. Moronne, P. Gin, S. Weiss, and A. P. Alivisatos, Semiconductor nanocrystals as fluorescent biological labels, *Science* **281**, 2013–2016 (1998).
70. M. Tomaselli, J. L. Yarger, M. Bruchez, R. H. Havlin, D. deGraw, A. Pines, and A. P. Alivisatos, NMR study of InP quantum dots: Surface structure and size effects, *J. Chem. Phys. B* **110**, 8861–8864 (1999).
71. X. G. Peng, J. Wickham, and A. P. Alivisatos, Kinetics of II-VI and III-V colloidal semiconductor nanocrystal growth: "Focusing" of size distribution, *J. Am. Chem. Soc.* **120**, 5343–5344 (1998).
72. C. B. Murray, D. J. Norris, and M. G. Bawendi, Synthesis and characterization of nearly monodisperse CdE (E = S, Se, Te) semiconductor nanocrystallites, *J. Am. Chem. Soc.* **115**, 8706–8715 (1993).
73. X. G. Peng, L. Manna, W. D. Yang, J. Wickham, E. Scher, A. Kadavanich, and A. P. Alivisatos, Shape controle of CdSe nanocrystals, *Nature* **404**, 59–61 (2000).
74. L. Spanhel, M. Haase, H. Weller, and A. Henglein, Photochemistry of colloidal semiconductors. 20. Surface modification and stability of strong luminescing CdS particles, *J. Am. Chem. Soc.* **109**, 5649–5655 (1987).

75. C. B. Murray, C. R. Kagan, and M. G. Bawendi, Self-organization of CdSe nanocrystallites into three-dimensional quantum dot superlattice, *Science* **270**, 1335–1338 (1995).
76. Y. Nakamura and I. Tagawa, An analysis of perpendicular magnetic recording using a newly-developed 2D-FEM combined with a medium magnetization model, *IEEE Trans. Magn.* **25**, 4159–4161 (1989).
77. J. R. Thomas, Preparation and magnetic properties of colloidal cobalt particles, *J. Appl. Phys.* **37**, 2914–2915 (1966).
78. E. Papirer, P. Horney, H. Balard, R. Anthore, C. Petipas, and A. Martinet, The preparation of a ferrofluid by decomposition of dicobalt octacarbonyl, *J. Colloid Interface Sci.* **94**, 207–228 (1983).
79. M. Li, H. Schnablegger, and S. Mann, Coupled synthesis and self-assembly of nanoparticles o give structures with controlled organization, *Nature* **402**, 393–395 (1999).
80. T. Morits, J. Reiss, K. Diesner, D. Su, and A. Chemseddine, Nanostructured crystalline TiO_2 through growth control and stabilization of intermediate structural building units, *J. Phys. Chem. B* **101**, 8052–8053 (1997).
81. E. Hao, B. Yang, S. Yu, M. Gao, and J. Shen, Formation of orderly organized cubic PbS nanoparticles domain in the presence of TiO_2, *Chem. Mater.* **9**, 1598–1600 (1997).
82. H. Matsumoto, T. Sakata, H. Mori, and H. Yoneyama, Preparation of monodisperse CdS nanocrystals by size selective photocorrosion, *J. Phys. Chem. B* **100**, 13781–13785 (1996).
83. P. C. Ohara, D. V. Leff, J. R. Heath, and W. M. Gelbart, Crystallization of opals from polydisperse nanoparticles, *Phys. Rev. Lett.* **75**, 3466–3469 (1995).
84. Z. Zhang and M. G. Lagally, Atomistic processes in the early stages of thin-film growth, *Science* **276**, 377–383 (1997).
85. J. R. Heath, P. J. Kuekes, G. S. Snider, and R. S. Williams, A defect-tolerant computer architecture: Opportunities for nanotechnology, *Science* **280** 1716–1721 (1998).
86. C. Zhou, M. R. Deshpande, M. A. Reed, L. Jones, and J. M. Tour, Nanoscale metal self-assembled monolayer metal heterostructures, *Appl. Phys. Lett.* **71**, 611–613 (1997).
87. G. Markovich, C. P. Collier, and J. R. Heath, Reversible metal–insulator transition in ordered metal nanocrystal monolayers observed by impedance spectroscopy, *Phys. Rev. Lett.* **80**, 3807–3810 (1998).
88. A. P. Alivisatos, K. P. Johnson, X. G. Peng, T. E. Wilson, C. J. Loweth, M. P. Bruchez, and P. G. Schultz, Organization of 'nanocrystal molecules' using DNA, *Nature* **382**, 609–611 (1996c).
89. S. W. Chen, R. S. Ingram, M. J. Hostetler, J. J. Pietron, R. W. Murray, T. G. Schaaff, J. T. Khoury, M. M. Alvarez, and R. L. Whetten, Gold nanoelectrodes of varied size: Transition to molecule-like charging, *Science* **280**, 2098–2101 (1998).
90. G. E. Thompson, R. C. Furneaux, G. C. Wood, J. A. Richardson, and J. S. Gode, Nucleation and growth of porous anodic films on aluminum, *Nature* **272**, 433–435 (1978).
91. H. Masuda and K. Fukuda, Ordered metal nanohole arrays made by a two-step replication of honeycomb structures of anodic alumina, *Science* **268**, 1466–1468 (1995).
92. M. Shiraki, Y. Wakui, T. Tokushima, and N. Tsuya, Perpendicular magnetic media by anodic oxidation method and their recording characteristics, *IEEE Trans. Magn.* **21**, 1465–1467 (1985).
93. S. A. Majetich and Y. Yin, Magnetization directions of individual nanoparticles, *Science* **284**, 470–473 (1999).
94. M. Saito, M. Kirihara, T. Taniguchi, and M. Miyagi, Micropolarizer made of the anodized alumina film, *Appl. Phys. Lett.* **55**, 607–609 (1989).
95. J. Li, C. Papadopoulos, and J. Xu, Nanoelectronics – Growing Y-junction carbon nanotubes, *Nature* **402**, 253–254 (1999).
96. D. S. Xu, D. P. Chen, Y. J. Xu, X. S. Shi, G. L. Guo, L. L. Gui, and Y. Q. Tang, Preparation of II-VI group semiconductor nanowire arrays by dc electrochemical deposition in porous aluminum oxide templates, *Pure Appl. Chem.* **72**, 127–135 (2000).
97. C. A. Huber, T. E. Huber, M. Sadoqi, J. A. Lubin, S. Manalis, and C. B. Prater, Nanowire array composites, *Science* **263**, 800–802 (1994)
98. C. G. Wu, and T. Bein, Conducting carbon wires in ordered, nanometer-sized channels, *Science* **266**, 1013–1015 (1994).
99. T. J. Beveridge, Bacterial S-layers, *Curr. Opin. Struct Biol.* **4**, 204–212 (1994).

100. K. W. Shenton, D. Pum, U. B. Sleytr, and S. Mann, Synthesis of cadmium sulphide superlattices using self-assembled bacterial S-layers, *Nature* **389**, 585–587 (1997).
101. K. Douglas, G. Devaud, and N. A. Clark, Transfer of biologically derived nanometer-scale patterns to smooth substrates, *Science* **257**, 642–644 (1992).
102. T. A. Winningham, H. P. Gillis, D. A. Choutov, K. P. Martin, J. T. Moore, and K. Douglas, Formation of ordered nanocluster arrays by self-assembly on nanopatterned Si(100) surfaces, Surf. *Science* **406**, 221–228 (1998).
103. T. Vossmeyer, E. DeIonno, and J. R. Heath, Light-directed assembly of nanoparticles, *Angew. Chem. Int. Ed. Eng.* **36**, 1080–1083 (1997).
104. W. Z. Li, S. S. Xie, L. X. Qian, B. H. Chang, B. S. Zou, W. Y. Zhou, R. A. Zhao, and G. Wang, Large-scale synthesis of aligned carbon nanotubes, *Science* **274**, 1701–1703 (1996).
105. Z. F. Ren, Z. P. Huang, J. H. Xu, P. B. Wang, M. P. Siegal, and P. N. Provencio, Synthesis of large arrays of well-aligned carbon nanotubes on glass, *Science* **282**, 1105–1107 (1998).
106. S. Fan, M. G. Chapline, N. R. Franklin, T. W. Tombler, A. M. Cassell, and H. Dai, Self-oriented regular arrays of carbon nanotubes and their field emission properties, *Science* **283**, 512–514 (1999).
107. S. M. Huang, L. M. Dai, and A. W. H. Mau, Patterned growth and contact transfer of well-aligned carbon nanotube films, *J. Phys. Chem. B* **103**, 4223–4227 (1999).
108. D. Bimberg, Quantum dots: Paradigm changes in semiconductor physics, *Semiconductors* **33**, 951–955 (1999).
109. Q. Xia, N. P. Kobayashi, T. R. Ramachandran, A. Kalburge, and P. Chen, Strain coherent InAs quiantum box islands on GaAs(100); size equalization, vertical self-organization, and optical properties, *J. Vac. Sci. Technol. B* **14**, 2203–2207 (1996).
110. Q. Xia, A. Madhukar, P. Chen, and N. P. Kobayashi, Vertically self-organized InAs quantum box islands on GaAs(100), *Phys. Rev. Lett.* **75**, 2542–2545 (1995).
111. D. Bimberg, M. Grundmann, and N. N. Ledentsov, *Quantum Dot Heterostructures* (Wiley, New York,1998).
112. L. Goldstein, F. Glas, J. Y. Marzin, M. N. Charasse, and G. Le Roux, Growth by molecular beam epitaxy and characterization of InAs/GaAs strained-layered superlattices, *Appl. Phys. Lett.* **47**, 1099--1101 (1985).

5

Structural Characterization of Nanoarchitectures

Periodic packing of nanocrystals is different from 3D packing of atoms in several aspects. First, to an excellent approximation atoms are spherical, while nanoparticles can be faceted polyhedra, thus, the 3D packing of particles can be critically affected by their shapes and sizes. Second, the sizes of atoms are fixed, but the size of nanoparticles can have a slight variation although their size distribution is very narrow. Finally, atomic bonding is due to the outshell electrons via ionic, covalence, metallic bonding, or the mixture, and in most cases the interatomic distance is fixed, while the bonding between nanoparticles is generated by the passivating surfactant with a controllable length, thus, the ratio of particle size to interparticle distance is adjustable. This is a parameter that is likely to determine the 3D packing of the nanoparticles.

There are many characterization methods that can be used to measure the properties of the self-assembly of nanocrystals. After the formation of self-assembly, XRD and TEM are the most common tools for structure analysis. In the stable colloidal solution, UV light absorption can be used to determine the polydispersity of the nanocrystals dispersed in the solution. This method applies to metal nanocrystals and semiconductor nanocrystals. The interparticle interaction in the self-assembly can be analyzed by impedance spectroscopy. This chapter reviews the techniques for structure characterization of self-assembled nanostructures.

5.1. X-RAY DIFFRACTION

5.1.1. Crystallography of Self-Assembly

X-ray diffraction is the first choice to examine the formation of crystalline assembly, provided the amount of the sample is sufficient. The diffraction spectrum at the high-angle range is directly related to the atomic structure of the nanocrystals, whereas the spectrum at the lower angle region is directly associated to the ordered assembly of nanocrystals.[1,2] By examining the diffraction peaks that are extinct in the spectrum, one may identify the crystallography of the packing. This analysis is based on an assumption that each particle is identical in size, shape, and even orientation (i.e., of the same X-ray scattering factor), so that the extinction rules derived from diffraction physics apply. In practice, however, a fluctuation in the size, orientation, or shape can easily abolish this hypothesis. This is the reason that a quantitative analysis of the low-angle diffraction spectrum is rather difficult.

XRD is the most powerful technique to evaluate the average interparticle distance, and it is unique for studying the *in situ* pressure and/or temperature induced phase transformation in nanocrystals.[3] Figure 5.1 gives a low-angle XRD pattern recorded for self-assembly of Au nanocrystals. The positions of

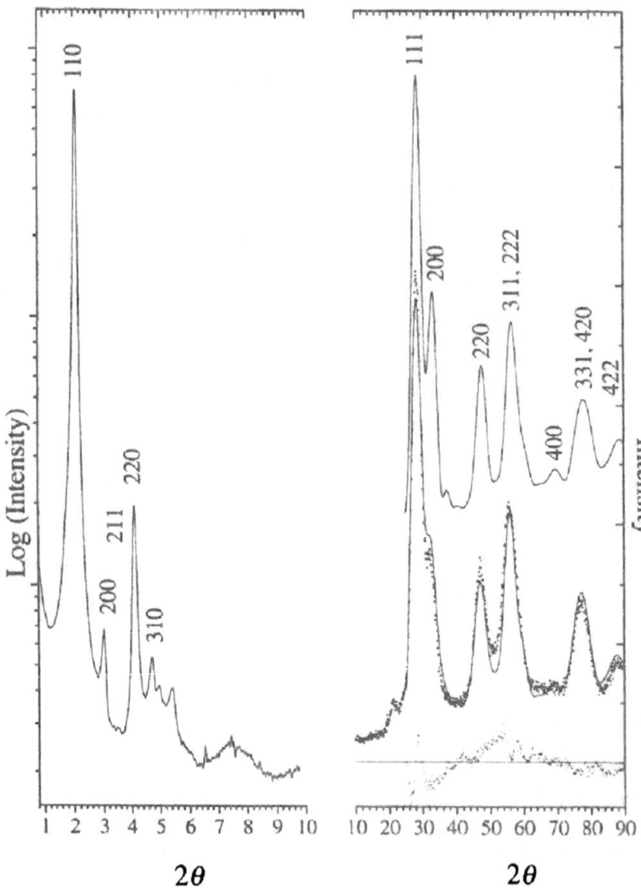

FIGURE 5.1. Low-angle XRD spectrum from Au nanocrystals self-assembled superlattices. The atomic structure of the Au nanocrystals is characterized by the scattering at high angles. (Courtesy of Whetten *et al.*[1])

the diffracting peaks can be indexed to be bcc, while their intensities deviate far from the theoretically expected values due to possibly the reasons just discussed.

An important result of low-angle XRD is determining the lattice constant(s) of the self-assembled structure, from which the core-to-core interparticle distance, D_{nn}, can be measured. On the other hand, mass spectroscopy is very powerful to precisely measure the mean size of the particles, D_{MS}. Therefore, the surface-to-surface distance as defined by the interdigitative surface adsorbed monolayers is $\Delta D = D_{nn} - D_{core}$. Figure 5.2(a) shows the dependence of ΔD on the size of the Au particles. Just as expected, the length of the thiolates remains constant although the particle size varies. Figure 5.2(b) shows a linear relationship between D_{nn} and D_{core}. The crystallography of the self-assembly depends strongly on the ratio between fully extended chain length (L) of the thiolates to the size of the metal core, which is $\xi = 2L/D_{core}$. A phase diagram for the crystallographic structure of the self-assembly of Au nanocrystals is plotted as a function of ξ in Fig. 5.2(c), which shows distinctive regions for self-assemblies with fcc, bcc and body-centered tetragonal (bct) structures.[4]

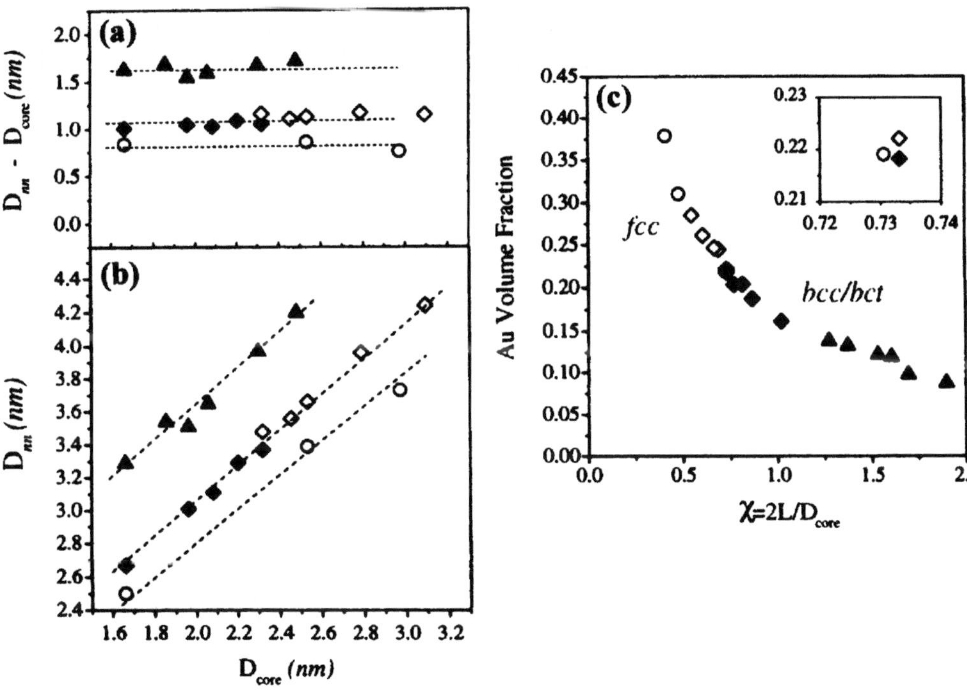

FIGURE 5.2. Collected structural parameters and phase diagram for Au nanocrystals passivated with thiolates of different lengths. The symbole code is the following: triangles = C12, diamonds = C6, circles = C4, filled symbols = bcc (or bct) structure, empty symbols = fcc structure observed. (a) Face-to-face distance and (b) nanocrystal-to-nanocystal distance as a function of core size. (c) The computed volume fraction of Au core and the corresponding packing structure as a function of $\xi = 2L/D_{core}$ (see the text). (Courtesy of Whetten *et al.*[4])

5.1.2. Structure of Nanocrystals

X-ray diffraction is a powerful tool for refining the atomic structure of nanoclusters.[5] This is possible if monodispersive nanocrystals can be prepared and separated at a reasonable quantity. If the particles are oriented randomly so that the entire assembly can be treated as a "polycrystalline" specimen composed of nanocrystals with identical structure but random orientations so that their scattering from each can be treated independently, the diffraction intensity $I(s)$ as a function of the diffraction vector length $s = 2 \sin \theta/\lambda$, where θ is the diffraction half-angle and λ is the wavelength of X-ray photons, is

$$I(s) = \sum_{i,j}^{N} f_i f_j \sin(2\pi s r_{ij})/(2\pi s r_{ij}) \tag{5.1}$$

where, r_{ij} is the distance between atom i and atom j in the cluster of a total N atoms, and f_i and f_j are their X-ray scattering factors. To make quantitative comparison of the calculated intensity profile with XRD spectrum, the contribution from the background created from the substrate used to support the nanocrystals must be subtracted; a Debye–Waller factor must also be introduced in the X-ray scattering factor to account for the thermal diffuse scattering effect.

The structure of nanocrystals can be refined by quantitative comparison of the theoretically calculated diffraction spectra for different nanocrystal models

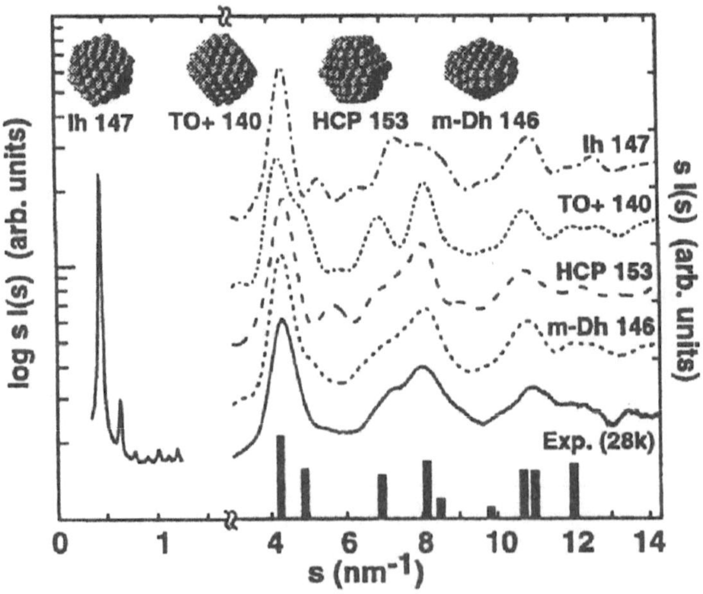

FIGURE 5.3. Experimental and calculated XRD ($\lambda = 0.15405$ nm) intensity plotted vs s. The small-s (left, log $sI(s)$) and large-s (right, $sI(s)$) for the separated 28k nanocrystal gold sample. The theoretical pattern were calculated for various relaxed atomistic structural models (denoted above each of the curves by the structural symbol followed by the number of atoms in the model shown at the top). (Courtsey of Cleveland *et al.*[6])

with the experimentally observed ones. Figure 5.3 shows a comparison between the experimentally observed XRD pattern from Au clusters of ~150 atoms and the theoretically calculated intensity for nanocrystals of different structures, including icosahedral (Ih), truncated octahedral (TO), hexagonal close-packed, and the Marks pentagonal decahedral (m-Dh). It is apparent that the m-Dh model fits the experimental data the best. This is a unique application of XRD for quantitative structure determination of particles with sizes >1.5 nm.[6]

The width of the diffraction lines is closely related to the size, size distribution and defects in nanocrystals. As the size of the nanocrystals decreases, the line width is broadened, resulting in inaccuracy in quantitative structural analysis of nanocrystals smaller than ~1 nm. Moreover, the data analysis is complicated by the presence of the substrate, which usually gives a broad diffuse scattering background that can seriously affect the accuracy of data quantification.

Mass spectroscopy is also very powerful for detecting the critical sizes of the smallest nanoclusters that can be synthesized and the magic number of atoms comprising the clusters.[7] This type of analysis is unique for clusters in which the atoms have not formed a well-defined crystal lattice, prohibiting the access of XRD. Determination of the structure of clusters of 1 nm in sizes or smaller is still a challenge.

5.2. SCANNING PROBE MICROSCOPY

Scanning probe microscopy (SPM) represents a group of techniques, such as STM, AFM, magnetic force microscopy and chemical force microscopy, etc., which have been extensively applied to characterize nanostructures.[8] A common characteristic of these techniques is that an atom sharp tip scans across the specimen surface and the images are formed by either measuring the current flowing through the tip or the force acting on the tip (longitudinal or transverse). SPM has impacted greatly the research in many fields because it can be operated in a variety of environmental conditions, in liquid, air, or gas, allowing direct imaging of inorganic surfaces and organic molecules. SPM not only provides the "eyes" for imaging nanoscale world, it also provides the "hands" for manipulation and construction of nanoscale objects. The discovery of SPM is a landmark contribution to nanotechnology.

STM is based on the vacuum tunneling effect in quantum mechanics.[9] The wave function of the electrons in a solid extends into the vacuum and decay exponentially. This is a barrier that prevents the passage of the electrons belonging to the solid. If a tip is brought sufficiently close to the solid surface, the overlap of the electron wave functions of the tip with that of the solid results in the tunneling of the electrons from the solid to the tip when a small electric voltage is applied (Fig. 5.4). Images are obtained by detecting the tunneling current when the bias voltage (V) is fixed while the tip is scanned across the surface, because the magnitude of the tunneling current (I_t) is extremely sensitive to the gap distance (x) between the tip and the surface

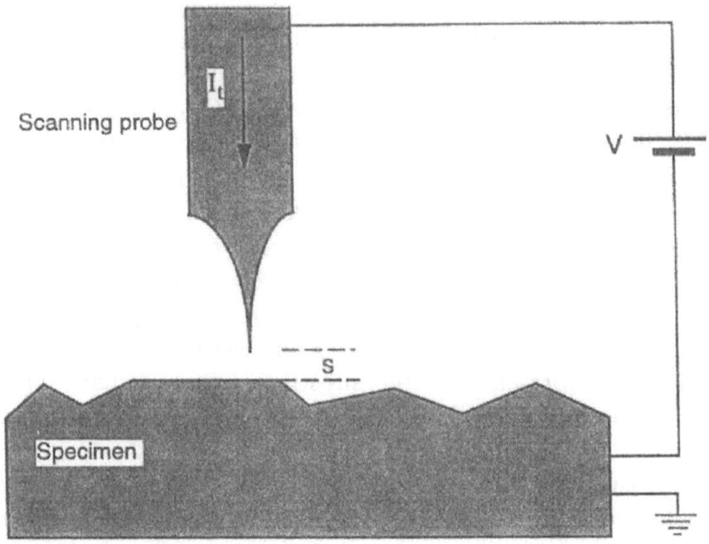

FIGURE 5.4. Schematic diagram showing the principle of a STM.

$$I_t \propto V \exp[-4\pi(2m\phi)^{1/2}x/h] \tag{5.2}$$

where m is the electron mass, h, Planck's constant, and ϕ the surface wave function. The exponential decay of the tunneling current as a function of the gap x is the basis of high resolution imaging. Based on $I_t \sim V$ curve measured experimentally, the surface electronic structure can also be derived. Therefore, STM is not only an imaging technique but also a spectroscopy technique.

It must be pointed out that STM works for conductive specimens. To image organic molecules which are most likely insulators, the molecules must be deposited on a conductive substrate. For non-conductive polymers, AFM can be an optimum choice.[10,11] AFM operates in an analogous mechanism except the signal is the force between the tip and the solid surface. The interaction between two atoms can be represented by a short-range repulsive force and a long-range attractive force. The repulsive force takes the dominant role when the two atoms are close enough, which can be approximated as a spring. By measuring the force acting on the tip, the compression of the spring is measured, which reflects the distance from the tip atom(s) to the surface atom, thus, images can be formed by detecting the force while the tip is scanned across the specimen. This technique overcomes the requirement of electric conductive specimen. A more generalized application of AFM is the scanning force microscopy, which measures the force of magnetic, electrostatic, friction, or molecular (van der Waals) interactions, allowing nanomechanical measurements.

Figure 5.5 shows an AFM image of the InAs islands grown by molecular beam epitaxy on GaAs substrate surface. The size of the InAs islands can be directly measured provided the image broadening from the tip is deconvoluted. Quantitative measurement shows that the average island size is 437 nm^2

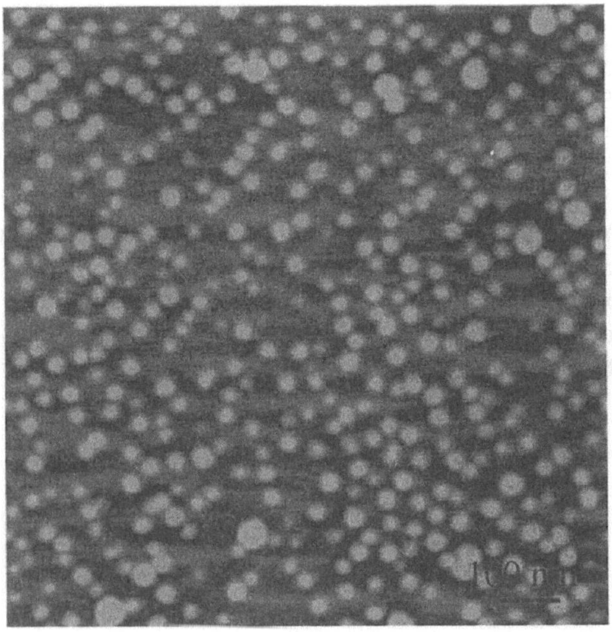

FIGURE 5.5. AFM image of InAs QDs grown on GaAs surface.

(diameter, 24 nm), with a standard deviation of 172 nm^2. The average area density is about $4.27 \times 10^{10} cm^{-2}$.

SPM imaging and spectroscopy provide nanoscale imaging and physical property measurements, which have been demonstrated as a powerful approach for nanotechnology. There is no question that the development of SPM has revolutionized our understanding of many fundamental phenomena. But one must realize the limitation of the SPM techniques. One of the most difficult questions is the interpretation of SPM images, because the image contrast strongly depends on the shape of the tip. A typical concern is the multi-contact problem, which means the tunneling current or contacting force may come from several contacting points because the tip may not be a single atom sharp. Therefore, caution must be exercised in quantitative interpretation of SPM images.

5.3. SCANNING ELECTRON MICROSCOPY

Scanning electron microscopy (SEM) is a powerful technique for imaging the surfaces of almost any materials. It is likely the most popular tool for material characterization. A modern SEM can furnish a resolution of 1 nm, relatively simple image interpretation, and large depth of focus, making it possible to directly image the 3D structure of nanomaterials.

SEM operates in a voltage of 100 V–30 kV and the electron probe can be as small as 1–3 nm with the use of a field emission source. The image resolution offered by SEM depends not only on the property of the electron probe, but also on the interaction of the electron probe with the specimen. Interaction of an incident electron beam with the specimen produces secondary electrons, with energies typically smaller than 50 eV, the emission efficiency of which sensitively depends on surface geometry, surface chemical characteristics, and bulk chemical composition.[12] Shown in Fig. 5.6 is a schematic illustration of the interaction volume of an incident electron probe with a semi-infinity bulk specimen at normal incidence. The X-ray signal generated is largely determined by the spatial distribution of the beam inside the bulk crystal, which is usually much larger than the size of the incident probe, strongly limiting the resolution of X-ray micro-analysis in SEM. Back scattered electrons, with an energy close to the incident

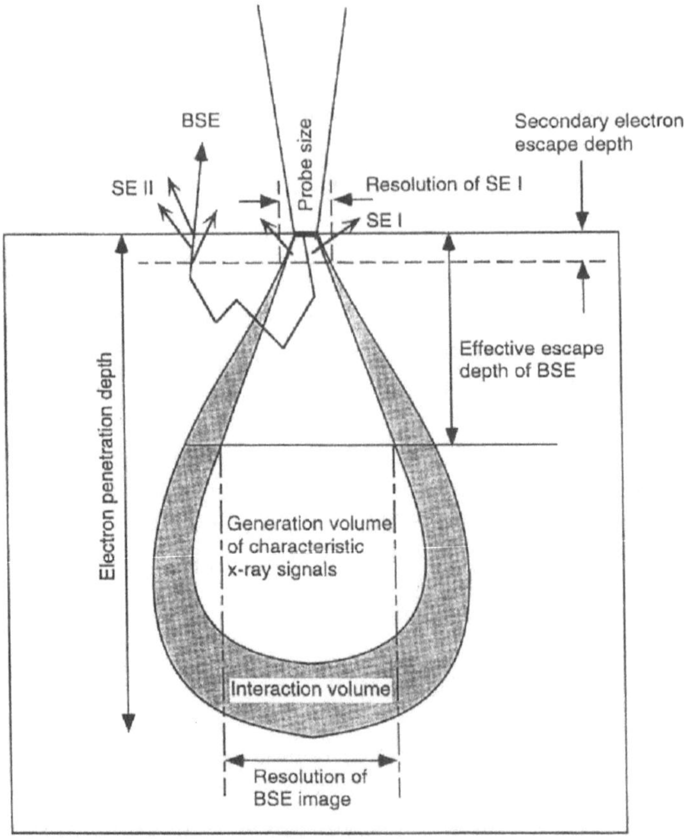

FIGURE 5.6. Interaction of an incident electron beam with a bulk specimen in SEM.

beam energy, come from a depth smaller than the depth of the X-ray generation volume. The secondary electrons can be emitted either in the region covered by the incident electron probe or in the extended region due to the backscattered electrons. The former can give a resolution close to the probe size because the secondary electrons that can escape the solid must be generated within a depth into the material smaller than the escape depth of the secondary electrons; and the latter is the result of multiple scattering and the lateral extension could be much larger than the probe size. The size and lateral extension of the interaction volume depend not only on the energy of the incident beam, but also on the solid material because of stopping power. The interaction volume can be relatively large in polymer specimens because of the low atomic number of the target atoms.

Figure 5.7 is an SEM image recorded from the aligned silica nanowires grown from a gallium droplet.[13] The nanowires self-assembled into bunches and aligned radially. The diameters of the nanowires are ~10 nm and rather uniform. The most interesting phenomenon is that a nanowire can self-split into multi-nanowires that form a bunch (see the upper-right hand corner of Fig. 5.7). The splitting is directly related to the structure of the silica glass, in which the tetrahedral unit of $[SiO_4]^{-4}$ is the basic structure, and tetrahedrons are interconnected by sharing corners. One tetrahedron can have at least three corners that can be shared, leading to the nucleation and replicating growth of nanowire branches.

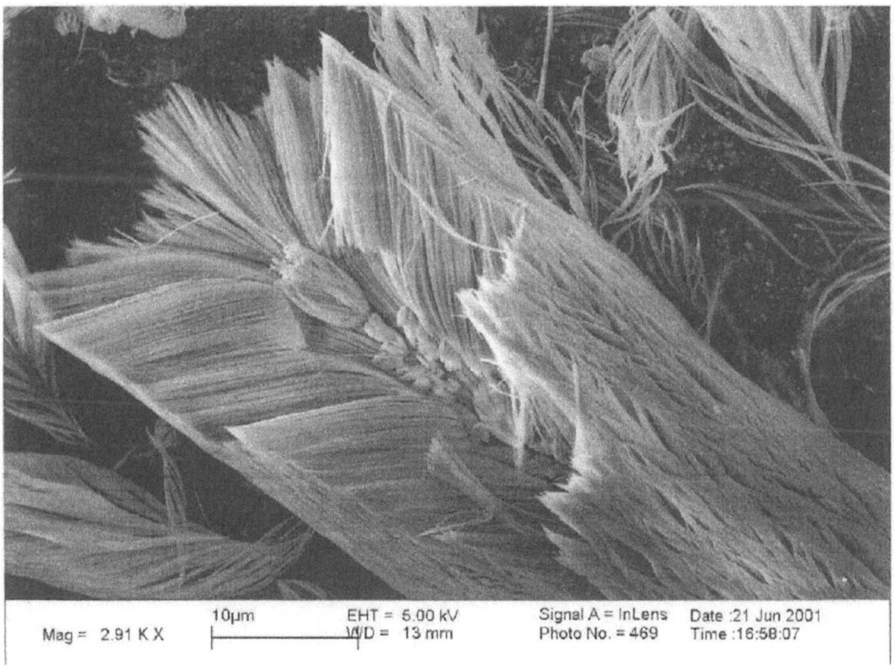

FIGURE 5.7. SEM image of silica nanowires grown on the surface of a gallium droplet.

Resolution that can be provided by a SEM is in the nanometer range. Figure 5.8 is a SEM of Au particles on a carbon film, where two particles separated by a few nm can be resolved, and 1 nm resolution is possible.

Inelastic excitation is inevitably involved in electron–specimen interaction; thus, radiation damage is the up most concern in SEM imaging of polymers. Radiation damage arises mainly because of the following three processes. First, knock-on damage is an effect of electron impact on the atoms in the solids, resulting in atom displacement and/or broken bonds. This effect usually increases as the accelerating voltage increases. Second, atomic ionization is caused by the energy transfer of the incident electron to the atomic electrons owing to inelastic interaction, resulting in X-ray emission. The repulsion among the ionized atoms induces the destruction of the local structure. Finally, the local temperature introduced by the incident beam can also be a factor of radiation damage.

Low-voltage SEM is an effective technique for imaging specimens that are sensitive to beam radiation. Low-voltage SEM has attracted a great deal of attention in recent years because of its application in polymers, biology, and material science. The resolution of low-voltage SEM is primarily limited by chromatic aberration. If the energy spread of the electron emission source is ΔE, the incident beam energy is E_0 and the chromatic aberration coefficient is C_c, the probe size limited by this effect is calculated by $d_c = \alpha\Delta E/E_0\ C_c$, where α is the beam convergence angle. The resolution of a SEM at 200 V can be as high as 2.5 nm, which is capable of resolving a lot of details on the nanoscale.

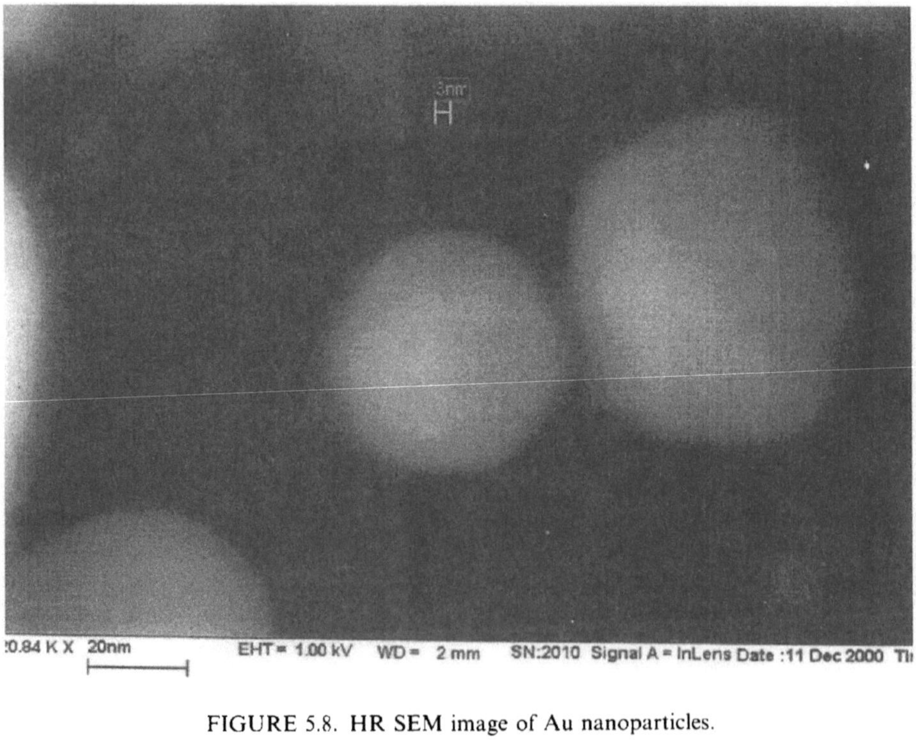

FIGURE 5.8. HR SEM image of Au nanoparticles.

5.4. TRANSMISSION ELECTRON MICROSCOPY

Transmission electron microscopy, as a high spatial resolution structure and chemical microanalysis tool, has been proven to be powerful for characterization of nanomaterials.[14] A modern transmission electron microscope gives one the capability to directly imaging atoms in crystalline specimens at resolutions close to 0.1 nm, smaller than interatomic distance. An electron beam can also be focused to a diameter smaller than ~0.3 nm, allowing quantitative chemical analysis from a single nanocrystal. This type of analysis is extremely important for characterizing materials at a length scale from atoms to hundreds of nanometers. In this section, we mainly focus on the application of TEM and associated techniques for characterizing nanocrystal self-assembled materials with an emphasis on analysis of particle shapes, crystallography of self-assembly and the interparticle interaction.[15,16] We first introduce the fundamental imaging principle of TEM.

5.4.1. Image Formation

For easy illustration, a TEM is simplified into a single-lens microscope, as given in Fig. 5.9 in which only a single objective lens is considered for imaging and the intermediate lenses, and projection lenses are omitted. This is because the resolution of the TEM is mainly determined by the objective lens.[17] The entrance

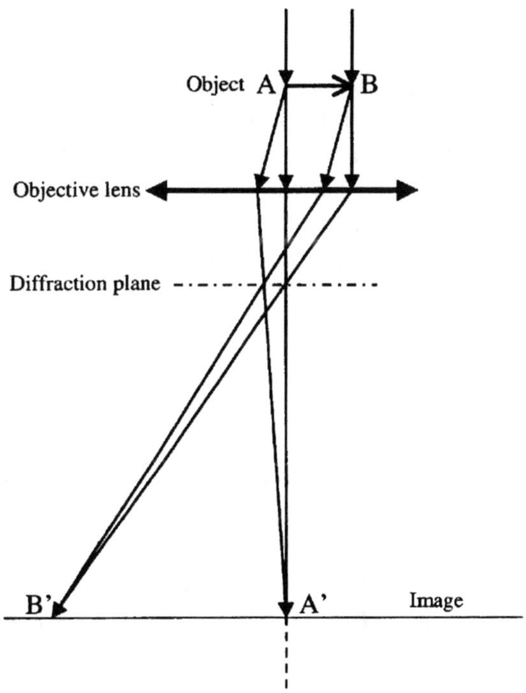

FIGURE 5.9. Abbe's theory of image formation in an one-lens TEM. This theory is for a general optical system in TEM.

surface of a thin foil specimen is illuminated by a parallel or nearly parallel electron beam. The electron beam is diffracted by the lattices of the crystal, forming the Bragg beams that are propagating along different directions. The electron–specimen interaction results in phase and amplitude changes in the electron wave that are determined by quantum mechanical diffraction theory.[18,19] For a thin specimen and high-energy electrons, the transmitted wave function at the exit face of the specimen can be assumed to be composed of a forward-scattered wave. The exit wave contains the full structural information of the specimen. Unfortunately, this wave will be transmitted nonlinearly by the optic system.

The non-near-axis propagation through the objective lens is the main source of nonlinear information transfer in TEM. The diffracted beams will be focused on the back-focal plane, where an objective aperture could be applied. An ideal thin lens brings the parallel transmitted waves to a focus on the axis in the back focal plane. Waves leaving the specimen in the same direction are brought together at a point on the back focal plane, forming a diffraction pattern. The electrons scattered to angle θ experience a phase shift introduced by the spherical aberration and the defocus of the lens, and this phase shift is a function of the scattering angle. The phase shift due to spherical aberration is caused by a change in focal length as a function of the electron scattering angle, and the phase shift owing to defocus is caused by the spherical characteristics of the emitted wave in free-space (e.g., the Huygens principle). The phase shift introduced by spherical aberration is due to the variation in focal length as a function of the electron scattering angle, resulting in a path length difference (Fig. 5.10).

The phase shift introduced by defocus can result in contrast reversal in TEM imaging. Figure 5.11 gives a pair of TEM images recorded from self-assembled Co nanocrystals. The nanocrystals show dark contrast in Fig. 5.11(a), whereas the contrast is reversed in Fig. 5.11(b) with a change in objective lens defocus. This example apparently shows that a TEM image has to be recorded under appropriate conditions to ensure the reliability of the information.

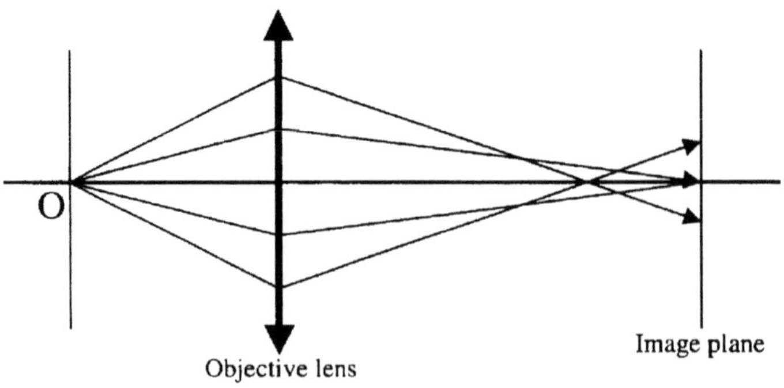

FIGURE 5.10. Schematic diagrams showing the phase shift introduced by spherical aberration. This nonlinear information transfer characteristics of the TEM gives rise the complexity in interpretation of TEM images.

FIGURE 5.11. A comparison of NCS of Co nanocrystals recorded under different defocus conditions, showing reversal in image contrast.

5.4.2. Contrast Mechanisms

Images in TEM are usually dominated by three types of contrast. First, diffraction contrast, which is produced due to the perturbation of local strain/defects/dislocation on the intensities of the Bragg reflected beams.[20] This contrast emphasizes the effect of defects on the amplitude of the transmitted wave and it is also called amplitude contrast. If the image is formed by selecting the central transmitted wave only, the image contrast is most sensitive to defect structures and the image resolution is limited to ~2 nm. For nanocrystals, most of the grains are defect-free in volume, while a high density of defects are localized at the surface or grain boundary, diffraction contrast can be useful for capturing strain distribution in nanocrystals of sizes larger than 15 nm.

Second, phase contrast is produced by the phase modulation of the incident electron wave when transmitting through the crystal potential.[17] This type of contrast is sensitive to the atom distribution in the specimen. When the electron goes through a crystal potential field, its kinetic energy is perturbed by the variation of the potential field, resulting in a phase shift with respect to the electron wave that travels in a space free of potential field. For a specimen of thickness *d*, the phase shift is

$$\Psi \approx \sigma V_{\rm p}(x, y) = \sigma \int_0^d {\rm d}z\, V(x, y, z) \tag{5.3}$$

where $\sigma = \pi/\lambda\ U_0$, U_0 is the acceleration voltage, and $V_p(x,y)$ the thickness-projected potential of the crystal. Equation (5.3) clearly shows that the phase contrast image is the result of the thickness-projected image along the beam direction (z-axis) of a 3D object. The 3D shape of the crystal can be revealed using the images recorded at least from two independent orientations.

Electron transmission through a thin crystal can be characterized by a phase modulation function, which is known to be the phase object approximation (POA). This approximation assumes that the electron wave is modulated only in phase but not in amplitude (e.g., phase contrast). Therefore, the electron wave at the exit face of a thin crystal is approximated to be

$$\Psi(x, y) = \exp[i\sigma V_p(x, y)] \tag{5.4}$$

If the incident beam travels along a low index zone-axis, the variation of $V_p(x, y)$ across atom rows is a sharp varying function because an atom can be approximated by a narrow potential well and its width is in the order of 0.2–0.3 Å. This sharp phase variation is the basis of phase contrast, the fundamental of atomic-resolution imaging in TEM.

Finally, mass-thickness or atomic number produced contrast. Atoms with different atomic numbers exhibit different powers of scattering. Specimens with a larger thickness-projected density, a strong dark contrast would be seen in conventional TEM image. This contrast is the dominant at low-magnification TEM images and is mostly used for imaging nanocrystal superlattices (NCSs).

5.4.3. Shapes of Nanocrystals

Determining the shape of nanocrystals is an important application of TEM. Using a combination of high-resolution lattice imaging and possibly electron diffraction, the crystal structure and facets of a nanocrystals can be determined. To determine the 3D shape of a nanocrystal, one requires at least two TEM images recorded from different orientations. The observed TEM images are approximately the 2D projection of the particle with the complication introduced by the lenses. The shape of some simple nanocrystals can be directly determined from the TEM image, but the shape of some of the particles relies on a comparison of the experimental images with the theoretically calculated ones. The Ih and decahedral nanocrystals are two typical cases. Figure 5.12 shows a group of calculated images for the two types of particles as a function of the particle orientation and objective lens defocus. The image simulation was carried out using the well-established multislice theory developed by Cowley and Moodie.[21] It is unambiguous that the image strongly depends on the orientation of the particle and the defocus of the objective lens. The easiest orientation for distinguishing the Ih and decahedral particles is along their fivefold symmetry axes.

Figure 5.13 shows comparisons between the simulated images, adapted from Buffat *et al.*[22] (left most column), with experimental images (second column) recorded from Ag nanocrystals, the Fourier transform of the experimental images (column 3) and the Fourier filtered images (right most column) for the different

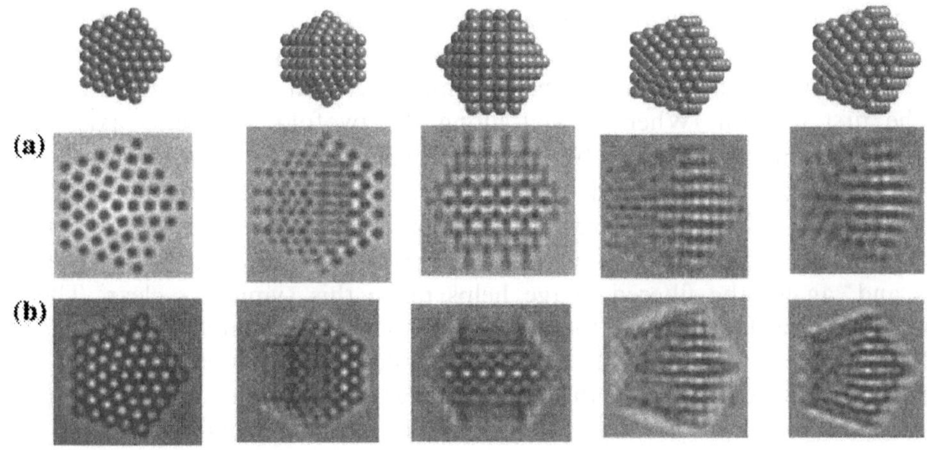

FIGURE 5.12. Theoretically simulated images for a decahedral Au particle at various orientations and at focuses of (A) $\Delta f = 42$ nm and (B) $\Delta f = 70$ nm, illustrating the contrast reversal in the two cases. (Courtesy of Drs. Ascencio and M. José-Yacamán).

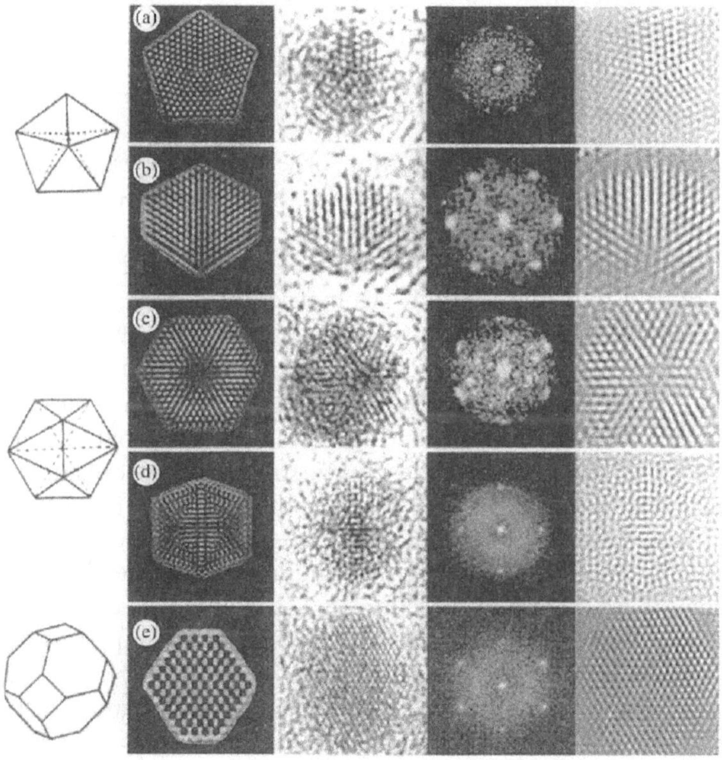

FIGURE 5.13. A series of simulated images (first column), experimental TEM images (second column), the corresponding Fourier transforms of the experimental images (third column) and the corresponding Fourier filtered images (fourth column). (a,b) A decahedron viewed along the fivefold and the twofold symmetry axes, respectively. (c,d) An icosahedraon viewed along the threefold and twofold symmetry axes, respectively. (e) An truncated octahedron.

shapes of particles, oriented along different crystal zone axes. A decahedral nanocrystal viewed along a fivefold multi-twin symmetry axis is shown in row a, the experimental image in the second column agrees well with the simulated image in the first column. When viewed along the twofold symmetry axis of the decahedral nanocrystal (row b), the experimental image again shows a good agreement with the simulated one. For an Ih nanocrystal oriented along the threefold axis (row c), the threefold symmetry of the image is apparent in the filtered image. Row d shows an Ih particle oriented along its twofold symmetry axis and, again, the filtered image helps make this symmetry clear. The TO morphology was fairly rare in these specimens but a few examples were found and one is shown in row e. Single twin nanocrystals (row f) were rare, and some distortion is evident between the filtered image (4th column) and the simulated image (1st column). The analysis was performed for 300 nanocrystals, among which about 170 were oriented well enough to uniquely determine their morphologies; icosahedron were predominant and accounted for almost 80% of the determinable nanocrystal cores.[23] Decahedron accounted for about 16%, single crystal and single twinned crystals followed with 3% and 1%, respectively. The remaining unidentified nanocrystal cores were due to their random orientations projected in the image.

5.4.4. Crystallography of Self-Assembly

Translation ordering and orientation ordering are both present in NCSs. The translation ordering is directly seen from the experimental images; whereas the orientational order requires diffraction study from the local region. Shown in Fig. 5.14 is a selected area electron diffraction pattern recorded from an Ag NCS with a size of ~4 nm, the sixfold symmetric diffraction pattern from the superlattice is apparent near the central transmitted spot, while the diffraction from the atomic lattices of Ag gives a ring pattern, indicating possibly no orientation order in this region. Before reaching this conclusion one must, however, consider the short-range ordering characteristics in particle orientations. The dark-field TEM imaging can be useful in solving this problem.

Dark-field TEM imaging is formed using the electrons that are diffracted at angles other than the (000) central transmitted spot. By selecting the electrons falling in a specific angular range, one may find the real space distribution of the particles which are oriented in a direction that is likely to give a Bragg reflection falling in the angular range selected by the small objective aperture for forming the image. Figure 5.15(a) is a bright-field image of a monolayer Ag NCS and the corresponding electron diffraction pattern. To map the orientation distribution of the Ag particles, the reflection from Ag atomic lattice, such as the (220) and (311) rings, is selected. Figure 5.10(c) and (d) shows two dark-field images recorded by selecting the electrons scattered to the angular ranges as indicated by circles c and d in the diffraction pattern, respectively, and the size of the circle approximately represents the size of the objective aperture used. By examining the intensity distribution in the volume of each particle, one finds the fact that only part of the particle gives rise the reflection selected by the objective aperture. This

FIGURE 5.14. An electron diffraction pattern recorded from an hcp packed Ag NCS, showing the sixfold symmetry of the superlattice reflections at low angle, while the high angle scattering from the atomic Ag lattices is a continuous ring pattern, indicating no average orientation ordering across a large specimen area (~4 μm).

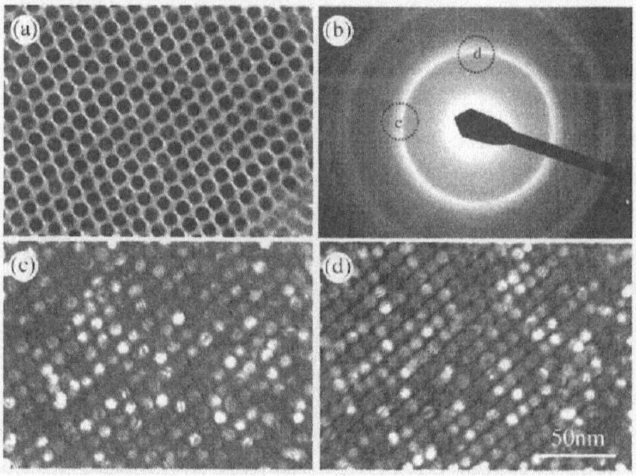

FIGURE 5.15. (a) Bright-field TEM image and (b) the corresponding electron diffraction pattern of self-assembled Co nanocrystals. (c,d) Dark-field TEM images recorded from the same region by selecting the electrons scattered to the angular ranges indicated by the c and d circles in (b), respectively, showing the short-range orientation ordering.

unambiguously proves that the particles are not single crystalline, rather they could have twins or multiple twins. By examining the intensity distribution across the entire image, one finds that the particles distributed in the regions, as indicated by A, B, C, and D, respectively, have the preferred particle orientations based on the similarity in contrast of each particle. A few examples are indicated by arrowheads in the image. The short-range ordering in particle orientation cannot be seen in the electron diffraction pattern since it is an average over all of the particles oriented differently.

In NCSs the sixfold projected symmetric distribution of the particles is usually seen, such as that shown in Fig. 5.14. This configuration could correspond to the $[111]_S$ of fcc (bcc) or the $[0001]_S$ of hcp, provided no other types of packing is possible. To uniquely determine the 3D crystallography of the packing, a 45° specimen tilting is required to change the imaging zone axis from $[111]_S$ (or $[0001]_S$) to another low index zone axis. This experiment is indispensable to pin down the answer.[23]

5.4.5. *Self-Assembly of Shape Controlled Nanocrystals*

If the nanocrystals can be taken as the building blocks, their 3D assembly is unavoidably affected by the particle shape.[24] TEM is unique in determining the relationship between particle orientations and the assembling crystallography of the ordered array. Figure 5.16(a) is a TEM image recorded from a Ag NCS deposited on a carbon substrate, from which the orientation of the particles and the arrays can be identified. The Ag nanocrystals have a TO shape [Fig. 5.16(b)] and they are oriented along the [110] of the Ag atomic lattice in the image, along which four {111} and two {100} facets are imaged edge-on. The unit cell of the NCS is also oriented along $[110]_S$ of fcc, where the subscript refers to the supercrystal. Therefore, the orientational relationship between the Ag particles and the nanocrystal lattice is $[110] \parallel [110]_S$ and $[001] \parallel [\bar{1}10]_S$. This registered orientational order can be used to determine the orientations of the nanocrystals in the unit cell of the superlattice, as shown in Fig. 5.16(c). The nanocrystals are packed into a fcc lattice, and the nanocrystals are arranged following a principle of *face-to-face*.[15] The image recorded from the same area at slightly large defocus condition gives the channels (the white dots in Fig. 5.16(b)) enclosed by the bundling distributed thiolates molecules on the surface of the nanocrystals. This model has been proved by the energy-filtered chemical imaging technique in TEM.[25]

The self-assembly of nanocrystals is strongly affected by the shape of the nanocrystals. The tetrahedral shaped nanocrystals have sharp corners and their assembly may have long-range translational symmetry but the orientational order is limited only to a short range.[26] Figure 5.17(a) gives the monolayer self-assembly of tetrahedral Ag nanocrystals. Two fundamental principles are usually observed in particle self-assembly. The most close-packing is the favorable configuration from the energy point of view. The passivation molecules are likely to form interdigitative bonds that link the nanocrystals. Two possible monolayer assembly models satisfying the two requirements are given in Fig. 5.12(b) and (c). The only difference between the two models is that half of the particles face down, while

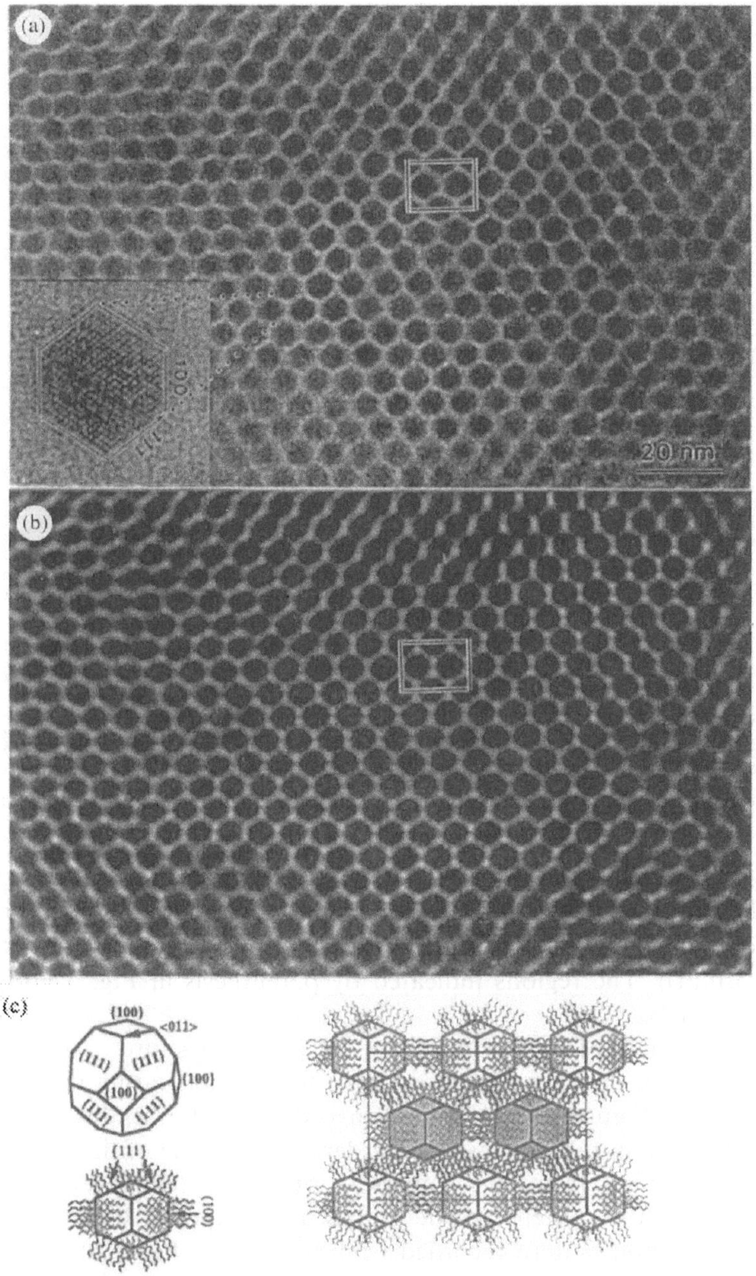

FIGURE 5.16. (a) TEM image of a fcc packed Ag nanocrystals whose shape is dominated by TO (see the inset HRTEM image of a single Ag nanocrystal). The image displays the orientational relationship between the nanocrystals and the assembled superlattice. (b) An TEM image recorded from the same specimen area under slightly defocus condition. The image in (a) is recorded at near focus condition and it is sensitive to the particle shape, and the image in (b) is most sensitive to the phases caused by a variation in the projected mass-thickness. (c) A structural model of the TO particle, and the $[110]_S$ projection of the unit cell as indicated in (a), where the gray pasted particles are located at different z heights. The channels enclosed by the bundled thiolates are presented.

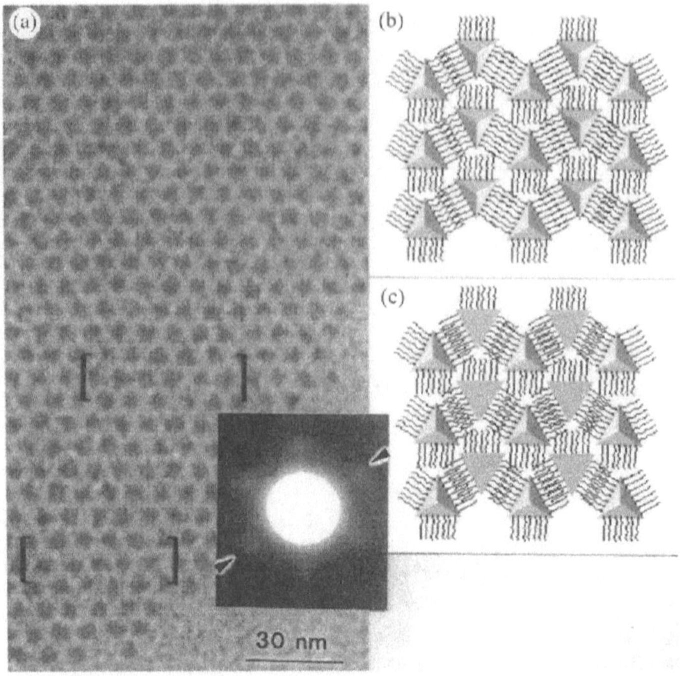

FIGURE 5.17. (a) TEM image of a monolayer self-assembly of tetrahedral Ag nanocrystals. (b,c) Two possible assemblies of tetrahedra with short-range orientational order. A small angle electron diffraction pattern from the monolayer assembly, reflecting the mirror symmetry (not threefold).

the projected structures of the two are identical. These models do not support the threefold symmetry of the packing due to the shape of the particles, in agreement with the mirror symmetry presented by the electron diffraction pattern (the inset in Fig. 5.17(a)). The regions indicated by parenthesis in Fig. 5.17(a) agree well with the models.

Self-assembly of the ϵ-Co NCs of the β-Mn structure[27] (ϵ-Co phase, space group $P4_132$)[28] is another typical example for demonstrating the role played by particle shape. Figure 5.18(a) shows an ordered self-assembly of the Co NCs, which exhibits long-range translation order. The packing of these shaped Co NCs is rather different from the spherical ones reported previously,[29] and is complicated by the existence of the polyhedral shapes of the NCs. A low-angle electron diffraction pattern recorded from the region [Fig. 5.18(b)] disproves the sixfold symmetry of the array. Two typical structures of the self-assembly are displayed in Fig. 5.18(c) and (d). From these enlarged images, it is apparent that three types of shapes, square-like, hexagon-like and square exist in the assembly. These different shapes of NCs result in the breaking of the normal sixfold symmetry in the superlattice assembly.

The unique feature of the Co NC shapes is easily seen in the HRTEM images (Fig. 5.19). By matching the interplanar distances and plane-to-plane angles through the Fourier transform of the images, the NC facets can be directly linked to the {221} and {310} planes. The (221) and (310) reflection rings are

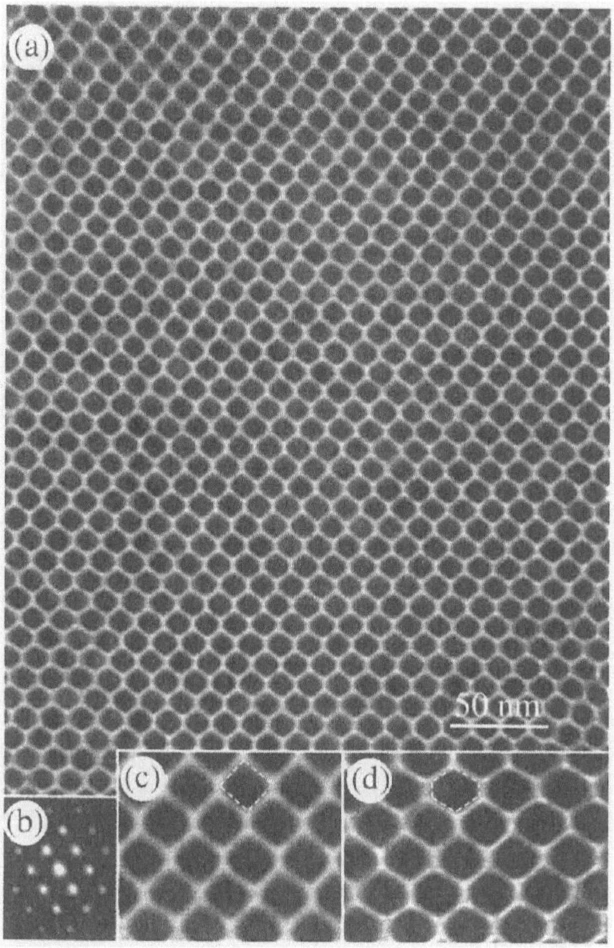

FIGURE 5.18. (a) Self-assembly of the 11 nm Co NCs. The thickness of the stacking is ~3 nanocrystals. (b) A low-angle electron diffraction pattern recorded from the area showing the symmetry of the self-assembly. (c,d) Two typical local structures introduced by the anisotropy shape of the NCs (Specimen courtesy of Dr. S. Sun).

almost inseparable in the electron diffraction pattern, but the angles among them are critical in identifying the facets. From the images shown in Fig. 3(a) and (b), where the electron beam is parallel to $[62\bar{1}]$, the NCs are partially enclosed by the $\pm(2\bar{1}2)$, $\pm(122)$ and $\pm(\bar{1}30)$ facets. When the electron beam is parallel to [001] [Fig. 3(c)], the $(\bar{3}\bar{1}0)$ and $(\bar{1}30)$ facets are imaged edge-on. Along the [120] direction [Fig. 5.18(d)], the $(2\bar{1}\bar{2})$ and $(2\bar{1}2)$ facets are imaged edge-on. These HRTEM images recorded from three different NC orientations are used to reconstruct the NC 3D shapes.

HRTEM images show that the ϵ-Co has a polyhedral shape bounded by $\pm(2\bar{1}2)$, $\pm(122)$, $\pm(\bar{1}30)$, $\pm(310)$, $\pm(2\bar{1}\bar{2})$, and $\pm(12\bar{2})$ facets. Using the information provided by the images, a structural model of the nanocrystal is given [Fig. 5.20(a)].[30] The projected shapes of the polyhedron along $[62\bar{5}]$, [120], and [001] are given in Fig. 5.20(b)–(d), respectively. These projections match well with the NC

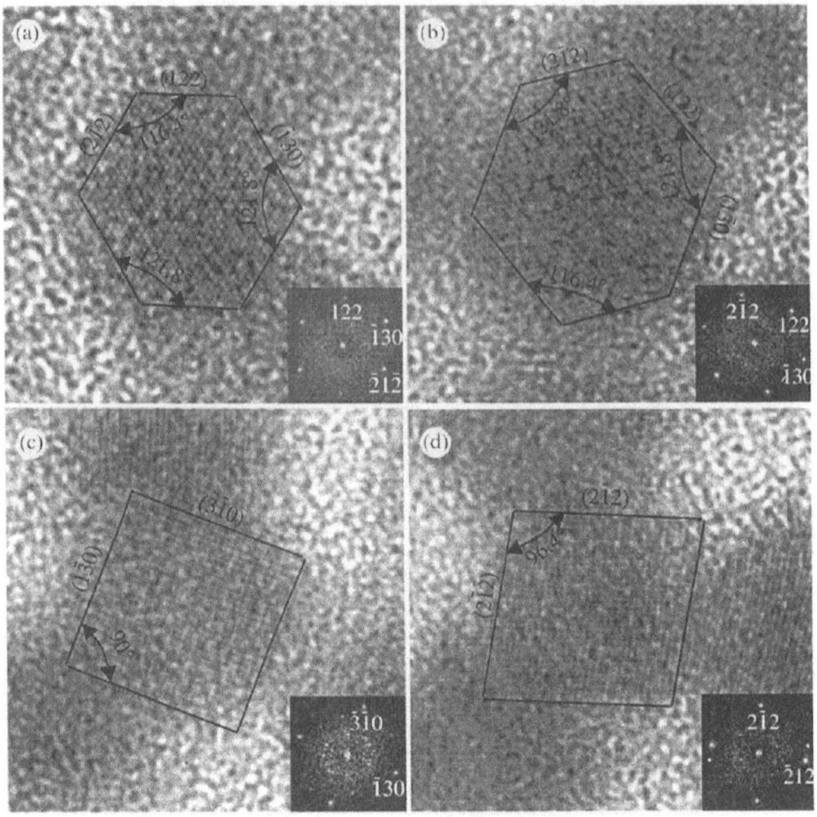

FIGURE 5.19. Typical HRTEM images of the Co nanocrystals oriented along (a, b) [$62\bar{2}$], (c) [001], and (d) [120], showing the {221} and {310} facets. The inserts are the Fourier transforms of the corresponding images.

shapes observed in the images displayed in Fig. 5.18(c) and (d). The self-assembly is constructed following the principle of the face-to-face for the same type of nanocrystal facets. For example, the ($\bar{1}30$) face for one particle is against the ($1\bar{3}0$) face of the other, and the ($2\bar{1}2$)/(122) face is to the ($\bar{2}1\bar{2}$)/($\bar{1}\bar{2}\bar{2}$). This configuration is observed in Fig. 5.18(c). If one uses the model given in Fig. 5.19(c) to form a 2D close packing, the configuration given in Fig. 5.20(f) is constructed, which is just the assembly observed in Fig. 5.18(d), where the NCs prefer to orient along [120]. Therefore, the two typical self-assemblies presented in Fig. 5.18(c) and (d) result from the ordered orientations of the Co NCs due to their anisotropic shape.

5.4.6. Defects in Self-Assembly

It is well known that defects and dislocations are created in solid materials to accommodate the local strain and deformation. Structural characterization of solid state materials is usually concentrated on the analysis of defects and interfaces. The as-synthesized nanocrystals are dispersed in liquid, in which the size selection according to mass is performed. NCSs are formed by depositing

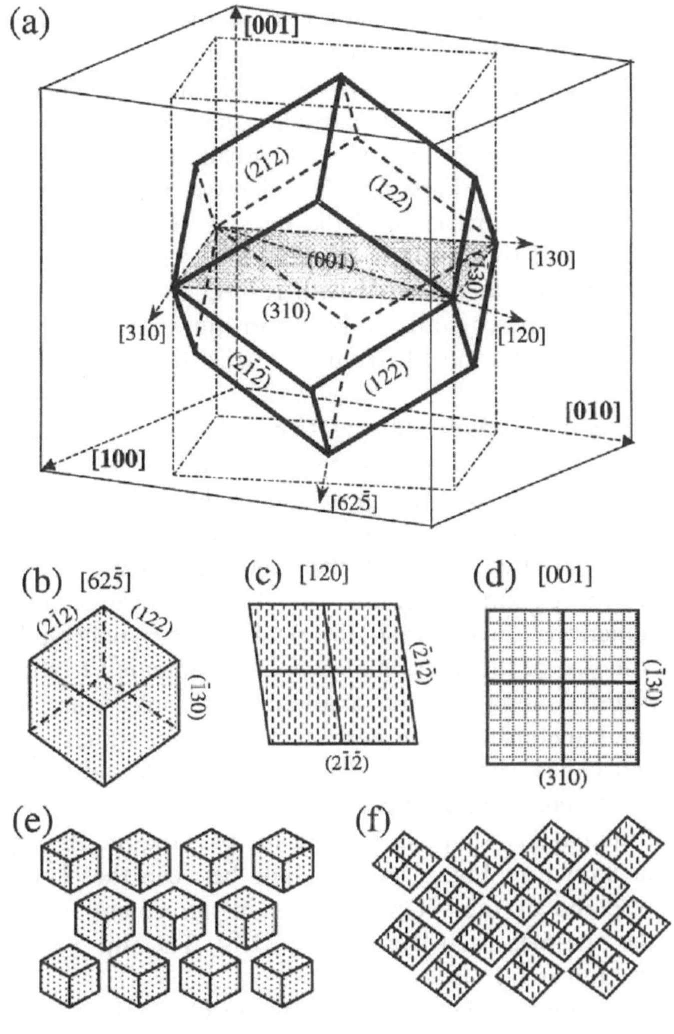

FIGURE 5.20. (a) Three-dimensional polyhedral shape of the Co nanocrystals constructed using the information provided by HRTEM images. (b–d) The projected shapes of the polyhedral model along [625̄], [120], and [001], respectively. (e,f) Monolayer close-packing of the polyhedra oriented in [625̄] and [120], respectively.

droplets of passivated nanocrystals on a solid state substrate via a slow drying process. The diffusion of the nanocrystals on the substrate is possibly driven by the hydraulic force as drying continue, thus, the wetting problem, the viscosity and the limited concentration of the nanocrystals determine the types of defects and their density. Our objective here is to illustrate the defect analysis in NCS using TEM, which, however, cannot be provided by XRD.

5.4.6.1. Slip Plane

Slip is a typical microstructure of metallurgy materials, and it is an important structure responsible at least in part to the deformation of materials.

This type of structure is created by slide part of the crystal parallel to the slip plane along a specific direction, namely the slip direction. This type of structure also occurs in NCS. Figure 5.21(a) shows a TEM image recorded from the self-assembled Au nanocrystals. The Au particles align along some lines, and the line direction forms an angle of $107 \pm 1°$, exhibiting a "twin" relationship between the A, B, and C domains. To examine the registered orientation relationship between the particles and the lines, a HRTEM image is recorded from the domain B and the result is given in Fig. 5.21(b). The orientations of the Au {111} fringes can be measured with respect to the line direction (or the $(111)_S$ plane of the supercrystal), it is surprising that the average angle is $17 \pm 2°$ clockwise as determined from the statistical plot inserted in Fig. 5.21(b).

The defect model is built starting from a perfect fcc structure oriented along $[110]_S$ (Fig. 5.22). If the nanocrystal assembly is made layer-by-layer parallel to the substrate, the stacking positions of the nanocrystals can be shifted due to existence of four equivalent positions, as marked by A–D in the upper-left corner. It must be pointed out that these positions may not be allowed in conventional atom packing because of the short-range and long-range interatomic interactions. In contrast, the packing of nanocrystals is primarily determined by the size of the interstitial positions compared to the available size of the nanocrystals. Among the four available positions, A and B or C and D can be taken simultaneously by the next stacking layer, which is represented by layer 3 in the figure. The slip of layer 3 to occupy the A and B positions results in the formation of densely packed crystal rows, as shown at the right-hand side of the figure. If the particle shape is TO and the packing model can be described by a similar one shown in Fig. 5.16, the angle between the $(1\bar{1}\bar{1})$ planes of Ag and the $(1\bar{1}\bar{1})_S$ plane of the NCS is 19.4°, in agreement with the sign and magnitude of the one measured from Fig. 5.21(b). The slip plane here is $(110)_S$ and the slip direction is $\frac{1}{8}[1\bar{1}2]$.

On the other hand, if the slip positions are C and D instead, the densely packed nanocrystal lines shown on the left-hand side are formed. The two sets of lines on the left- and the right-hand sides form a twin structure and the angle between the two is calculated to be 109.5° theoretically, in excellent agreement with the 107° measured experimentally. The slip plane here is $(110)_s$ and the slip direction is $\frac{1}{8}[1\bar{1}2]$. The twin image is formed as a result of the two equivalent slip directions.

5.4.6.2. *Twin and Stacking Fault*

Twins and stacking faults are important microstructures in material's deformation. These types of planar defects are also observed in NCSs, as shown in Fig. 5.23. The presence of twin structure is directly reflected by the Fourier transform of the image, exhibiting mirror symmetry. The twin and stacking planes are $(111)_s$ as indicated by arrowheads. The Ag nanocrystals have the TO shape and the NCS has the fcc packing. The formation of these planar defects can be understood as follows.

(a)

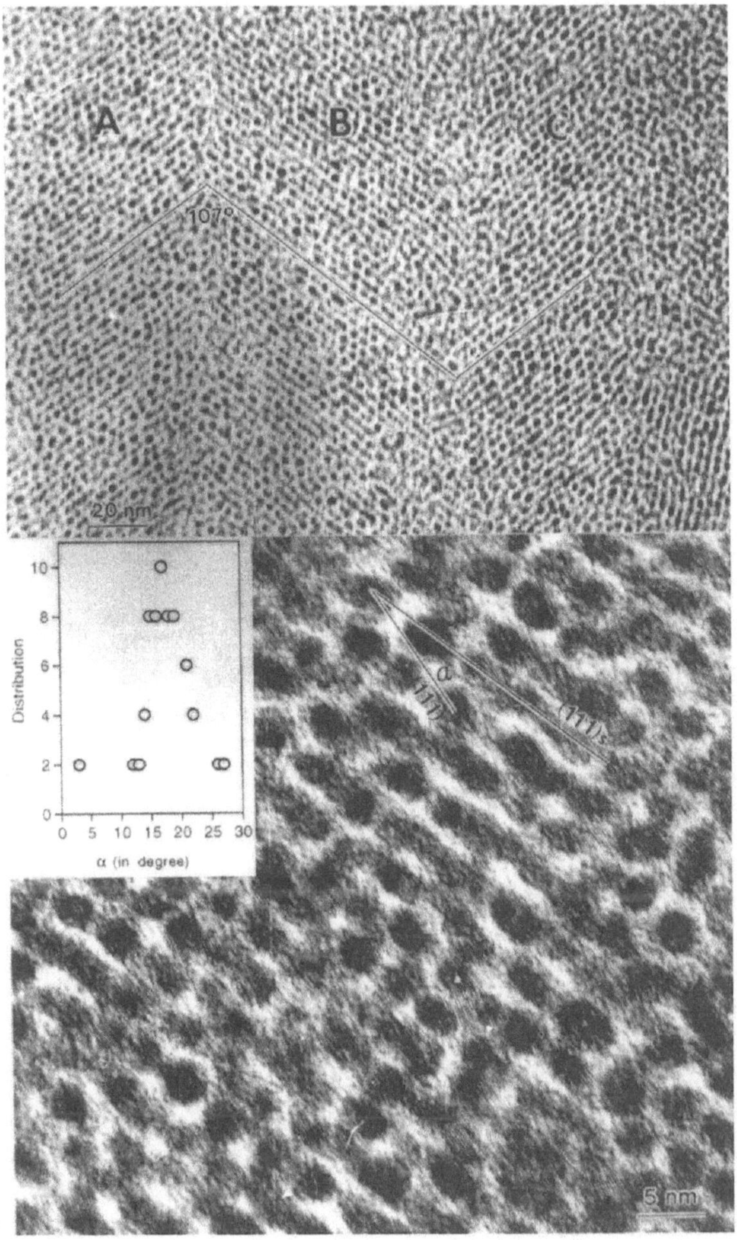

(b)

FIGURE 5.21. (a) TEM image of a self-assembled structure of Au nanocrystals, exhibiting "twin" symmetry. (b) An enlarged image of area B showing the single crystallinity of the gold particles and their ordered orientations with respect to the particle assembling plane. A statistical plot is given for the measured angles between (111) of Au and $(111)_S$ of NCS. The particle shapes are likely to be dominated by TO, but the assembling could be affected by the size fluctuation of nanocrystals.

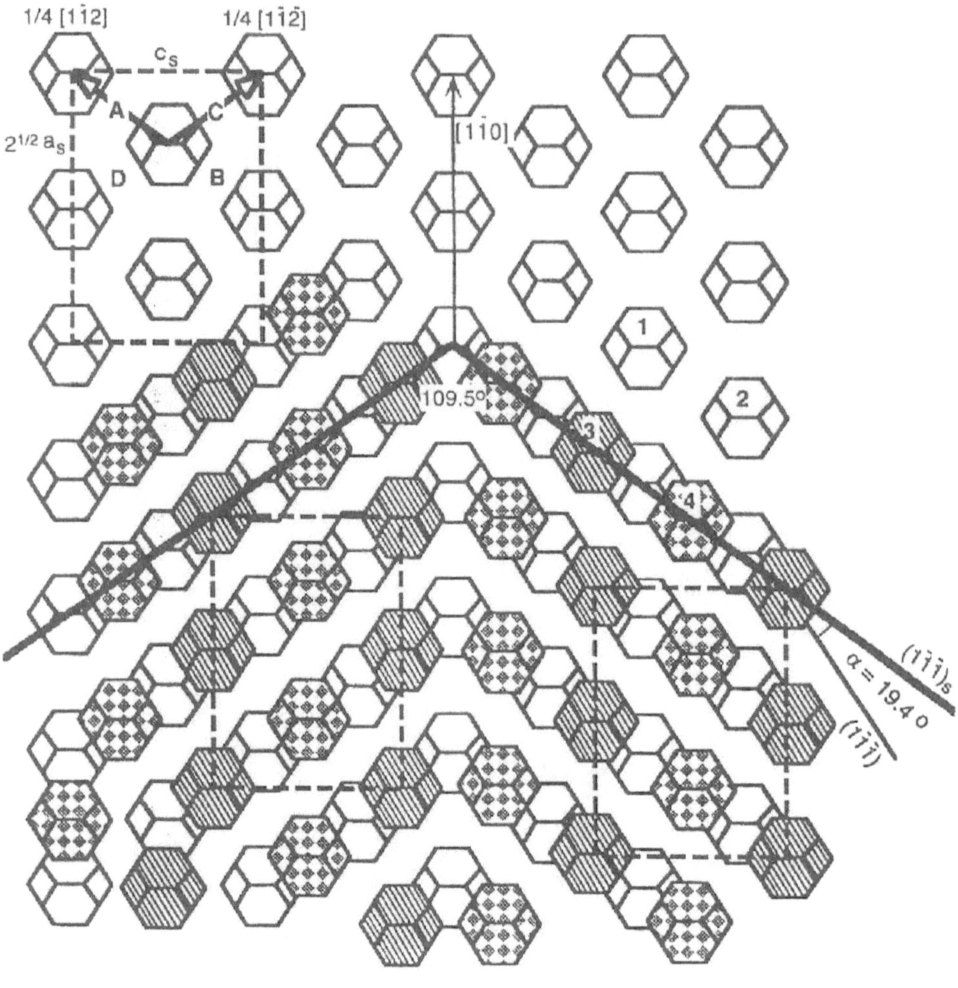

FIGURE 5.22. (a) A model of the twin structure constructed by an fcc NCS oriented along $[110]_S$ with the introduction of slip plane and different slip directions. The first two layers, labeled by 1 and 2, preserve the fcc model, while the third and fourth layers are stacked on the top with static displacements in the $(110)_S$ plane (see the text).

From the study of Ag nanocrystals with a TO shape, the intermolecular bonds are formed by the bundling of the surfactants on the {111} and {100} facets of the TO Ag particles, thus, the {111} facets can be interconnected naturally with either the {111} facets or the {100} facets, the latter results in a rotation of the superlattice, leading to the formation of a $\{111\}_S$ twin. A model for this case is shown in the inset in Fig. 5.23, in which a common layer shared by both sides of the twinned crystals is proposed, which serves as an intermediate step for rotating the particle orientations so that the face-to-face molecular bonding between particles still holds. By the same token, if the particle sites are shifted to modulate the interparticle distance, a stacking fault is formed. This

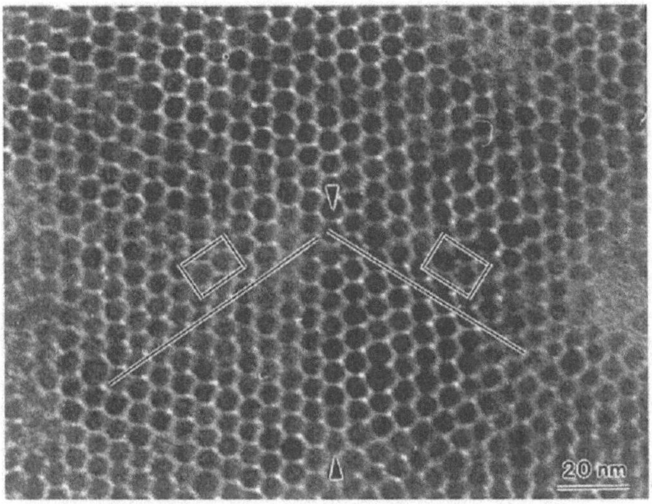

FIGURE 5.23. TEM image of an fcc Ag NCS with twin and stacking fault planar defects, where the models for the twin and stacking fault are given. The average particle size is 6 nm.

modulation is limited to only one row of nanoparticles but still preserves the face-to-face packing.[15]

The formation of twin and stacking fault can also be understood in light of the theoretical calculation by Luedtke and Landman.[31] The thiolates are bundled on the facets of the Ag nanocrystals, the assembling of the nanocrystals can be achieved by a "gear" model, in which the bundled thiolates are the "teeth" of the gear and the two nanocrystals are assembled simply by filling the space, resulting in a rotation in the particle orientation. This type of assembling is simply a geometrical matching between the nanocrystals, while the van der Waals molecular interaction dominates the bundling of the thiolates belonging to the same particle. The twin model is thus given in Fig. 5.24, in which the two $\{111\}_S$ planes of the NCSs forming the twin has an angle of 109.4°, while the (111) faces of the nanocrystals on the left-hand side is not exactly parallel to the (100) faces of those on the right-hand side. This again suggests the erected bundling of the thiolates on the nanocrystal surfaces.[32]

Multiply-twinned NCS with a fivefold symmetry can also be found [Fig. 5.25(a)], which is similar to the projection of a decahedral particle along the fivefold axis. Amazingly, by accounting seven particles along each side of the "pentagon", a gap is created at the low-left side, near where a stacking fault is being created (see the arrowhead). As we discussed earlier, strain must be present in decahedral particle to fill the gap in the atomic scale. This rule also applies to NCS although the Ag nanoparticles have the TO shape and are packed as fcc. A corresponding structure model of the multiply-twinned NCS is given in Fig. 5.25(b), where the particles are packed following the rule of face-to-face. This model is proposed for a 2D projected multiply-twinned NCS without considering the thickness variation across the assembling for a 3D decahedral

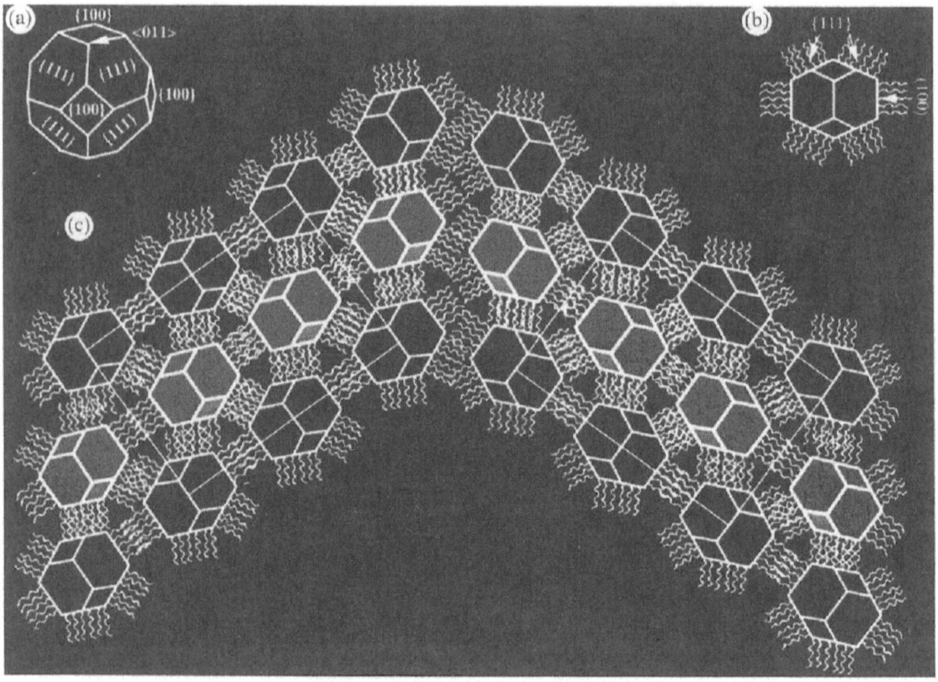

FIGURE 5.24. The "gear"-matching model for the formation of stacking fault and twins by matching of the thiolates bundles.

NCS. From the image contrast, this nanocrystal assembly is likely to be a thin, platelet nanocrystal structure.

The atomic-scale decahedral particle model was first proposed by Ino,[33] who found that the total energy of the particle is minimized by creating twins because of the increased portion of {111} plane and the lower interface energy. This is possible only if the particle is smaller than a critical value. Our example shown above clearly shows that this can also be the case for NCSs.

5.4.6.3. Dislocations and Lattice Distortion

The assembling of nanocrystals is complicated largely by their shapes and in some cases, the unit cell of the superlattice is distorted due to the non-uniformity in particle sizes and their shapes. Figure 5.26 shows the co-existence of twins, stacking faults as well as a partial dislocation in the self-assembly. The "strain" induced by the dislocations and defects results in relative misalignments among the NCs. These stacking faults and twins totally disturb the local ordering of the self-assembly. The creation of defects is a major problem in practical applications of NCSs, because the ordering of the structure is disturbed, limiting the growth of a large ordered structure. The solution based self-assembly process is not easy to control. The solution to this technical difficulty remains to be discovered.

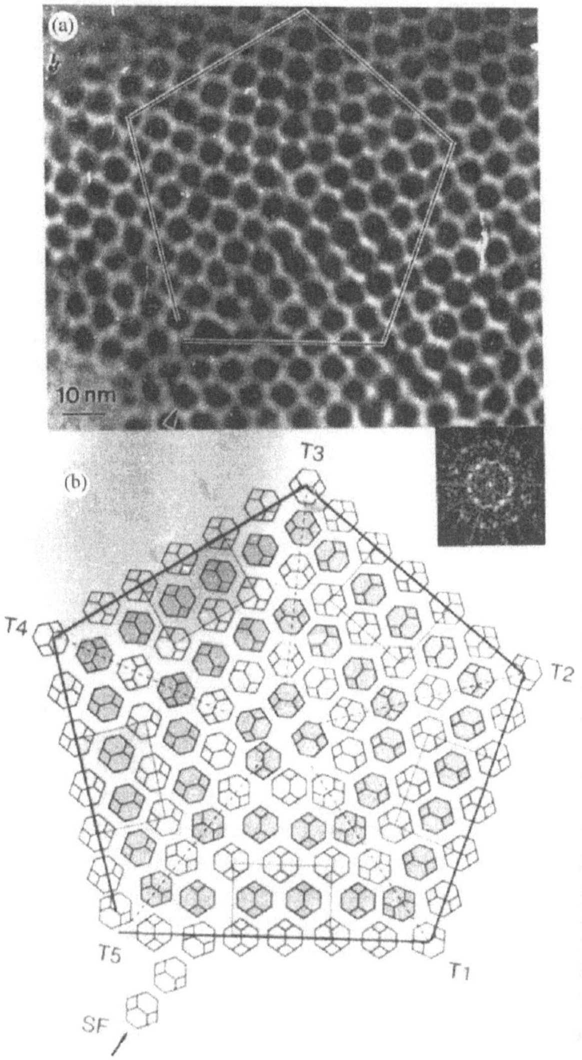

FIGURE 5.25. TEM image of an fcc self-assembled Ag nanocrystal array showing a fivefold multiply-twinned structure. The inset is a Fourier transform of the image. (b) A structure model of the Ag nanocrystal aggregate comprised faceted TO particles, where a stacking fault (SF) starts to be formed at the low-left corner.

5.4.7. *Chemical Imaging of Self-Assembled Structures*

High resolution TEM is a powerful technique for identifying the crystal structure of nanophase materials, but it may not be sensitive enough to detect a small change in atomic number. To precisely determine the content of a particular element in a specimen, energy dispersed X-ray spectroscopy and electron

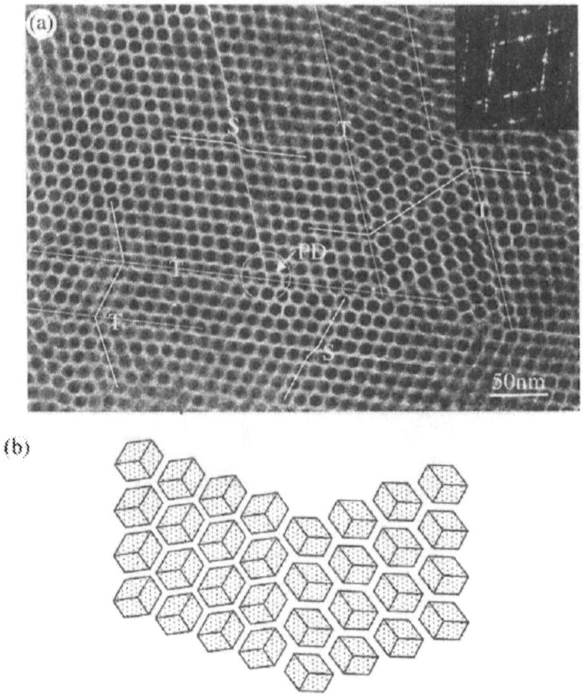

FIGURE 5.26. Twins (T), stacking faults (S) and partial dislocations (PD) observed in self-assembly of Co nanocrystals. The inset is a Fourier transform of the image showing the twin reflections.

energy-loss spectroscopy can be very useful. Under the impact of an incident electron, the electrons bound to the atoms may be excited either to a free electron state or to a unoccupied energy level with a higher energy. The quantum transitions associated with these excitations will result in emission of photons (e.g. X-rays) and Auger electrons. These inelastic scattering signals are the finger-prints of the elements that can provide quantitative chemical and electronic structural information.[19,34] Two analytical techniques are most commonly used in chemical microanalysis in TEM, EDS,[35] and EELS.[36] The size of the incident electron probe can be as small as 2–3 nm, giving the feasibility of acquiring EDS and EELS spectra from individual nanocrystals. The data can be used quantitatively for determining the chemical compositions of nanoparticles. EDS is mainly sensitive to heavy elements due to the fluorescence effect arising from Auger emission for light elements, whereas EELS is most adequate for light elements. Both EDS and EELS can be applied, in complement, to determine the chemical composition. The detection limit for EDS can be as high as 1%, and that for EELS can be as high as 5% depending on elements.

Figure 5.27 is an eds spectrum acquired from La-Ce-O nanocrystals, clearly indicating the presence of La, Ce, O, and P in the nanocrystals. quantitative analysis using the intensities of the X-ray lines gives the composition of La : Ce : O : P = 8.0 : 16.3 : 73.6 : 2.0. eds analysis can achieve a spatial resolution of sub-nanometer using a nanoelectron probe. this is a powerful, effective and

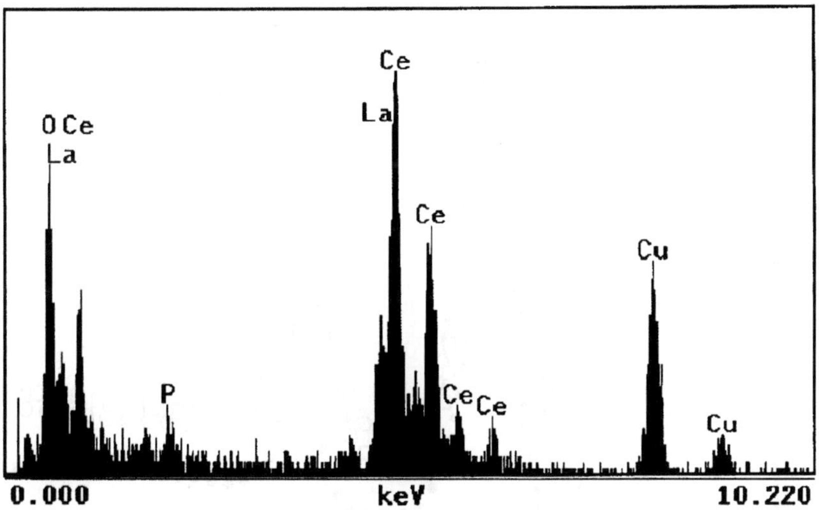

FIGURE 5.27. EDS spectrum acquired form a La–Ce–O nanocrystals showing the X-ray lines at distinct energies. The copper signals came from the grid used to support the sample.

reliable technique for determination of the contents of heavier elements in nanomaterials.

Figure 5.28 is an EELS spectrum acquired from as-synthesized CoO nanocrystals, which is clearly different from the spectrum from the Co_3O_4 standard sample. This is a "finger-print" for identifying the cation valence states and electronic structure of nanocrystals.[37,38] After annealing the CoO nanocrystal in oxygen, the received spectrum shows an almost identical shape as that of Co_3O_4, indicating the phase transformation from CoO to Co_3O_4. Again, the EELS analysis can be carried out with sub-nanometer spatial resolution.

Chemical analysis in TEM can achieve high spatial resolution with the use of either the small electron probe or the energy filtering system. The energy-filtered TEM (EF-TEM) relies on the principle of EELS and the images (or diffraction patterns) are formed by electrons with specific energy losses.[39] If the electrons which have excited the carbon K ionization edge of carbon, for example, are selected for forming the image, the image contrast is approximately proportional to the thickness of projected carbon atoms in the specimen, providing a direct chemical map of carbon. The thiolate molecules are composed of mainly carbon, thus the EF-TEM of the carbon K edge can give the distribution of the thiolates around the nanocrystals.[40] For this analysis, Ag NCSs are deposited on an amorphous SiO_x substrate and the effects from the substrate can be removed by processing the experimental images acquired pre- and post-edge.

The EF-TEM was performed for the Ag NCS oriented along $[110]_s$, which is the optimum orientation for imaging thiolate distribution between the particles. The EF-TEM image acquired using the carbon K edge from an Ag NCS gives an interesting contrast feature [Fig. 5.29(a)]. The projected carbon density between the particles shows a contrast pattern that is the strongest between the A and B

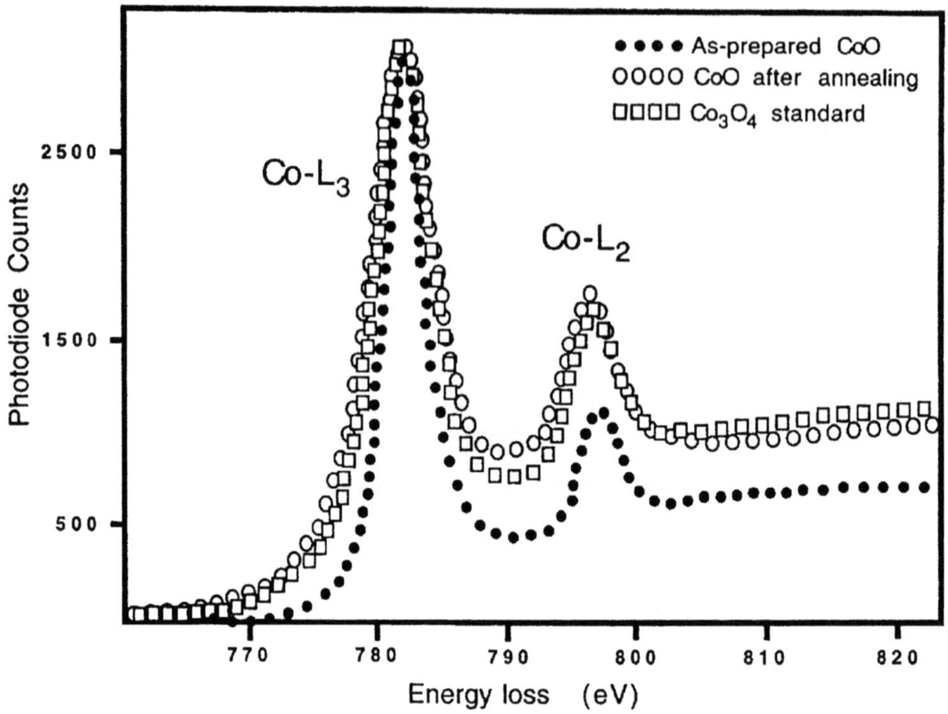

FIGURE 5.28. EELS spectra acquired from Co_3O_4 standard sample, the As-synthesized CoO nanocrystals, and the CoO nanocrystals after annealing in oxygen. Through the change in the Co-L edges one concludes that the CoO has been transformed to Co_3O_4 after annealing.

types of particles, while the contrast is lower between the A and C or B and C types of particles. To interpret this phenomenon, the $[110]_S$ projection of the NCS was first constructed, and the result is shown in Fig. 5.29(b). From the structural point of view, the molecular bonds tend to parallelly align the facets on which they are tethered. For the nanocrystals A and B assembled by facing the $\{100\}$ faces, in addition to the carbon density contributed by the interdigitated thiolates passivated on the $\{100\}$ facets (which are edge-on while viewed along $[110]_S$), the thiolates passivated on the four $\{111\}$ planes (not edge-on) also contribute to the projected carbon density although the $\{111\}$ faces are at an angle with the projection direction. Therefore, the projected density of the thiolate molecules between particles A and B is expected to be higher than that between A and C (or B and C) if the size of $\{111\}$ faces is the same as the that of $\{100\}$ as well as the density of the thiolate passivation is the same on both $\{111\}$ and $\{100\}$. With consideration of the resolution of the EF-TEM of $\sim$2 nm, the channels formed by the bundled thiolates may not be resolved in this type of images.

5.4.8. In Situ Structural Transformation

In situ study of the temperature induced phase transformation, structural, and chemical evolution of nanocrystals is important for understanding the

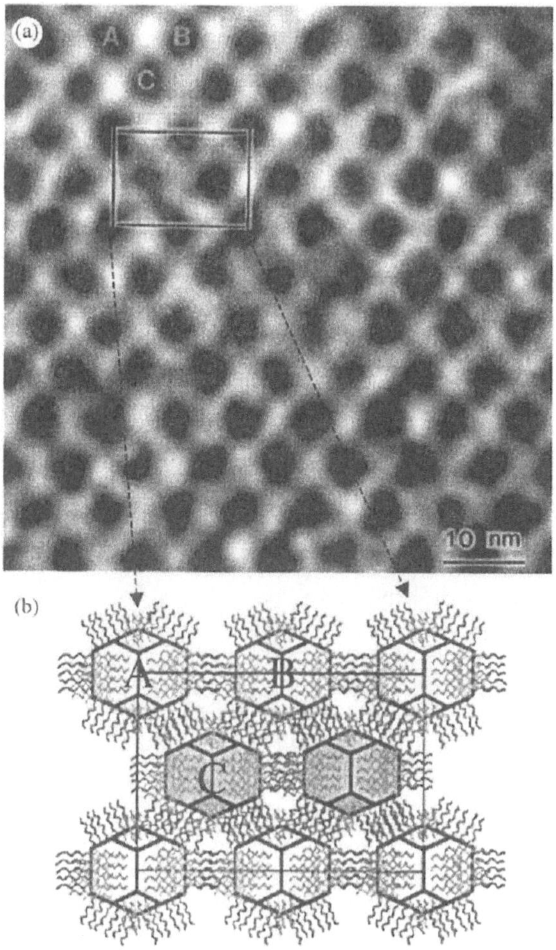

FIGURE 5.29. (a) Energy-filtered TEM image formed by electrons that have excited the K ionization edge of carbon, and the image contrast corresponds to the thickness projected density of carbon around the nanocrystals. The NCS was supported by silica substrate. (b) Structural model of the molecular passivation on the crystal surfaces, forming the interparticle molecular bonding.

structure and structural stability of nanomaterials. TEM is an ideal approach for conducting this type of experiments, in which a specimen can be cooled down to the liquid nitrogen or liquid helium temperatures or heated to 1000°C. The *in situ* process can be recorded at TV rate for exhibiting the time and temperature-dependent phenomena.

Figure 5.30 shows a group of TEM images recorded *in situ* from Ag NCS at different temperatures. The ordered assembly of Ag nanocrystals of ~5 nm in core size preserves the order to temperature as high as 500°C, much higher than the expected temperature. This is likely due to the strengthened binding force owing to the formation of the interdigitative molecular bonds of the adsorbed molecules on the nanocrystal surfaces. Monolayer self-assembly has been found

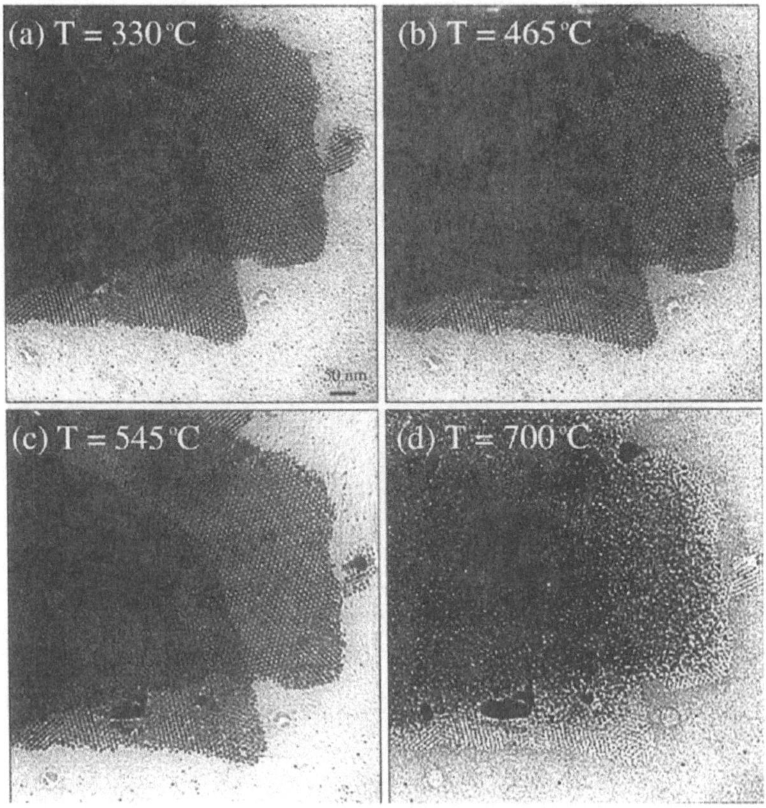

FIGURE 5.30. *In situ* TEM observation of structural damage of self-assembled Ag nanocrystals at different temperatures.

to exhibit the highest thermal stability,[41] possibly because of the strong effect from the carbon substrate.

TEM can also be applied to study the phase transformation of nanocrystals. FePt nanocrystals, for example, have the chemically disordered fcc lattice structure (A1, $FM\bar{3}m$) and the chemically ordered phase ($L1_0$, P4/mmm). Phase transformation occurs by annealing the nanocrystals to ~550°C. Figure 5.31(a) is a low-magnification TEM image showing a typical morphology of the product, where the monodisperse FePt nanocrystals with the size of ~6 nm in their diameters self-assemble to form a 2D arrangement of nanocrystals on a supporting carbon film. A select-area electron diffraction pattern [Fig. 5.31(c)] indicates that the as-synthesized FePt nanocrystals are of the chemically disordered A1–FePt phase. To obtain the chemically ordered ferromagnetic $L1_0$–FePt phase, *in situ* thermal annealing is applied to the as-synthesized FePt nanocrystals.[42] Shown in Fig. 5.31(b) is a TEM image of the FePt nanocrystals isothermally treated at 530°C for 1 h. The corresponding electron diffraction pattern is shown in Fig. 5.31(d), in which the extra reflection rings, such as (110), (120), (112), etc., appear, indicating the occurrence of phase transformation from chemically

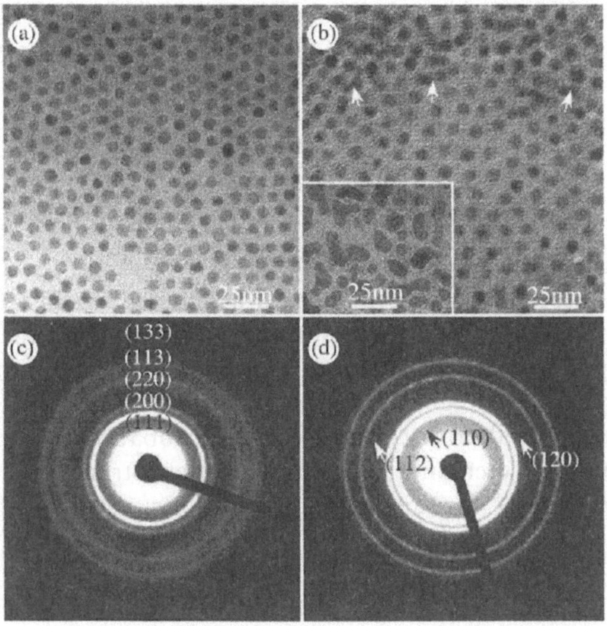

FIGURE 5.31. TEM image of the as-synthesized FePt nanocrystals (a) and the corresponding select area electron diffraction pattern (c). (b,c) TEM images of the $Fe_{52}Pt_{48}$ nanocrystals after annealing at 530°C for 1 h and corresponding select area electron diffraction pattern, respectively. The inset in (b) is a TEM image of the nanocrystals post annealing at 600°C for 1 h.

disordered A1 fcc phase to chemically ordered $L1_0$ face centered tetragonal (fct) phase. The inset in Fig. 5.31(b) shows a TEM image taken from the specimen undergoing an isothermal treatment at 600°C for 1 h, after which most of nanocrystals have coalesced with only some nanocrystals in the monolayer region staying intact. Figure 5.32 shows a HRTEM image of the coalesced FePt nanocrystal that consists of two originally separated nanocrystals (grains), in which the grain boundary still can be distinguished, as marked by the white

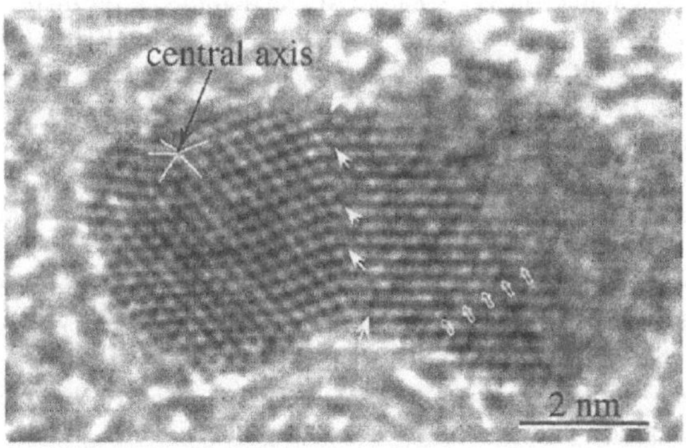

FIGURE 5.32. Coalesced FePt nanocrystal that consists of two original FePt nanocrystals, and one of them is of decahedron shape.

arrowheads. The white open arrowheads labeled on the grain on the right-hand side show compositional modulation resulting from chemical ordering due to annealing.[43]

5.5. SUMMARY

Structural analysis is important for understanding the fundamental mechanisms and processes in forming self-assembled nanostructures, and the knowledge derived is critical for controlling and applying the science and technology developed from nanostructures. This chapter describes the main tools for structural analysis, including XRD, SPM, SEM, and TEM. The application of TEM was extensively elaborated because it is probably the most popular technique used for characterizing self-assembled nanostructures. This chapter should serve as the basis for understanding the structure of self-assembled nanocrystals.

REFERENCES

1. R. L. Whetten, J. T. Khoury, M. M. Alvarez, S. Murthy, I. Vezmar, Z. L. Wang, P. W. Stephens, C. L. Cleveland, W. D. Luedtke, and U. Landman, Nanocrystal gold molecules, *Adv. Mater.* **8**, 428–433 (1996).
2. C. B. Murray, C. R. kagan, and M. G. Bawendi, Self-organization of cdse nanocrystallites into 3-dimensional quantum-dot superlattices, *Science* **270**, 1335–1338 (1995).
3. A. P. Alivisatos, Semiconductor clusters, nanocrystals, and quantum dots, *Science* **271**, 933–937 (1996).
4. R. L. Whetten, M. N. Shafigullin, J. T. Khoury, T. G. Schaaff, I. Vezmar, M. M. Alvarez, and A. Wilkinson, Crystal structures of molecular gold nanocrystal arrays, *Acc. Chem. Res.* **32**(5), 397–406 (1999).
5. B. D. Cullity, *Elements of X-ray Diffraction*, 2nd edition (Addison-Wesley, New York, 1958).
6. C. L. Cleveland, U. Landman, T. G. Schaaff, M. N. Shafigullin, P. W. Stephens, and R. L. Whetten, Structural evolution of smaller gold nanocrystals: The truncated decahedral motif, *Phys. Rev. Lett.* **79**, 1873–1876 (1997).
7. M. M. Alvarez, J. T. Khoury, T. G. Schaaff, M. Shafigullin, I. Vezmar, and R. L. Whetten, Critical sizes in the growth of Au clusters, *Chem. Phys. Lett.* **266**, 91–98 (1997).
8. R. Wiesendanger, *Scanning Probe Microscopy and Spectroscopy: Method and Applications* (Cambridge University Press, Cambridge, 1994).
9. G. Binnig, H. Rohrer, Ch. Gerber, and E. Weibel, Surface studies by scanning tunneling microscopy, *Phys. Rev. Lett.* **49**, 57–60 (1982).
10. G. Binning, C. F. Quate, and Ch. Gerber, Atomic force microscope, *Phys. Rev. Lett.* **56**, 930–933 (1986).
11. H. J. Guntherodt, D. Anselmetti, and E. Meyer (eds.), *Forces in Scanning Probe Methods*, NATO ASI series E, Applied Sciences, vol. 286 (Kluwer Academic, New York, 1995).
12. J. I. Goldstein, D. E. Newbury, P. Echlin, D. C. Joy, A. D. Romig Jr., C.E. Lyman, C. Fiori, and E. Lifshin, *Scanning Electron Microscopy and X-ray Microanalysis*, 2nd edition (Plenum Press, New York, 1992) Chapter 3 and 4.
13. Z. W. Pan, Z. R. Dai, C. Ma, and Z. L. Wang, Molten gallium as a catalyst for the large-scale growth of highly aligned silica nanowires, *J. Am. Chem. Soc.*, **124**(8), 1817–1822 (2002).
14. Z. L. Wang (ed.) *Characterization of Nanophase Materials* (Wiley-VCH, 1999).

15. Z. L. Wang, Structural analysis of self-assembling nanocrystal superlattices, *Adv. Mater.* **10**, 13–30 (1998).
16. Z. L. Wang, Transmission electron microscopy of shape-controlled nanocrystals and their assemblies (Invited review article), *J. Phys. Chem. B* **104**(6), 1153–1175 (2000).
17. P. Buseck, J. Cowley, and L. Eyring (ed.) *High-Resolution Transmission Electron Microscopy – Theory and Applications* (Oxford University Press, 1989).
18. J. M. Cowley, *Diffraction Physics*, 3rd edition (North-Holland Physics Publishing, Amestadam, 1995).
19. Z. L. Wang, *Elastic and Inelastic Scattering in Electron Diffraction and Imaging* (Plenum Press, New York, 1995), Chapters 2–5.
20. D. B. Williams and C. B Carter, *Transmission Electron Microscopy* (Plenum Press, 1996).
21. J. M. Cowley and A. F. Moodie, The scatterin of electrons by atoms and crystals. 1. A new theoretical approach, *Acta Cryst.* **10**, 609–619 (1957).
22. P.-A. Buffat, M. Flueli, R. Spycher, P. Stadelmann, and J.-P. Borel, Crystallographic structure of small gold particles studied by high-resolution electron-microscopy, *Faraday Discuss.* **92**, 173–187 (1991).
23. S. A. Harfenist, Z. L. Wang, M. M. Alvarez, I. Vezmar, and R. L. Whetten, Hexagonal close packed thin films of molecular Ag nanocrystal arrays, *Adv. Mater.* **9**(10), 817–822 (1997).
24. S. A. Harfenist, Z. L. Wang, M. M. Alvarez, I. Vezmar, and R. L. Whetten, Highly oriented molecular Ag-nanocrystal arrays, *J. Phys. Chem. B* **100**, 13904–13910 (1996).
25. Z. L. Wang, S. A. Harfenist, R. L. Whetten, J. Bentley, and N. D. Evans, Bundling and interdigitation of passivated thiolate molecules in self-assembled nanocrystal superlattices, *J. Phys. Chem. B* **102**, 3068–3072 (1998).
26. Z. L. Wang, S. A. Hurfenist, I. Vazmar, R. L. Whetten, J. Bentley, N. D. Evans, and K. B. Alexander, Superlattices of self-assembled tetrahedral Ag nanocrystals, *Adv. Mater.* **10**, 808–812 (1998).
27. D. P. Dinega and M. G. Bawendi, A solution-phase chemical approach to a new crystal structure of cobalt, *Angew Chem. Int. Edit.* **38**, 1788–1991 (1999).
28. C. B. Shoemaker, D. P. Shoemaker, T. E. Hopkins, and S. Yidepit, Refinement of the structure of β-manganese and of a related phase in the Mn–Ni–Si system, *Acta Crystallogr. B* **34**, 3573–3576 (1978).
29. S. Sun and C. B. Murray, Synthesis of monodisperse cobalt nanocrystals and their assembly into magnetic superlattices, *J. Appl. Phys.* **85**, 4325–4330 (1999).
30. Z. L. Wang, Z. R. Dai, and S. Sun, Polyhedral shapes of cobalt nanocrystals and their effect of ordered nanocrystal assembly, *Adv. Mater.* **12**, 1944–1946 (2000).
31. W. D. Luedtke and U. Landman, Structure, dynamics, and thermodynamics of passivated gold nanocrystallites and their assemblies *J. Phys. Chem. B* **100**, 13323–13329 (1996).
32. Z. L. Wang, Self-assembled superlattices of size and shape selected nanocrystal-interdigitative and gear molecular assembling models, *Mater. Charact.* **42**, 101–109 (1999).
33. S. Ino, Epitaxial growth of metals on rocksalt faces cleaved in vacuum. II. Orientation and structure of gold particles formed in ultrahigh vacuum, *J. Phys. Soc. Jpn.* **21**, 346–362 (1966).
34. Z. L. Wang and Z. C. Kang, *Functional and Smart Materials–Structural Evolution and Structure Analysis* (Plenum Press, New York, 1998), Chapter 6.
35. J. I. Goldstein, D. E. Newbury, P. Echlin, D. C. Joy, A. D. Romig, C. E. Lyman, C. Fiori, and E. Lifshin, *Scanning Electron Microscopy and X-ray Microanalysis, A Text for Biologists, Materials Scientists and Geologists* (Plenum Press, New York, 1992).
36. R. F. Egerton, *Electron Energy-Loss Spectroscopy in the Electron Microscope*, 2nd edition (Plenum Press, New York, 1996).
37. Z. L. Wang, J. S. Yin, W. D. Mo, and Z. J. Zhang, In situ analysis of valence state conversion in transition metal oxides using electron energy-loss spectroscopy, *J. Phys. Chem. B* **101**, 6793–6798 (1997).
38. Z. L. Wang, J. Bentley, and N. D. Evans, Maping the valence states of transition metal elements using energy-filtered transmission electron microscopy, *J. Phys. Chem. B* **103**, 751–753 (1999).
39. L. Reimer (ed.) *Energy Filtering Transmission Electron Microscopy*, Springer Series in Optical Science (Springer, New York, 1995).

40. Z. L. Wang, S. A. Harfenist, R. L. Whetten, J. Bentley, and N. D. Evans, Bundling and interdigitation of passivated thiolate molecules in self-assembled nanocrystal superlattices, *J. Phys. Chem. B* **102**, 3068–3072 (1998).
41. S. A. Harfenist and Z. L. Wang, High temperature stability of passivated silver nanocrystal superlattices, *J. Phys. Chem. B* **103**, 4342–4345 (1999).
42. Z. R. Dai, Z. L. Wang, and S. H. Sun, Shapes, Multiply twins and surface structures of monodispersive FePt magnetic nanocrystals, *Surf. Sci.* **505**, 325–335 (2002).
43. Z. R. Dai, Z. L. Wang, and S. H. Sun, Phase Transformation, coalescence and twinning of monodisperse FePt nanocrystals, *Nanoletters* **1**, 443–447 (2001).

6

Fabrication of Nanoarchitectures Using Lithographic Techniques

6.1. FABRICATION TECHNIQUES AND NANOLITHOGRAPHY

To make nanostructures of materials, two major approaches can be taken: non-lithographic and lithographic methodologies. Examples of non-lithographic methods are discussed in Chapters 2–4, and 10 of this book, among which QD synthesis has recently gained strong momentum. Other non-lithographic methods include a mix-and-grow procedure or using exchange reactions. Nanodomains of organic thin film materials have been produced by mixing desired components in solution (or vapor), then soaking solid substrates in the mixture.[1] Similar nanodomains may be obtained by immersing a pure thin film in a solution containing another adsorbate.[2] Figure 6.1 shows nanodomains of long-chain alkanethiols produced using the mix-and-adsorb approach. The domain sizes and distribution are determined by the interplay of surface reaction kinetics and thermodynamics. Typically, nanoislands of various sizes coexist, as visualized in Fig. 6.1.

In contrast to non-lithographic methods, the nanostructures produced using lithographic approaches are not at the mercy of thermodynamics and kinetics. Instead, the features are determined by the designs of lithographic masks, or by the trajectory of the probes. Figure 6.2 lists lithographic techniques reported to produce various feature sizes. Micropatterns can be readily produced using well-known techniques. Frequently reported methods of microfabrication include photolithography[3–6] and micromachining.[7] These techniques produce microfeatures as small as 1 μm. Recent advances in electron- and ion-beam lithography,[8–10] and microcontact printing (μCP)[11,12] have broken the wavelength barrier to produce patterns as small as 300 nm. To distinguish between micro and nanolithography, we refer to techniques that are capable of producing feature sizes of 100 nm (0.1 μm) or smaller as nanolithography.

This chapter is organized by first introducing X-ray, electron- and ion-beam lithography, which utilize masks or focused beams (Section 6.2). Nanoparticle

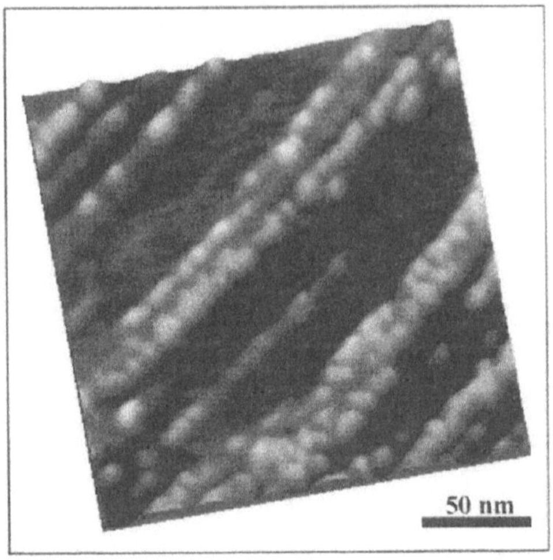

FIGURE 6.1. Nanoislands of $CH_3(CH_2)_{17}SH$ (bright areas) in a $CH_3(CH_2)_9S/Au$ self-assembled monolayer (dark area).

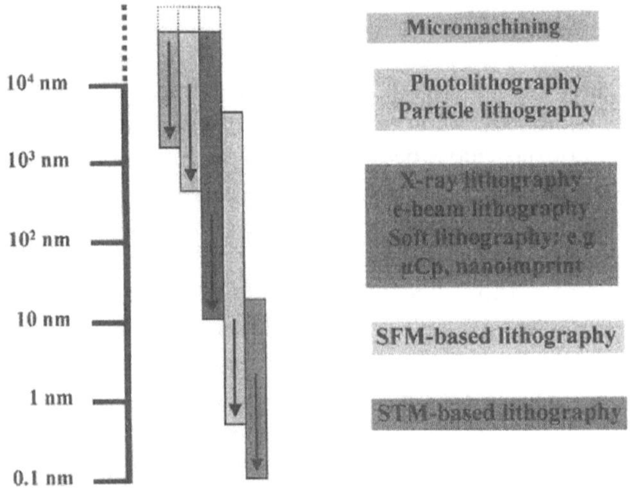

FIGURE 6.2. Summary of lithographic techniques and their highest resolution.

lithography will then be discussed (Section 6.3), followed by scanning probe lithography (Section 6.4).

6.2. X-RAY, ELECTRON, AND ION-BEAM LITHOGRAPHY

Over the past 15 years, the search for techniques to produce nanoscale features has been very active, driven by the need to produce smaller and more powerful devices and circuits in the semiconductor industry. Moore's law predicts that the dimensions MOS devices is decreased by a factor of 1/2 every three years, therefore, the feature size will be around 70 nm in year 2010. In optical lithography, short wavelengths have been employed, such as deep UV (F_2 laser with $\lambda = 157$ nm)[13–15] and extreme UV lithography (EUV, with radiation of 13.4 nm).[16] For further miniaturization, soft X-ray ($\lambda = 0.5$–4 nm), electron-beam and ion-beam-based lithography have come into play.[17] Less expensive approaches include nonconventional lithographies such as nanoimprint[18–21] and microcontact printing.[11,12]

6.2.1. *X-Ray Lithography*

In conventional photolithography, a projection configuration is used, where a collimated light passes through a mask and illuminates the resist surfaces. Depending on the optics used, various magnifications may be projected onto the resists. There are, however, no ideal materials available for X-ray projection. Therefore, shadow mask techniques were developed, which consist of a thin membrane of SiC, Si_3N_4, or Si, and heavy metals such as Ta, W, or Au to absorb X-ray photons.[17,22] These masks are produced using microfabrication/processing and electron-beam lithography. In the United States, a standard mask format has been established by NIST, which consists of a very flat 4″ wafer containing the designed features.[17,23] Such masks are kinematically mounted and positioned ~10 μm above resists.[17]

Frequently used X-ray sources include synchrotron X-ray (e.g., the advanced light source in Lawrence Berkeley Laboratory) or laser-induced plasma sources.[17] The resolution of proximity X-ray lithography is limited by the Fresnel diffraction and by the scattering of photoelectrons in the resist materials such as polymethylmethacrylate (PMMA). At present, standard X-ray lithography has produced features as small as 70 nm, while phase-shifting masks have reached the 40–70 nm region. Soft X-ray is capable of exposing relatively thick resists with a high aspect ratio. PMMA resists with submicron features and 100 μm depth are reported.[15,17,24–26] Although X-ray lithography is still at the research stage, many researchers predict that it is the most promising tool in meeting the high spatial precision and high throughput criteria of manufacturers.

6.2.2. *Electron- and Ion-Beam Lithography*

Figure 6.3 illustrates the fundamental approaches in electron- and ion-beam lithography and subsequent pattern transfer. In electron-beam lithography, a focused electron-beam may be achieved by converting a commercial SEM or TEM instruments, which operate at 50–100 kV.[27–29] The electron-beam has very high energy, and is focused on a tiny spot (a few nanometers). The spot can be moved by the electron optics following a programmed trajectory. The resolution is not determined by the spot size, but rather by electron scattering, which impacts both the exposed and nearby areas (proximity effect). The correction and improvement of proximity effect has been intensely studied.[30,31] The advantages of this

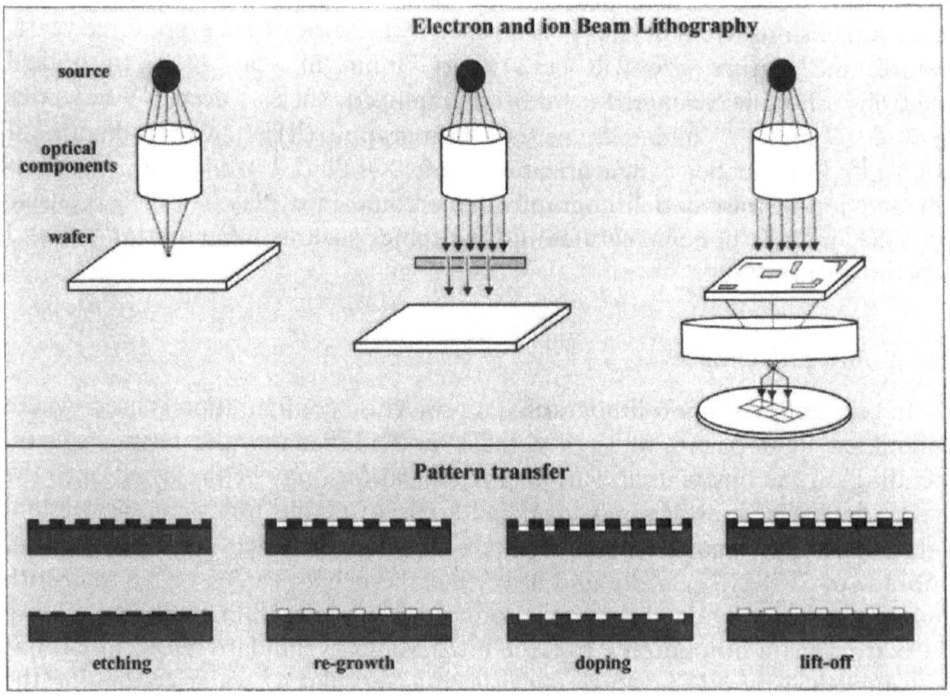

FIGURE 6.3. Three basic approaches adopted by electron- and ion-beam lithography, and subsequent methods for pattern transfer.

configuration include bypassing masks, and the simplicity in employing home-designed features. The shortcoming is its serial nature, that results in low throughput as judged by the standards of industry. Electron-beam lithography with focused electron-beams has been widely employed to produce masks that are used for X-ray lithography and nonconventional lithography such as microcontact printing and nanoimprint. In addition, electron-beams are also used for hole drilling, and for surface modification of various inorganic materials as well as organic thin films.[15,17]

The focused beam configuration can also be applied to ion-beam lithography, where beam of positive ions (instead of negatively charged electrons) such as H^+ or He^+ are produced by plasma ionization of gas molecules, and then extraction of ions. The ions may be focused by electrostatic lens to a "point." Analogous to electron-beam lithography, the ions may be moved by ion optics following a programmed trajectory. Resolution and speed are similar to electron-beam lithography, although ion-beams suffer far less scattering effect. Taking advantage of the high resolution (20 nm) and high energy (up to hundreds of kiloelectron volt), focused ion-beam is used in the integrated circuit (IC) industry to ion mill transistors and contacts.[17,32–35]

For manufacturing IC, high throughput is required. Parallel approaches are used which involve the use of masks. In masked ion-beam lithography, a mask (often a 1X stencil mask) is posted in close proximity to the resist, while a well-collimated ion-beam illuminates the resist surface through the mask to transfer the feature to the resist coatings. Typical ion energies range from 70 to 150 keV. For high-resolution applications, ion-projection lithography is used. Ion optical components are adopted after the ions pass through the mask to form a reduced image on the resist (demagnification 3–10×). Stencil masks utilize single crystal Si membranes (~2.5 μm thickness) since mask fabrication may be achieved using in the standard Si processing procedures. IMS and Siemens are two well-known companies in the technology development of ion projection lithography. At present, a 100 nm feature size with less than 10 nm distortion can be reached using commercial setups.[17,32–35]

6.3. NANOPARTICLE LITHOGRAPHY

To produce periodic arrays of nanostructures, a relatively inexpensive and simple method of nanoparticle (or nanosphere) lithography has attracted much attention.[36–41] Monodispersed latex particles are simple to synthesize and purify. For particle lithography of inorganic materials, the procedures are illustrated in Fig. 6.4. Particles can assemble into ordered, and closely packed 2D and 3D structures under carefully controlled conditions. These assemblies are used as templates, where the void space is filled with the materials of interest. Particles can then be removed by either calcination or solvent dissolution. Arrays of nanostructures are produced, which include 2D and 3D arrays of metals, metal oxides and silica, porous membranes of polyurethane, as well as 3D opal or reversed opal structures of photonic band-gap materials such as SiO_2 and TiO_2.[36–42]

Similar methods may be adapted to produce 2D arrays of protein nanostructures, as shown in Fig. 6.4. First the protein and latex are mixed to desired concentrations. The mixture is then deposited on the substrate and allowed to dry. Guided by the assembly of latex spheres, an ordered structure is formed, consisting of close-packed latex particles decorated by proteins in the void spaces. The latex template is then rinsed away with deionized water, resulting in periodic arrays of protein nanostructures on the substrate.

Periodic arrays of nanostructures of proteins have been successfully produced. Figure 6.5 illustrates the production of nanostructures of bovine serum albumin (BSA). The periodicity of the BSA nanoarrays is 206 nm. Most protein molecules retain their activity because specific antibodies such as rabbit-anti-BSA immunoglobulin G (IgG) can bind to the adsorbed BSA particles. Compared with X-ray, electron- and ion-beam lithography, nanosphere lithography is simpler and less expensive. One can easily adapt this method to produce periodical nanoarrays in a research laboratory. In comparison to scanning-probe lithography (SPL) (discussed later in the chapter), nanoparticle lithography is able to produce nanostructures with higher throughput. Figure 6.6 shows a large area of BSA nanoarrays with a periodicity of 800 nm. Most of the latex particles in the template were removed except for several bright spots.

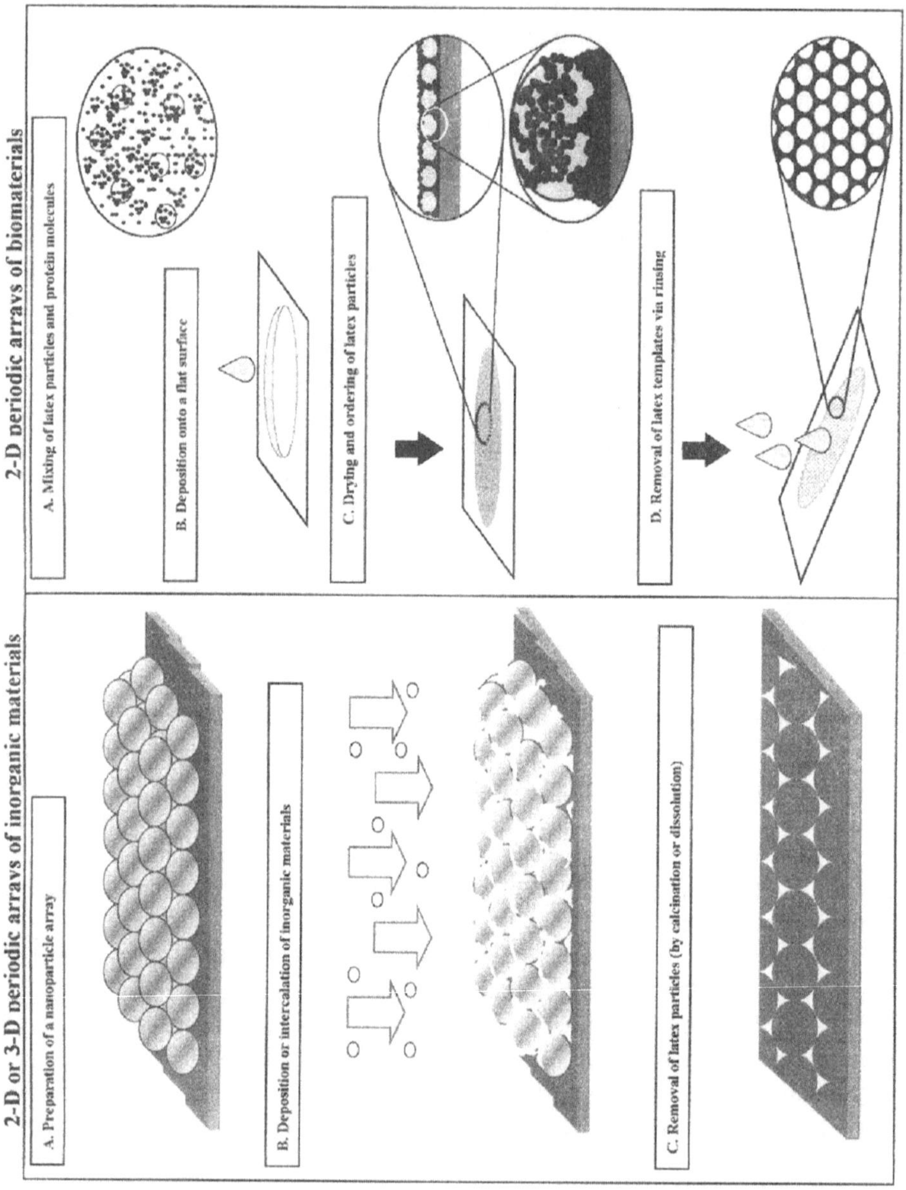

FIGURE 6.4. Schematic diagrams of nanoparticle lithography for inorganic materials (left) and biomolecules (right).

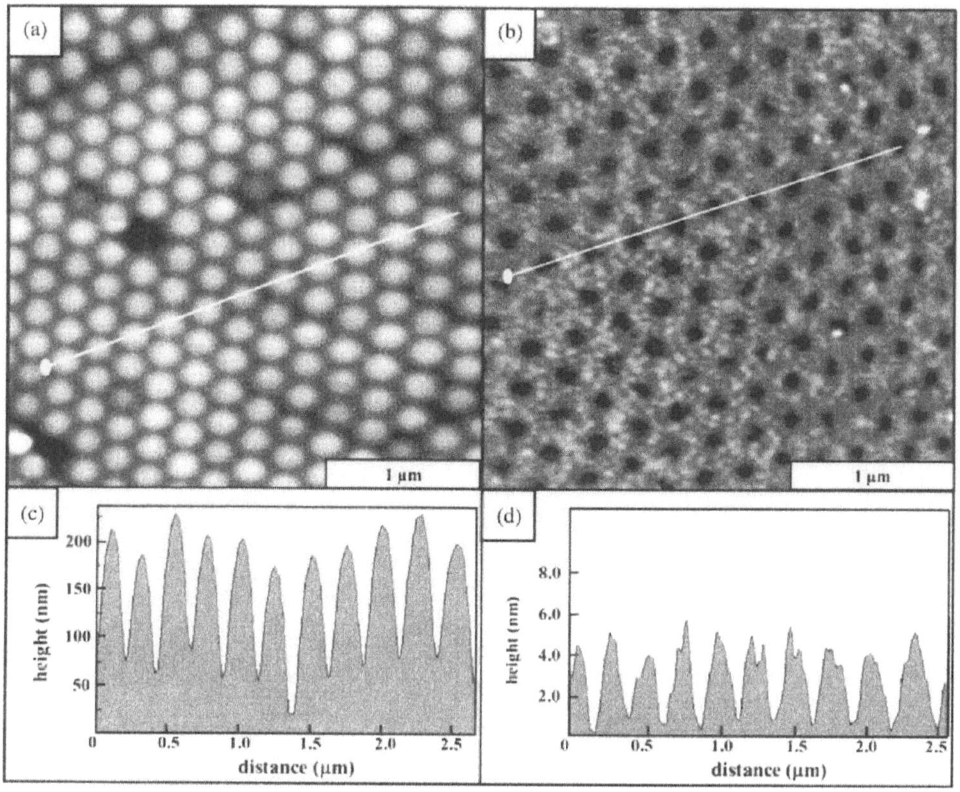

FIGURE 6.5. (a) A 3×3 μm^2 AFM topographic image of latex–BSA mixture deposited on mica. The latex particles have a diameter of 204 ± 6 nm. (b) Same system after removal of latex spheres. Periodic arrays of proteins are produced. (c) and (d) are cursor profiles as indicated in (a) and (b) respectively.

6.4. SCANNING PROBE LITHOGRAPHY

6.4.1. Resolution and Nanofabrication

Scanning probe microscopy such as STM and AFM are well known for their capability to visualize surfaces of materials with the highest spatial resolution.[43–46] In terms of structural characterizations, AFM can attain a lateral resolution of 0.1 Å and vertical resolution of 0.05 Å.[43,46] STM resolution is intrinsically five times higher than AFM.[43,46] These high resolutions are reached for crystalline systems. For noncrystalline surfaces or soft-and-sticky surfaces, it is more difficult to achieve high resolution.

Despite the structural complexity of SAMs and macromolecules, high-resolution images have been obtained using both AFM and STM.[47–55] Figure 6.7 includes four images which demonstrate the high resolution attainable by AFM and STM. In the case of an alkane-thiol SAM [Fig. 6.7(a) and (b)], defects such as steps, etch pits (gray spots) and single layer islands (bright spots) are clearly visible. More importantly, the c(4 × 2) superlattice with respect to the basic $(3\sqrt{2} \times 3\sqrt{2})R30°$

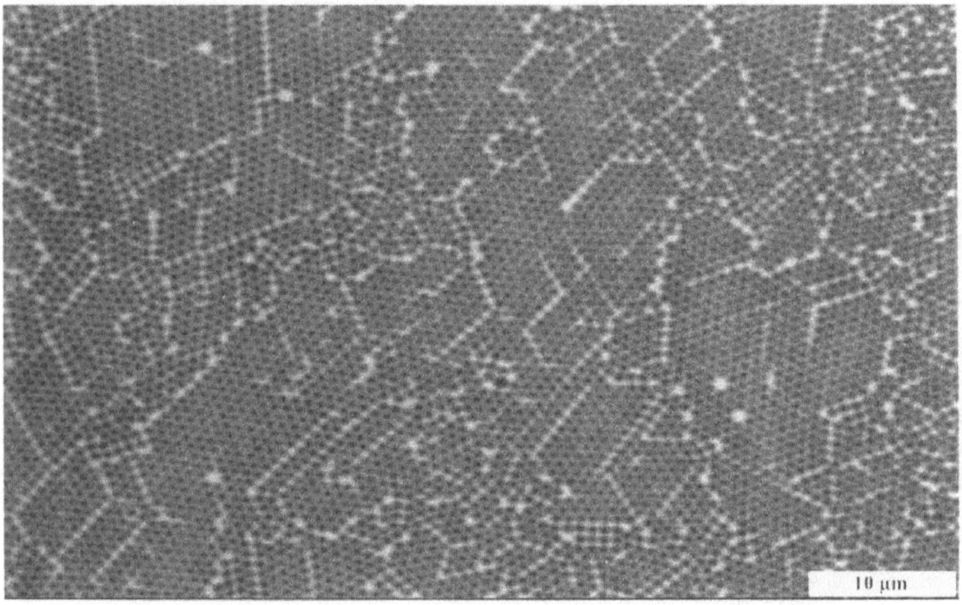

FIGURE 6.6. An optical micrograph of latex and BSA mixture deposited on a mica surface. The periodicity is 802 ± 6 nm. The scale bar is 10 μm.

structure was resolved for the first time using AFM. STM intrinsically has much higher resolution than AFM. In Fig. 6.7(b), the periodicity and single molecular defects at the domain boundary of a SAM are resolved within the same image.

In the case of immobilized proteins on a planar surface [Fig. 6.7(c)], individual lysozyme (LYZ) molecules are visible from the AFM topograph, from which the orientation of the proteins may be extracted.[47,56] Detailed structural features of nonplanar and large biosystems, such as bacteria and cells, can also be revealed using AFM.[57,62] Figure 6.7(d) is an AFM topograph of a cylindrical *E. coli* bacterium covered by IgG molecules. Individual IgG can be resolved in the high-resolution images (see the zoom-in of image shown if Fig. 6.7(d)).[56,63]

The fact that molecules within SAMs can be resolved indicates that the tip–SAM interaction in AFM imaging and the tunneling electrons in STM imaging are localized to molecular dimensions. Taking advantage of the sharpness of the tips, and strong and localized tip–surface interactions, SPM has also been used to manipulate atoms on metal surfaces, and to fabricate nanopatterns of metal and semiconductor surfaces.[64–67] These successful examples catalyze an emerging field of SPL.[67] Despite the fact that SPL cannot produce large numbers of patterns as rapidly as photolithography or electron lithography, SPL research still attracts tremendous attention among researchers because of its unparalleled spatial precision and its capability to fabricate and image *in situ*. More importantly for researchers in nanomaterial science and biotechnology, SPL is the only approach which promises to be able to position individual molecules.

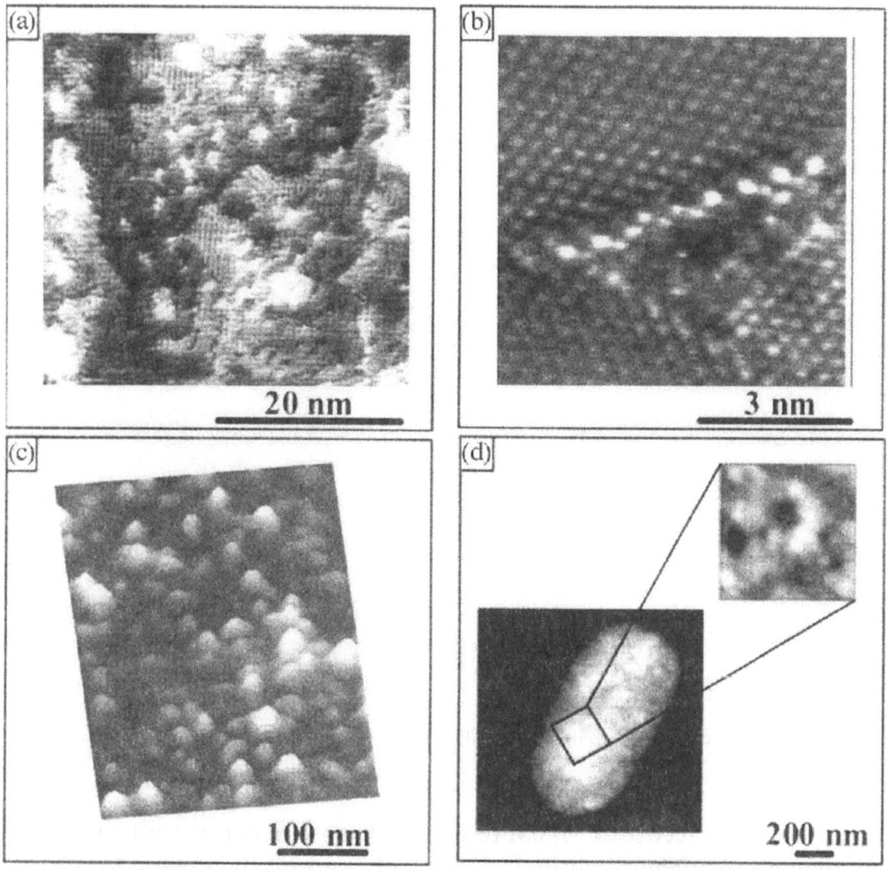

FIGURE 6.7. (a) An AFM topograph of a decanethiol SAM. (b) An STM topograph of the same monolayer. (c) LYZ immobilized on a carboxylic acid-terminated SAM. (d) An *E. coli* bacterium covered by IgG molecules.

The achievement of high resolution is the foundation of SPM-based molecular manipulation and nanolithography. In principle, by enhancing these local interactions such as atomic force, density of tunneling electrons or electrical field strength, one should be able to break chemical bonds selectively. The detailed methodology in controlling these local interactions is the key to obtaining sharp patterns with high spatial precision. Various approaches in controlling the local interactions have been reported, which include: AFM-based lithography such as tip-catalyzed surface reactions,[68,70] dip-pen nanolithography (DPN);[71–73] and STM-based lithography such as single molecule manipulation,[74,75] tip-assisted electrochemical etching and field-induced desorption.[76,77]

6.4.2. *STM-Based Nanolithography*

STM tips typically are atomically sharp and are composed of W, Pt–Rh, or Pt–Ir wires. The tip–surface distance is regulated by feedback electronics to

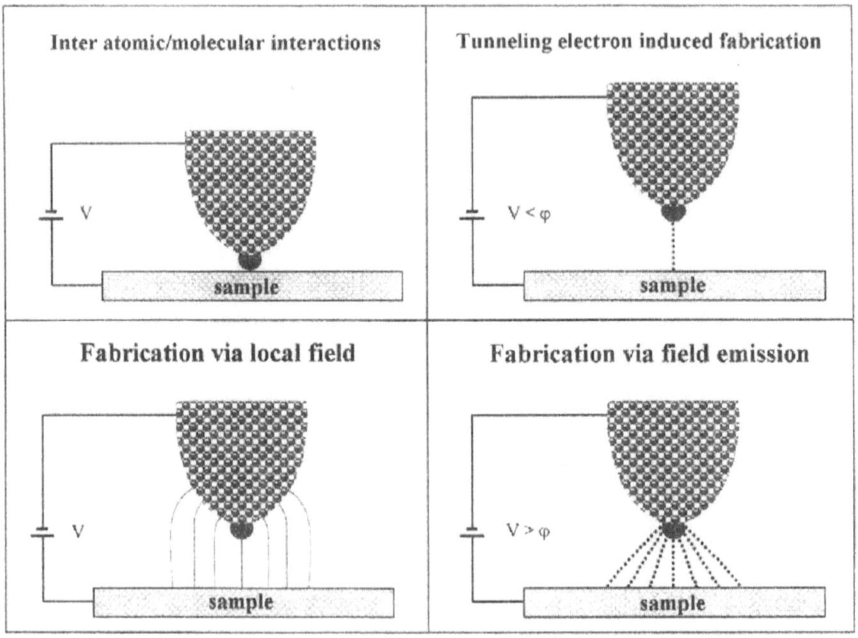

FIGURE 6.8. Four possible mechanisms of STM lithography or tip manipulation of molecules.

maintain a constant feedback signal (e.g., tunneling current or tip height). To go beyond imaging to achieve manipulation or fabrication, the targeted molecule or molecules must be broken free from the rest of the surface materials. In other words, local bonds must be weakened or broken. Experimentally, one can adjust the tip–surface separation, bias voltage, or tunneling current. The tunneling current depends on the separation distance, following a negative exponential relationship. Four possible mechanisms of STM manipulation of molecules are summarized in Fig. 6.8, following the example of Ho.[75] In principle, individual molecular manipulation is best achieved using interatomic or intermolecular forces between the tip and surface species. Fabrication involving breaking or formation of chemical bonds is best carried out using tunneling current, which is highly local (e.g., 1 nm or smaller). Field emission is not as local as the other approaches, and is mostly used to clean the tip.

Using atomic forces, Eigler and coworkers[74,78,79] first demonstrated the precision of STM manipulation at 4 K by advancing an STM tip to the very close proximity of the surface atoms or small molecules. The overlap of the wave function between the tip and surface species resulted in strong interatomic or intermolecular interactions between the tip and the selected atom or molecule. The tip can then slide or pick up the atoms or molecules and move them one-by-one to the designated locations. The atoms that were successfully moved by STM tips included Cu, Fe, Xe, Co, and the small molecules included CO and O_2.[74,78–84] These artificially engineered structures provide a unique opportunity for observation and

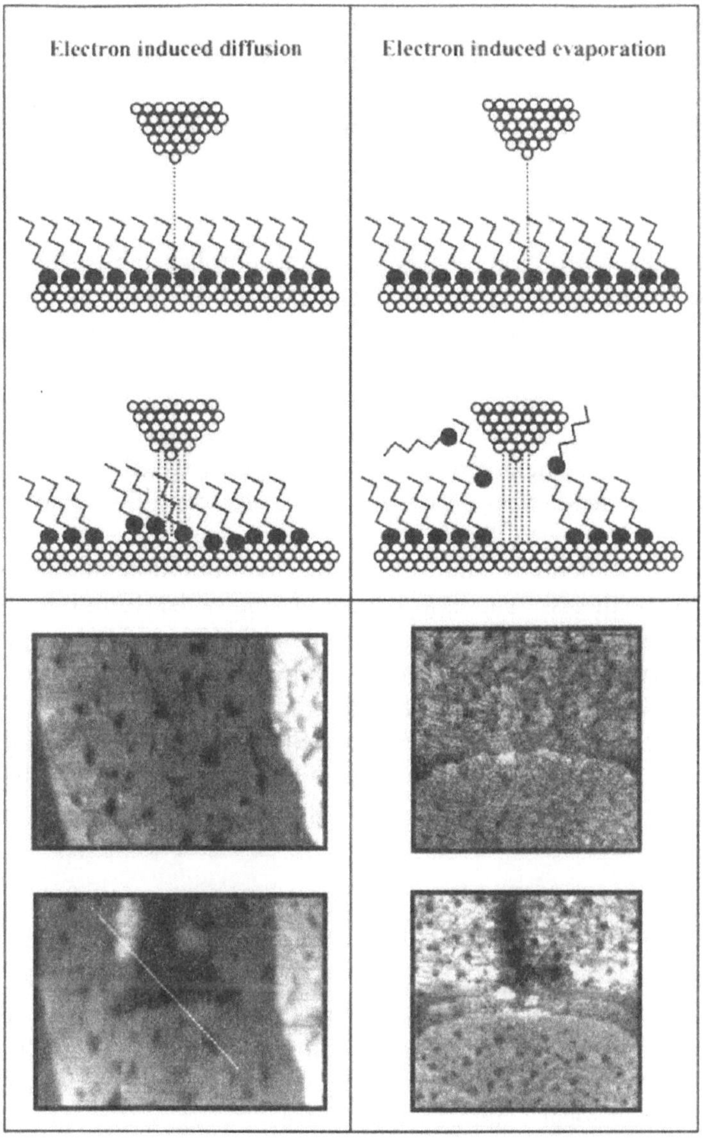

FIGURE 6.9. Two examples of STM-based nanofabrication. Tip induced diffusion of gold atoms is illustrated in the schematic diagram on the left. An 75 × 50 nm^2 STM topograph below the scheme was taken at 1.5 V and 6.6 pA, in which a triangular area was produced at 80 pA. The dimension corresponds well with the single atomic layer of gold. Tunneling electron induced desorption of thiols is illustrated in the schematic diagram on the right. The STM-topograph below was acquired at 2 V and 4 pA, within which an L-shaped area was produced at 80 pA. The depth of the patterns is 0.75 nm, consistent with the desorption of thiols.

investigation of size-dependent quantum behavior such as quantum corrals and quantum mirages.[79-84] Larger molecules such as porphyrins have also been positioned by Jung *et al.*[85] following a similar protocol, at room temperature.

Tunneling current has been used to induce local chemical reactions, that is, the breaking and formation of bonds. For small molecules, one can understand such processes as placing electrons in the antibonding orbital or removing electrons from the binding orbital, which cause the breaking of the chemical bonds. Ho[75] was successful in breaking O-O, and Si-Si bonds on surfaces, as well as in forming C-Fe bonds. Rieder and coworkers[86–89] combined tunneling current and forces to break C-I bonds, and to direct two phenyl radicals to form diphenyl molecules.

Less localized lithography may also be accomplished by using an STM tip as an electrode to induce local electrochemical reactions such as the oxidation of adsorbates[76,77] or the substrate itself in the case of conductive or semiconductive materials.

The local electrical field under the tip provides another means of manipulation. During STM imaging, the electrical field typically ranges from 10^6 to 10^7 V/cm. The field strength may be increased to $\sim 10^8$ V/cm, which is comparable to the field experienced by valence electrons. Si-Si may be dissociated under high electrical fields.[90]

It is more difficult to understand the manipulation mechanism for large molecules. Nevertheless, successful examples may be found in the literature including porphyrins,[85] and hydrocarbon chains.[47] Figure 6.9 shows two examples of STM fabrication of decanethiols on gold. In the first example, tunneling current did not perturb the SAM, but caused gold atoms to diffuse underneath the adsorbates. Only gold atoms under the tip within the first layer of the thin film are displaced. In the second case, the tunneling electrons broke the sulfur–gold chemisorption and thus caused the thiols to detach.[47]

6.4.3. AFM-Based Nanolithography

AFM-based lithography is a very active area of research because of the flexibility and simplicity of the technique. AFM tips may be used to carry catalysts to selectively induce surface reactions, or as a pen to attach molecules on surfaces in DPN and its derivative techniques. AFM tips may also be used as an electrode to direct local oxidation on surfaces.

Local force provides another means for AFM fabrication. Figure 6.10 shows cartoons of three AFM-based nanofabrication techniques – nanoshaving, nanografting, and nanopen reader and writer (NPRW).[47] In these methods, the surface structure is first characterized under a very low force or load. Fabrication locations are then selected, normally in flat regions, for example, Au(111) plateau areas. Then nanopatterns are produced under a high force. In nanoshaving, the AFM tip exerts a high local pressure at the contact. This pressure results in a high shear force during the scan, which causes the displacement of SAM adsorbates. In nanografting, the SAM and the AFM cantilever are immersed in a solution containing a different thiol. As the AFM tip plows through the matrix SAM, the thiol molecules in the solution adsorb onto the newly exposed gold surface. In NPRW, the tip is pre-coated with desired molecules and these molecules are

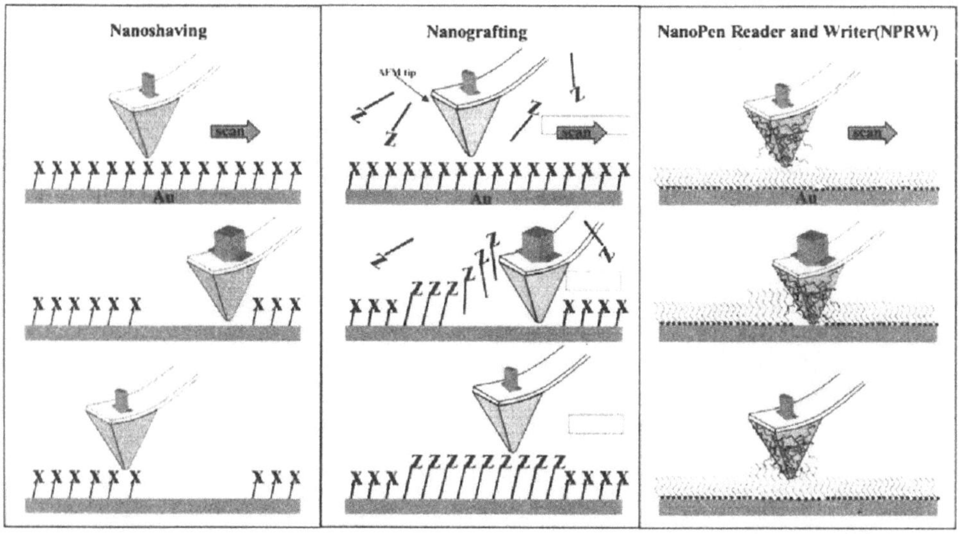

FIGURE 6.10. Schematic diagrams of three AFM-based lithography methods. The top and bottom rows represent imaging before and after fabrication, respectively, while the middle rows depict the fabrication process.

transferred under high force to the exposed substrate. The nanostructures are then characterized at reduced loads.

Compared with other fabrication techniques such as photolithography, and microcontact printing, nanografting and NPRW have several advantages, as illustrated in Fig. 6.11.[47,91,92] First, an edge resolution of 1 nm is routinely obtained. The smallest feature shown in Fig. 6.11 is 2×4 nm^2, consisting of 32 thiol molecules. Second, nanostructures can be characterized *in situ* and with high resolution. Third, the setup is able to produce complicated patterns automatically. In Fig. 6.11, various chain lengths and functionalities are demonstrated. Multiple components can be produced by varying the solution or tip coatings. Fourth, one can quickly change and/or modify the fabricated patterns *in situ* without changing the mask or repeating the entire fabrication procedure. An example of increasing the separation of two parallel lines is shown in Fig. 6.11.

Applications of AFM lithography are promising in the areas of nanoelectronics and nanobiotechnology. Preliminary success in producing nanopatterns of biosystems such as proteins and DNA have been demonstrated.[63,93] The size of protein patterns ranges from 5 nm to 1 μm.[63,93] The strategy is to first pattern SAMs using the SPL-based lithography described above. These SAM nanostructures serve as templates for subsequent fabrication processes such as patterning proteins by selective adsorption. Selectivity of protein adsorption can be achieved with knowledge of the variation in protein affinity towards different SAMs.[56,94–96]

The results of two proof-of-feasibility experiments are shown in Fig. 6.12, in which IgG and LYZ are positioned onto nanopatterns.[63,93] Figure 6.12(b) displays a 40×40 nm^2 pattern containing high coverage of IgG, produced via

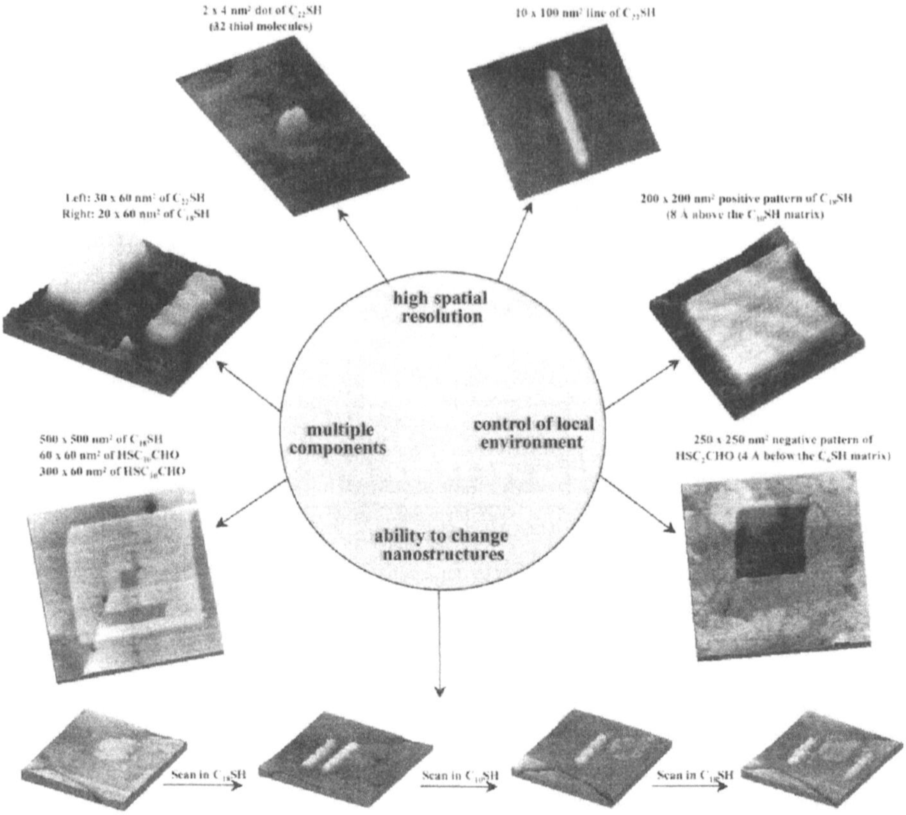

FIGURE 6.11. Key advantages of nanografting and NPRW are illustrated using actual AFM images of SAMs.

covalent immobilization on the CHO-terminated nanosquare. Figure 6.12(e) shows two smaller nanopatterns, a 150 nm line and a 100×150 nm^2 rectangle, fabricated by electrostatic immobilization of LYZ on HOOC-terminated nanopatterns.

Individual protein particles within the patterns can be resolved from the AFM images shown in Fig. 6.12(b) and (e). The corresponding cursor profiles shown in Fig. 6.12(c) and (f) reveal that the immobilized protein molecules exhibit different heights. IgG has a Y-shaped structure, while LYZ molecules are ellipsoidal with dimensions about $4.5 \times 3.0 \times 3.0$ nm^3 according to X-ray crystallographic studies.[97] The variation in heights is consistent with the fact that electrostatic and covalent interactions are not specific, and often result in various protein orientations with respect to the surface. The smallest protein pattern reported contains only three LYZ molecules (the line pattern in Fig. 6.12(e)).

The activity of the immobilized proteins may be studied by monitoring the specific antibody–antigen binding processes using the same AFM tip.[63,93] Some of

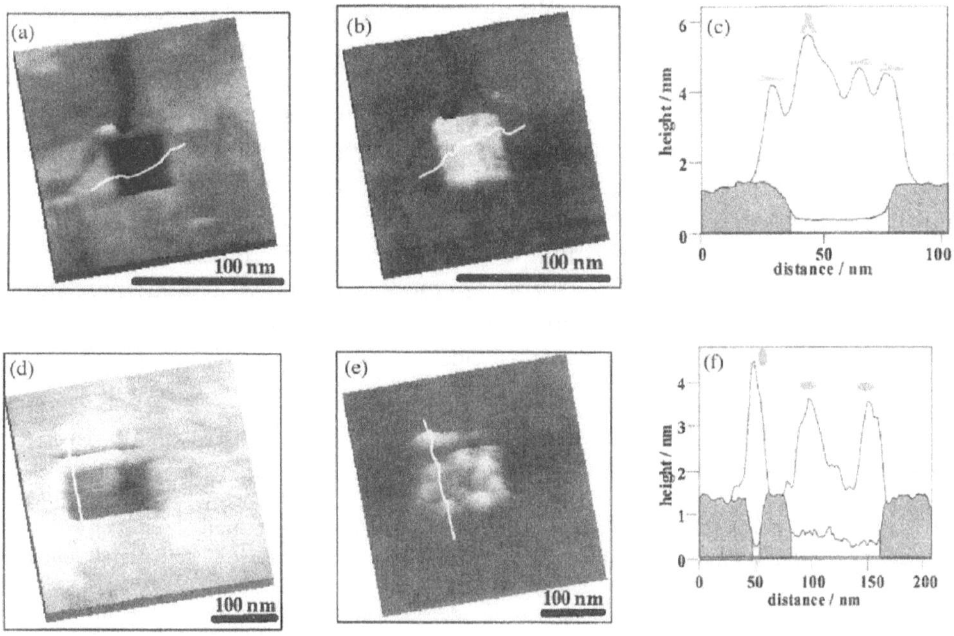

FIGURE 6.12. Production of nanopatterns of IgG and LYZ. (a) A 150 × 150 nm^2 topographic image with a 40 × 40 nm^2 pattern of 3-mercapto-1-propanal in a decanethiol SAM matrix. (b) The same area after IgG adsorption via covalent binding. (c) The combined cursor profile of images (a) and (b). (b) A 150 nm line and a 100 × 150 nm^2 rectangle, fabricated with $HOOC(CH_2)_2SH$ in a decanethiol matrix. (e) The same area imaged after LYZ immobilization via electrostatic adsorption. (f) The combined cursor profiles of images (d) and (d).

the proteins within the nanopatterns are found to be active as they can be recognized by specific antibodies.[63,93] The immobilized antigens exhibit various orientations, thus only the portion with the antibody binding sites exposed to solution can be recognized by the specific antibodies.

6.5. CONCLUDING REMARKS

Various nanolithography techniques are highlighted in this chapter, which represent an engineering approach to position molecules in two and three dimensions. The approaches are complementary to the "bottom-up" material synthesis methods (some of which are discussed in this book), because they provide a means to fabricate ultra-small electronic components, sensing elements, and scaffolds for biomaterial engineering. There are pros and cons among these approaches. Because of the high throughput, X-ray, electron- and ion-beam lithography are, in most situations, the choices of semiconductor industry as the future technology for manufacturing IC and devices. Soft lithography such as microcontact printing and nanoimprint provide simpler and less expensive means to produce nanostructures of organic and biological molecules. Nanoparticle lithography is specialized in producing periodical arrays of inorganic as well as biological materials.

SPL exhibits the highest spatial precision, and is still a research technique at present. This approach is mostly used to address fundamental scientific issues such as size dependent or quantum properties of selected systems. One concern with SPL is the serial nature of the process and low throughput. Two approaches are in progress to improve the throughput: automating the nanofabrication,[98–100] and/or using parallel probes.[101–104] The unique advantage of spatial precision makes SPL a promising tool for future biotechnology, especially in engineering biochips, sensors, and molecular based devices.

REFERENCES

1. P. E. Laibinis, R. G. Nuzzo, and G. M. Whitesides, Structure of monolayers formed by coadsorption of 2 normal- alkanethiols of different chain lengths on gold and its relation to wetting, *J. Phys. Chem.* **96**, 5097–5105 (1992).
2. C. E. D. Chidsey, Free-energy and temperature-dependence of electron-transfer at the metal-electrolyte interface, *Science* **251**, 919–922 (1991).
3. J. Y. Huang, D. A. Dahlgren, and J. C. Hemminger, Photopatterning of self-assembled alkalinethiolate monolayers on gold—a simple monolayer photoresist utilizing aqueous chemistry, *Langmuir* **10**, 626–628 (1994).
4. M. J. Tarlov, D. R. F. Burgess, and G. Gillen, UV photopatterning of alkanethiolate monolayers self-assembled on gold and silver, *J. Am. Chem. Soc.* **115**, 5305–5306 (1993).
5. E. W. Wollman, D. Kang, C. D. Frisbie, I. M. Lorkovic, and M. S. Wrighton, Photosensitive self-assembled monolayers on gold—photochemistry of surface-confined aryl azide and cyclopentadienylmanganese- tricarbonyl, *J. Am. Chem. Soc.* **116**, 4395 (1994).
6. S. P. A. Fodor, J. L. Read, M. C. Pirrung, L. A. T. Stryer, and D. Solas, Light-directed, spatially addressable parallel chemical synthesis, *Science* **251**, 767–773 (1991).
7. N. L. Abbott, A. Kumar, and G. M. Whitesides, Using micromachining, molecular self-assembly, and wet etching to fabricate 0.1–1-mm–m-scale structures of gold and silicon, *Chem. Mater.* **6**, 596–602 (1994).
8. J. A. M. Sondag-Huethorst, H. R. J. Van-Helleputte, and L. G. J. Fokkink, Generation of eelectrochemically deposited metal patterns means of electron-beam (nano) lithography of self-assembled monolayer resists, *Appl. Phys. Lett.* **64**, 285 (1994).
9. R. C. Tiberio *et al.*, Self-assembled monolayers electron-beam resist on GaAs, *Appl. Phys. Lett.* **62**, 476–478 (1993).
10. K. K. Berggren, A. Bard, J. L. Wilbur, J. D. Gillaspy, A. G. Helg, J. J. McClelland, S. L. Rolston, W. D. Philips, M. Prentiss, and G. M. Whitesides, Microlithography by using neutral metastable atoms and self-assembled monolayers, *Science* **269**, 1255–1257 (1995).
11. A. Kumar, N. L. Abbott, E. Kim, H. A. Biebuyck, and G. M. Whitesides, Patterned self-assembled monolayers and mesoscale phenomena, *Acc. Chem. Res.* **28**, 219–226 (1995).
12. Y. N. Xia and G. M. Whitesides, Use of controlled reactive spreading of liquid alkanethiol on the surface of gold to modify the size of features produced by microcontactprinting, *J. Am. Chem. Soc.* **117**, 3274 (1995).
13. T. M. Bloomstein, M. Rothschild, R. R. Kunz, D. E. Hardy, R. B. Goodman, and S. T. Palmacci, Critical issues in 157 nm lithography, *J. Vac. Sci. Technol., B* **16**, 3154–3157 (1998).
14. T. M. Bloomstein, M. W. Horn, M. Rothschild, R. R. Kunz, S. T. Palmacci, and R. B. Goodman, Lithography with 157 nm lasers, *J. Vac. Sci. Technol., B* **15**, 2112–2116 (1997).
15. F. Cerrina, X-ray imaging: Applications to patterning and lithography, *J. Phys. D: Appl. Phys.* **33**, R103–R116 (2000).

16. C. W. Gwyn, R. Stulen, D. Sweeney, and D. Attwood, Extreme ultraviolet lithography, *J. Vac. Sci. Techol. B* **16**, 3142–3149 (1998).
17. Y. Chen and A. Pepin, Nanofabrication: Conventional and nonconventional methods, *Electrophoresis* **22**, 187–207 (2001).
18. Z. N. Yu, S. J. Schablitsky, and S. Y. Chou, Nanoscale GaAs metal–semiconductor–metal photodetectors fabricated using nanoimprint lithography, *Appl. Phys. Lett.* **74**, 2381–2383 (1999).
19. M. M. Alkaisi, R. J. Blaikie, and S. J. McNab, Low temperature nanoimprint lithography using silicon nitride molds, *Microelectro. Engi.* **57–58**, 367–373 (2001).
20. W. Zhang and S. Y. Chou, Multilevel nanoimprint lithography with submicron alignment over 4 in. Si wafers, *Appl. Phys. Lett.* **79**, 845–847 (2001).
21. S. Y. Chou, Nanoimprint lithography and lithographically induced self-assembly, *MRS Bull.* **26**, 512–517 (2001).
22. C. K. Malek, K. H. Jackson, W. D. Bonivert, and J. Hruby, Masks for high aspect ratio X-ray lithography, *J. Micromech. Microeng.* **6**, 228–235 (1996).
23. S. Tsuboi, Y. Tanaka, T. Iwamoto, H. Sumitani, and Y. Nakayama, Recent progress in 1X X-ray mask technology: Feasibility study using ASET-NIST format TaXN X-ray masks with 100 nm rule 4 Gbit dynamic random access memory test patterns, *J. Vac. Sci. Technol. B* **19**, 2416–2422 (2001).
24. G. Feiertag, W. Ehrfeld, H. Lehr, A. Schmidt, and M. Schmidt, Accuracy of structure transfer in deep X-ray lithography, *Microelectron. Eng.* **35**, 557–560 (1997).
25. W. Ehrfeld, V. Hessel, H. Lowe, C. Schulz, and L. Weber, Materials of LIGA technology, *Microsyst. Technol.* **5**, 105–112 (1999).
26. W. Ehrfeld and A. Schmidt, Recent developments in deep X-ray lithography, *J. Vac. Sci. Technol. B* **16**, 3526–3534 (1998).
27. C. Vieu, F. Carcenac, A. Pepin, Y. Chen, M. Mejias, A. Lebib, L. Manin-Ferlazzo, L. Couraud, and H. Launois, Electron-beam lithography: Resolution limits and applications, *Appl. Surf. Sci.* **164**, 111–117 (2000).
28. Y. Chen, D. Macintyre, and S. Thoms, A study of electron forward scattering effects on the footwidth of T-gates fabricated using a bilayer of PMMA and UVIII, *Microelectron. Eng.* **53**, 349–352 (2000).
29. Y. Chen, D. Macintyre, and S. Thoms, Electron-beam lithography process for T- and Gamma-shaped gate fabrication using chemically amplified DUV resists and PMMA, *J. Vac. Sci. Technol. B* **17**, 2507–2511 (1999).
30. G. Owen, Proximity effect correction in electron-beam lithography, *Opt. Eng.* **32**, 2446–2451 (1993).
31. C. N. Berglund, N. I. Maluf, J. Ye, G. Owen, R. Borwning, and R. F. W. Pease, Spatial correlation of electron-beam mask errors and the implications for integrated-circuit yield, *J. Vac. Sci. Technol. B* **10**, 2633–2637 (1992).
32. J. S. Huh, M. I. Shepard, and J. Melngailis, Focused ion-beam lithography, *J. Vac. Sci. Technol. B* **9**, 173–175 (1991).
33. X. Xu, A. D. Dellaratta, J. Sosonkina, and J. Melngailis, Focused ion-beam induced deposition and ion milling as a function of angle of ion incidence, *J. Vac. Sci. Technol. B* **10**, 2675–2680 (1992).
34. J. Melngailis, Focused ion-beam lithography, *Nucl. Instrum. Methods Phys. Res. Sect. B – Beam Interactions with Materials and Atoms* **80–81**, 1271–1280 (1993).
35. J. Melngailis, A. A. Mondelli, I. L. Berry, and R. Mohondro, A review of ion projection lithography, *J. Vac. Sci. Technol. B* **16**, 927–957 (1998).
36. C. L. Haynes and R. P. Van Duyne, Nanosphere lithography: A versatile nanofabrication tool for studies of size-dependent nanoparticle optics, *J. Phys. Chem. B* **105**, 5599–5611 (2001).
37. A. Stein, Sphere templating methods for periodic porous solids, *Microporous Mesoporous Mater.* **44**, 227–239 (2001).
38. S. H. Park, D. Qin, and Y. Xia, Crystallization of mesoscale particles over large areas, *Adv. Mater.* **10**, 1028–1038 (1998).
39. O. D. Velev, T. A. Jede, R. F. Lobo, and A. M. Lenhoff, Porous silica via colloidal crystallization, *Nature* **389**, 447–448 (1997).

40. O. D. Velev and A. M. Lenhoff, Colloidal crystals as templates for porous materials, *Curr. Opin. Colloid Interface Sci.* **5**, 56–63 (2000).
41. O. D. Velev, P. M. Tessier, A. M. Lenhoff, and E. W. Kaler, Nanostructured porous materials templated by colloidal crystals: From inorganic oxides to metals, *Abstracts of Papers of the American Chemical Society* **219**, 425-PHYS (2000).
42. J. Cizeron and V. Colvin, Preparation of nanocrystalline quartz under hydrothermal conditions, *Abstracts of Papers of the American Chemical Society* **218**, 545-INOR (1999).
43. G. Binnig, Force microscopy, *Ultramicroscopy* **42**, 7–15 (1992).
44. G. Binnig, C. Gerber, E. Stoll, T. R. Albrecht, and C. F. Quate, Atomic resolution with atomic force microscope, *Surf. Sci.* **189**, 1–6 (1987).
45. G. Binnig and H. Rohrer, Scanning tunneling microscopy – from birth to adolescence, *Angew. Chem Int. Ed. Engl.* **26**, 606–614 (1987).
46. F. Ohnesorge and G. Binnig, True atomic-resolution by atomic force microscopy through repulsive and attractive forces, *Science* **260**, 1451–1456 (1993).
47. G.-Y. Liu, S. Xu, and Y. Qian, Nanofabrication of self-assembled monolayers using scanning probe lithography, *Acc. Chem. Res.* **33**, 457–466 (2000).
48. S. Xu, P. E. Laibinis, and G.-Y. Liu, Accelerating self-assembly on gold – a spatial confinement effect, *J. Am. Chem. Soc.* **120**, 9356–9361 (1998).
49. G. E. Poirier, Characterization of organosulfur molecular monolayers on Au(111) using scanning tunneling microscopy, *Chem. Rev.* **97**, 1127 (1997).
50. G. E. Poirier, E. D. Pylant, and J. M. White, Crystalline structures of pristine and hydrated mercaptohexanol self-assembled monolayers an Au(111), *J. Chem. Phys.* **105**, 2089–2092 (1996).
51. G. E. Poirier and M. J. Tarlov, The c(4×2) Superlattice of *N*-Alkanethiol monolayers self-assembled an Au(111), *Langmuir* **10**, 2853–2856 (1994).
52. P. E. Poirier, E. D. Pylant, and J. M. White, Crystalline structures of pristine and hydrated mercaptohexanol self-assembled monolayers on Au(111), *J. Chem. Phys.* **105**, 2089 (1996).
53. P. E. Poirier and M. J. Tarlor, The c(4×2) Superlattice of *N*-alkanethiol monolayers self-assembled an Au(111), *Langmuir* **10**, 2853 (1994).
54. P. E. Poirier, M. J. Tarlor, and H. E. Rushmeier, Two-dimensional liquid phase and Pxv3 phase of alkanethiol self-assembled monolayer on Au(111), *Langmuir* **10**, 3383 (1994).
55. H. J. Butt, K. Seifert, and E. Bamberg, Imaging molecular defects in alkanethiol monolayers with an atomic-force microscope, *J. Phys. Chem.* **97**, 7316–7320 (1993).
56. K. Wadu-Mesthrige, N. A. Amro, and G.-Y. Liu, Immobilization of proteins on self-assembled monolayers, *Scanning* **22**, 380–388 (2000).
57. N. A. Amro, L. P. Kotra, K. Wadu-Mesthrige, A. Bulchev, S. Mobashery, and G.-Y. Liu, Structural basis of the *Escherichia coli* outer-membrane permeability, *Proc. SPIE* **3607**, 108–122 (1999).
58. N. A. Amro, L. P. Kotra, K. Wadu-Mesthrige, A. Bulchev, S. Mobashery, and G.-Y. Liu, High-resolution atomic force microscopy studies of the *Escherichia coli* outer membrane: The structural basis for permeability, *Langmuir* **16**, 2789–2796 (2000).
59. A. M. Belcher, P. K. Hansma, E. L. Hu, G. D. Stucky, and D. E. Morse, Proteins controlling crystal phase, orientation and morphology in biocomposite materials, *Abstracts of Papers of the American Chemical Society* **214**, 56-MTLS (1997).
60. C. M. Kacher, I. M. Weiss, R. J. Stuart, C. F. Schmidt, P. K. Hansma, M. Radmacher, and M. Fritz, Imaging microtubules and kinesin decorated microtubules using tapping mode atomic force microscopy in fluids, *Eur. Biophys. J. Biophys. Lett.* **28**, 611–620 (2000).
61. S. Kasas, N. H. Thomson, B. L. Smith, P. K. Hansma, J. Miklossy, and H. G. Hansma, Biological applications of the AFM: From single molecules to organs, *Int. J. Imaging Syst. Technol.* **8**, 151–161 (1997).
62. B. L. Smith, D. R. Gallie, H. Le, and P. K. Hansma, Visualization of poly(A)-binding protein complex formation with poly(A) RNA using atomic force microscopy, *J. Struct. Biol.* **119**, 109–117 (1997).
63. K. Wadu-Mesthrige, N. A. Amro, J. C. Garno, S. Xu, and G.-Y. Liu, Fabrication of nanometer-sized protein patterns using atomic force microscopy and selective immobilization, *Biophys. J.* **80**, 1891–1899 (2001).

64. P. Avouris, Manipulation of matter at the atomic and molecular levels, *Acc. Chem. Res.* **28**, 95–102 (1995).
65. I. W. Lyo and P. Avouris, Field-induced nanometer-scale to atomic-scale manipulation of silicon surfaces with the STM, *Science* **253**, 173–176 (1991).
66. B. C. Stipe, M. A. Bezaei, W. Ho, S. Gao, M. Persson, and B. I. Lundqvist, Single-molecule dissociation by tunneling electrons, *Phys. Rev. Lett.* **78**, 4410–4413 (1997).
67. R. M. Nyffenegger and R. M. Penner, Nanometer-scale surface modification using the scanning probe microscope: Progress since 1991, *Chem. Rev.* **4**, 1195 (1997).
68. B. J. McIntyre, M. Salmeron, and G. A. Somorjai, Nanocatalysis by the tip of a scanning tunneling microscope operating inside a reactor cell, *Science* **265**, 1415–1418 (1994).
69. B. J. McIntyre, M. Salmeron, and G. A. Somorjai, Spatially (nanometer) controlled hydrogenation and oxidation of carbonaceous clusters by the platinum tip of a scanning tunneling microscope operating inside a reactor cell, *Catal. Lett.* **39**, 5–17 (1996).
70. W. T. Muller, D. L. Klein, T. Lee, J. Clarke, P. L. Mceuen, and P. G. Schultz, A strategy for the chemical synthesis of nanostructures, *Science* **268**, 272–273 (1995).
71. R. D. Piner, S. Hong, and C. A. Mirkin, Improved imaging of soft materials with modified AFM tips, *Langmuir* **15**, 5457–5460 (1999).
72. R. D. Piner and C. A. Mirkin, Effect of water on lateral force microscopy in air, *Langmuir* **13**, 6864–6868 (1997).
73. R. D. Piner, J. Zhu, F. Xu, S. H. Hong, and C. A. Mirkin, "Dip-pen" nanolithography, *Science* **283**, 661–663 (1999).
74. D. M. Eigler and E. K. Schweizer, Positioning single atoms with a scanning tunneling microscope, *Nature* **244**, 524–526 (1990).
75. W. Ho, Inducing and viewing bond selected chemistry with tunneling electrons, *Acc. Chem. Res.* **31**, 567 (1998).
76. C. B. Ross, L. Sun, and R. M. Crooks, Scanning probe lithography. 1. Scanning tunneling microscope induced lithography of self-assembled *N*-alkanethiol monolayer resists, *Langmuir* **9**, 632 (1993).
77. J. K. Schoer, F. P. Zamborini, and R. M. Crooks, Scanning probe lithography. 3. Nanometer-scale electrochemical patterning of Au and organic resists in the absence of intentionally added solvents or electrolytes, *J. Phys. Chem.* **100**, 11086–11091 (1996).
78. P. Zeppenfeld, C. P. Lutz, and D. M. Eigler, Manipulating atoms and molecules with a scanning tunneling microscope, *Ultramicroscopy* **42–44**, 128–133 (1992).
79. M. F. Crommie, C. P. Lutz, and D. M. Eigler, Confinement of electrons to quantum corrals on a metal surface, *Science* **262**, 218–220 (1993).
80. M. F. Crommie, C. P. Lutz, and D. M. Eigler, Confinement of electrons to quantum corrals on a metal-surface, *Science* **262**, 218–220 (1993).
81. M. F. Crommie, C. P. Lutz, D. M. Eigler, and E. J. Heller, Waves on a metal-surface and quantum corrals, *Surf. Rev. Lett.* **2**, 127–137 (1995).
82. M. F. Crommie, C. P. Lutz, D. M. Eigler, and E. J. Heller, Quantum corrals, *Physica D* **83**, 98–108 (1995).
83. E. J. Heller, M. F. Crommie, C. P. Lutz, and D. M. Eigler, Scattering and absorption of surface electron waves in quantum corrals, *Nature* **369**, 464–466 (1994).
84. H. C. Manoharan, C. P. Lutz, and D. M. Eigler, Quantum mirages formed by coherent projection of electronic structure, *Nature* **403**, 512–515 (2000).
85. T. A. Jung, R. R. Schlittler, J. K. Gimzewski, H. Tang, and C. Joachim, Controlled room-temperature positioning of individual molecules: Molecule flexure and motion, *Science* **271**, 181–184 (1996).
86. S. W. Hla, L. Bartels, G. Meyer, and K. H. Rieder, Inducing all steps of a chemical reaction with the scanning tunneling microscope tip: Towards single molecule engineering, *Phys. Rev. Lett.* **85**, 2777–2780 (2000).
87. S. W. Hla, G. Meyer, and K. H. Rieder, Inducing single-molecule chemical reactions with a UHV-STM: A new dimension for nano-science and technology, *Phys. Rev. Lett.* **2**, 361–366 (2001).
88. G. Meyer, L. Bartels, and K. H. Rieder, Atom manipulation with the STM: Nanostructuring, tip functionalization, and femtochemistry, *Comput. Mater. Sci.* **20**, 443–450 (2001).

89. J. J. Schulz, R. Koch, and K. H. Rieder, New mechanism for single atom manipulation, *Phys. Rev. Lett.* **84**, 4597–4600 (2000).
90. E. S. Snow and P. M. Campbell, Afm fabrication of sub-10-nanometer metal-oxide devices with in situ control of electrical-properties, *Science* **270**, 1639–1641 (1995).
91. S. Xu and G. Y. Liu, Nanometer-scale fabrication by simultaneous nanoshaving and molecular self-assembly, *Langmuir* **13**, 127–129 (1997).
92. S. Xu, S. Miller, P. E. Laibinis, and G.-Y. Liu, Fabrication of nanometer scale patterns within self-assembled monolayers by nanografting, *Langmuir* **15**, 7244–7251 (1999).
93. K. Wadu-Mesthrige, S. Xu, N. A. Amro and G.-Y. Liu, Fabrication and imaging of nanometer-sized protein patterns, *Langmuir* **15**, 8580–8583 (1999).
94. W. Norde, M. Giesbers, and H. Pingsheng, Langmuir–Blodgett films of polymerized 10,12-pentacosadionic acid as substrates for protein adsorption, *Colloids Surf. B: Biointerfaces* **5**, 255 (1995).
95. J. Buijs, D. W. Britt, and H. Vladimer, Human growth hormone adsorption kinetics and conformation on self-assembled monolayers, *Langmuir* **14**, 335 (1998).
96. N. Patel, M. C. Davies, M. Hartshorne, R. J. Heaton, C. J. Roberts, S. J. B. Tendler, and P. M. Williams, Immobilization of protein molecules onto homogeneous and mixed carboxylate-terminated self-assembled monolayers, *Langmuir* **13**, 6485 (1997).
97. C. F. Blake, D. F. Koenig, G. A. Mair, A. C. T. Morth, D. C. Phillips, and V. R. Sarma, Structure of hen egg-white lysozyme, *Nature* **206**, 757–761 (1965).
98. S. Cruchon-Dupeyrat, S. Porthun, and G. Y. Liu, Nanofabrication using computer-assisted design and automated vector-scanning probe lithography, *Appl. Surf. Sci.* **175–176**, 636–642 (2001).
99. C. Baur, B. C. Gazen, B. Koel, T. R. Ramachandran, A. A. G. Requicha, and L. Zini, Robotic nanomanipulation with a scanning probe microscope in a networked computing environment, *J. Vac. Sci. Technol. B* **15**, 1577–1580 (1997).
100. T. R. Ramachandran, C. Baur, A. Bugacov, A. Madhukar, B. E. Koel, A. A. G. Requicha, and C. Gazen, Direct and controlled manipulation of nanometer-sized particles using the non-contact atomic force microscope, *Nanotechnology* **9**, 237–245 (1998).
101. S. H. Hang and C. A. Mirkin, A nanoplotter with both parallel and serial writing capabilities, *Science* **288**, 1808–1811 (2000).
102. S. H. Hong, J. Zhu, and C. A. Mirkin, Multiple ink nanolithography: Toward a multiple-pen nano-plotter, *Science* **286**, 523–525 (1999).
103. G. Binnig and H. Rohrer, Scanning tunneling microscopy (Reprinted from IBM Journal of Research and development, vol. 30, 1986), *IBM J. Res. Development* **44**, 279–293 (2000).
104. M. Despont, *et al.*, VLSI-NEMS chip for parallel AFM data storage, *Sens. Actuators, A: Physical* **80**, 100–107 (2000).

7

Chemical and Photochemical Reactivities of Nanoarchitectures

7.1. REDOX POTENTIALS OF NANOMATERIALS

7.1.1. Quantum Size Confinement Effect

The chemical, electrochemical, and photochemical reactivities of nanometarials are primarily determined by the redox potentials of the materials. The effect of size is reflected in changes in redox potential of nanomaterials since the electronic energy levels shift with size, known as the quantum size confinement effect, which will be discussed in detail in Chapter 8. The shift in redox potential in turn affects the reactivities of the nanomaterial.

The most apparent effect of size on the redox potentials is the shift with changing size. The shift or enhancement in redox potential will change the energetics of the electrons and holes and thereby their reactivities. Quantization effectively increases the band gap of the semiconductor, therefore, the electrons and holes will have more negative and more positive redox potentials, respectively, at the lowest respective quantum level in the conduction band and valence band. The distribution of the confinement energy or increase in band gap, ΔE, between the electrons and holes depend on their effective masses. This distribution can be estimated from the following expression:[1]

$$\Delta E_e \cong (E_{ex} - E_g^{bulk})/(1 + m_e^*/m_h^*) \tag{7.1}$$

and

$$\Delta E_h \cong E_{ex} - E_g^{bulk} - \Delta E_e \tag{7.2}$$

where ΔE_e and ΔE_h are the increased redox potentials for electrons and holes, respectively; E_{ex} and E_g^{bulk} are the effective band gap of the nanoparticle and band gap of the bulk semiconductor, respectively; m_e^* and m_h^* are the effective masses of the electron and hole, respectively. The enhancement of redox potentials in colloidal semiconductor nanoparticles have been experimentally verified in a number of systems, including HgSe, PbSe, CdSe ($R < 2.5$ nm),[2] TiO_2,[3–6] PbS, CdS, CdTe,[6] and ZnO.[4,5]

Another consequence of the quantum confinement is slower hot electron relaxation, which in turn facilitates hot electron transfer to redox acceptors in solution.[7–12] This has been demonstrated mostly in relatively clean 2D quantum well systems,[7,8,10–12] consistent with theoretical expectations.[13] It has been suggested that hot electron transfer could produce much higher photon conversion efficiencies.[14] Direct observation of the slower relaxation in confined 3D systems has usually been made difficult by surface effects that increase the electron relaxation rate due to trapping.[15] The surface characteristics are expected to play an important role in the photochemical and photoelectrochemical properties, as discussed next.

7.1.2. Surface Effect

The surface plays a critical role in the properties and reactivities of nanomatreials because of their exceedingly large surface-to-volume ratio and thereby a high density of surface states introduced by defects and dangling bonds. Many of the surface atoms are more reactive than bulk atoms and thus surface reactivity is enhanced for nanomaterials relative to bulk. This enhanced reactivity is usually useful for surface reactions in catalysis, photochemical, and photoelectrochemical processes. However, the surface reactivity also tends to reduce the thermal as well as photo stability of the materials. Therefore, the surface reactivity can be useful or harmful, depending on the applications of interest. If the chemical or photochemical reactivities can be used for producing useful products or degrading wastes or pollutants, the enhancement is desirable and useful. On the other hand, if the reactivities result in undesirable by-products or cause degradation of the nanomaterials, the enhancement is clearly undesirable or harmful. Understanding and controlling the reactivities of nanomaterials are, therefore, critical to their applications in various technological areas. It is indeed one of the major challenges in the design of catalysts or photocatalysts to satisfy the desired surface reactivity while maintain material stability. This is particularly important for catalytic reactions since the catalysts need to be used repeatedly.

7.1.3. Effects of Light

Several important areas of applications of nanomaterials involve their interaction with light. These include photochemical, photoelectrochemical, and photocatalytic reactions. The effect of light with above band gap excitation is to produce very reactive electrons and holes in the semiconductor materials that subsequently react with species near or on the surface of the nanomaterial. The chemical reactions involving the photogenerated electrons are photoreduction reactions, while reactions involving photogenerated holes are photooxidations. A large percentage of the initially created charge carriers are quickly trapped by surface trap states (on the timescale of a few hundreds of fs to a few tens of ps).

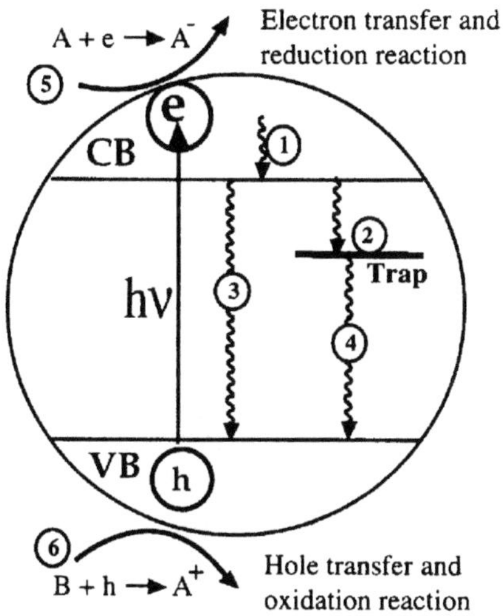

FIGURE 7.1. Schematic of different relaxation, transfer and reaction pathways of photogenerated charge carriers in semiconductor nanoparticles. (1) Electronic cooling with the CB, (2) electronic trapping by trap states, (3) bandedge electron–hole recombination, (4) trapped electron-hole recombination, (5) electron transfer and reduction reaction with an electron acceptor, and (6) hole transfer and oxidation reaction with an hole acceptor or electron donor. The electron and hole transfer reactions usually compete with their trapping and recombination.

Dynamic properties of charge carrier trapping and relaxation will be discussed in detail in Chapter 8. Both trapped and free carriers can participate in reactions with species on or near the surface. The tapped carriers are less energetic than free carriers. Electron or hole transfer across the interface region is a critical step in the overall reaction processes. Trapping and transfer of free electrons are competing processes and often occur on ultrafast timescales. Another competing process is electron–hole recombination. Electron transfer can take place following trapping as well, but on longer timescales, ns or longer. Similar events take place for the hole. However, the timescale for hole transfer and trapping can be different from that for the electron. As shown schematically in Fig. 7.1, the different processes involving photoexcited charge carriers are illustrated.

7.2. PHOTOCHEMICAL AND CHEMICAL REACTIONS

7.2.1. Photochemical Reactions

Colloidal semiconductor nanomaterials have been considered as catalysts and photocatalysts for various inorganic and organic chemical reactions. In particular, the potential for organic functional group transformation has been actively explored for organic synthesis.[16–19] As catalysts or photocatalysts, semiconductor nanomaterials, notably TiO_2, afford some advantages, including low cost, non-toxicity (environmentally benign), and high turnover number.[20] Photocatalytic organic reactions that have been investigated include oxidation,[20–27] reduction,[17,28,29] isomerization,[30,31] condensation,[32–35] dimerization, and polymerization.[36–43] For example, one of the most studied photooxidation reactions on semiconductor nanoparticles is dehydrogenetion of alcohols to aldehydes or ketones.[21–27] The oxidation is initiated by electron transfer from the alcohol to the hole. Reduction of H^+ by the photoinduced electron results in formation of hydrogen.

$$\frac{1}{2}H_2 + R_2CHO^- \xleftarrow{e^-} R_2CHOH \xrightarrow{h^+} R_2CHOH^+ \longrightarrow R_2C{=}O + H^+ + \frac{1}{2}H_2 \quad (7.3)$$

Similarly, an amine can be photooxidized in the presence of oxygen[44] and sulfides can be oxidized into sulfones.[45] It has been demonstrated that *cis*-2-butane on illuminated TiO_2, ZnS or CdS generates a mixture of *cis–trans* isomers.[30,31,46,47] 1,2-diarycyclopropanes have been shown to undergo *cis–trans* isomerization using ZnO as a photocatalyst.[48]

In the absence of oxygen, cations or radical cations generated from organic substrates can lead to polymerization. For example, illumination of colloidal TiO_2 in 1,3,5,7-tetraethylcyclotetrasiloane (TMCTS) initiates a ring-opening polymerization producing poly(methylsiloxane) (PMS).[39] A cationic reaction mechanism was proposed that includes cation generation ($TMCTS^+$) through hole transfer ($TMCTS + h^+$) followed by ring-opening. Synthesis of polypyrrole films on illuminated TiO_2 particles has also been reported.[40] Polymerization of methacrylic acid can be accomplished on CdS, CdS/HgS, or CdS/TiO_2 particles.[49] The polymerization was believed to be initiated by a radical generated through oxidation of the monomer.[41–43]

It has also been demonstrated that photooxidation of some small molecules on semiconductor nanoparticles can lead to the formation of biologically important molecules such as amino acids, peptide ologomers, and nucleic acids.[32–35] These condensation reactions typically have very low yields, however, they have interesting implications in the origin and evolution of biological molecules.

In addition to photooxidation, photoreduction based on semiconductor nanoparticles have also been explored for synthesis of organic molecules.[28,29] For example, photoinduced reduction of *p*-dinitrobezene and its derivatives on TiO_2 particles in the presence of a primary alcohol has been found to lead to the formation of benzimidoles with high yields.[50] Benzimidoles are of interest for biological and chemical applications.[51–53] The above mentioned examples demonstrate that semiconductor photocatalysis is potentially useful for organic and biochemical synthesis. A major practical challenge is to conduct the photocatalytic reactions at a large synthetic scale. Another difficulty to resolve is competitive absorption of light by the reactants and products. Competitive adsorption of products and reactants on the particle surface is also an issue that needs to be better addressed.[19]

7.2.2. Chemical Reactions without Light

Many semiconductor and metal nanoparticles are efficient catalysts for chemical reaction without light.[54,55] Similar to photocatalysis, nanoparticles often exhibit new and improved catalytic properties over their bulk counterparts. As mentioned earlier, the high density of surface states and extremely large surface area of nanoparticles are fundamentally important for catalytic as well as photocatalytic reactions. The environment or surface characteristics of the nanoparticles critically influence their catalytic activities. For example, core/shell iron oxide/iron nanoparticles encapsulated in mesoporous alumina has been found to be active towards formation of carbon nanotubes by a CVD process.[56] Pt and Pd nanoparticles encapsulated in dendrimers have shown electrocatalytic activity towards O_2 reduction[57] and high catalytic activities towards hydrogenation reactions of unsaturated alkenes in aqueous solution.[58] Interestingly, the hydrogenation rate can be controlled by using dendrimers with different generations, with the lower generation exhibiting the highest catalytic activity. It is believed that sterically crowded terminal groups of higher generation dendrimers hinder substrate penetration and thereby lower catalytic reaction rates.[58] The longer distance between the substrate and the particle surface for higher generation dendrimers may also contribute towards the lower rates.

The unique chemical reactivities of nanoparticles have been exploited for many years for commercial applications where small metal particles dispersed on oxide supports are used as high surface area catalysis for heterogeneous reactions. Examples include supported Ag nanoparticles used for the epoxidation of ethylene glycol production, around 10 million tons a year worldwide,[59] and Rh, Pd, and Pt particles used for the treatment of automobile exhaust in catalytic converters.[60] The metal particles in the size range of 1–100 nm are the active sites for adsorption and reaction. The reactivity and selectivity of these metal catalysts are strongly dependent on the particle size and size distribution (metal loading), metal–support interaction, and surface characteristics (chemical modification). The optimal particle size for high reactivity depends on the nature of the chemical reactions and other reaction conditions such as pressure and temperature.[61,62]

Nanoparticles can also be directly involved in chemical reactions and many have shown enhanced chemical reactivities in comparison to their bulk counterparts or microparticles. For example, ZnO nanoparticles have shown greatly enhanced chemical reactivity in their reactions with CCl_4, SO_2 and paraoxon.[63] The enhancement was thought to be due to morphological differences in that smaller particles posses a higher concentration of reactive sites. Similar enhanced reactivity has been demonstrated in other semiconductor nanoparticles including CaO[64,65] and MgO.[65–68] The increase in density of reactive sites could be due to both defects as well as intrinsic surface states. At high temperatures (400–500°C), these MgO and CaO nanoparticles have been found to be effective destructive adsorbents for toxic substances such as organophosphourus compounds and chlorocarbons.[65] The destruction capability for chlorocarbons can be greatly enhanced by placing a monolayer of transition metal oxide, for example Fe_2O_3, on the nanoparticle surface, which is likely due to a catalyzed Cl^-/O^{2-} exchange in the presence of Fe_2O_3.

7.3. PHOTOELECTROCHEMICAL REACTIONS

Another important area of applications of semiconductor nanoparticles is in photoelectrochemical reactions. Similar to photochemical reactivities, the photoelectrochemical properties of nanoparticles are often quite different from those of their bulk counterparts. For example, structurally controlled generation of photocurrents has been demonstrated for double-stranded DNA-cross-linked CdS nanoparticle arrays upon irradiation with light.[69] The electrostatic binding of $[Ru(NH_3)_6]^{3+}$ to the ds-DNA units provides tunneling routes for the conduction band electrons and thus results in enhanced photocurrents. This could be useful for DNA sensing applications.

Photoelectrochemical behavior has been demonstrated in a number of semiconductor nanoparticle films, including CdS and CdSe,[70-72] ZnO,[73] TiO_2,[74-78] Mn-doped ZnS,[79,80] WO_3,[81] SnO_2/TiO_2 composite,[82] and TiO_2/In_2O_3 composite.[83] The large band gap semiconductors, e.g. TiO_2 and ZnO, often require sensitization with dye molecules so that photoresponse can be extended to the visible region of the spectrum.[74,79,84] Some of the same films, for example, CdS and CdSe, were found to exhibit no corresponding solid state photovoltaic behavior.[72] It has been proposed that the electron/hole separation is determined by kinetic differences in charge injection into an electrolyte rather than by a built-in space charge layer in the semiconductor. The films can behave as either "n" or "p" type with respect to direction of photocurrent flow by changes in the semiconductor surface properties and/or the electrolyte.

Electrochemical reactions of semiconductor and metal nanoparticles without light have also been actively explored over the last decade. These will be discussed in detail mainly in Chapter 10. One example to be discussed here is the use of nanoparticles as catalysts in fuel cells. Fuel cells have attracted a great deal of attention recently as a possible alternative, environmentally friendly energy technology.[85] Fuel cells operate based on the principles of catalytic electrochemistry. One of the electrodes used in solid polymer membrane fuel cells is a layered structure composed of a porous carbon substrate covered by a catalyst layer. The catalyst layer is usually platinum nanoparticles supported on carbon microparticles.[86,87] Simaultaneous incorporation of TiO_2 nanoparticles with Pt nanoparticles in polymer electrolyte membranes have been found to improve the cell performance.[88] As shown in Fig. 7.2, the TiO_2 nanoparticles were believed to enhance the back-diffusion of water produced by the Faradic reaction at the cathode by the hygroscopic property, resulting in efficient humidification of the polymer electrolyte membrane of the anode side dried by the electroosmotic drag. The cathode potential was found improved distinctively. This was attributed to elimination of the short-circuit reaction of crossover gases in the cathode catalyst

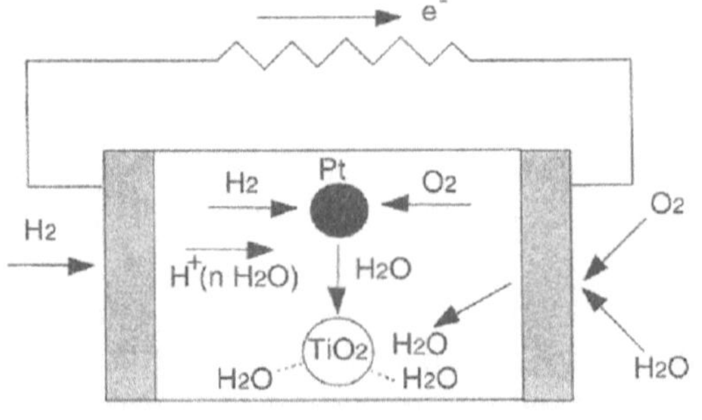

FIGURE 7.2. Schematic operation concept of a polymer electrolyte fuel cell using self-humidifying Pt–oxide–polymer electrolyte membrane. Adapted with permission from Fig. 1 of Ref. 88.

layer, resulting in a small non-Faradic consumption of H_2 and no disturbance of reactant O_2 diffusion by the produced water vapor.[88] Carbon coated cobalt nanoparticles activated by electric arc discharge have also been found to show activity as catalysts for O_2 reduction in H_2/O_2 polymer electrolyte fuel cells.[89]

7.4. PHOTOCATALYSIS AND ENVIRONMENTAL APPLICATIONS

7.4.1. Small Inorganic Molecules

Photocatalysis based on semiconductors plays an important role in chemical reactions of both small inorganic molecules and large organic, or biological molecules. The photocatalytic reactivities are strongly dependent on the nature and properties of the photocatalysts, including pH of the solution, particle size, and surface characteristics.[16] These properties are sensitive to the methods of preparation.[16] Impurities or dopants can significantly affect these properties as well as reactivities. For example, it has been shown that selectively doped quantum-sized particles have a much greater photoreactivity as measured by their quantum efficiency for oxidation and reduction than their undoped counterparts.[90] A systematic study of the effects of over 20 different metal ion dopants on the photochemical reactivity of TiO_2 colloids with respect to both chloroform ($HCCl_3$) oxidation and carbon tetrachloride (CCl_4) reduction has been conducted.[90,91] A maximum enhancement of 18-fold for CCl_4 reduction and 15-fold for $CHCl_3$ oxidation in quantum efficiency for Fe(III)-doped TiO_2 colloids have been observed. The mechanism for the observed enhancement is yet to be well understood.[92]

A large number of small inorganic molecules are sensitive to photooxidation on semiconductor surfaces. Examples include ammonia,[93,94] nitrogen,[95] oxygen,[96,97] ozone,[98] nitric oxide and nitrogen dioxide,[99–101] nitrates and nitrites,[102–104] cyanide,[94,105–107] chromium,[108,109] copper,[110,111] gold,[112,113] silver,[108,112,114] mercury,[108,115] iron species,[110,116] and halide ions.[117–119] Removal or destruction of toxic metals or metal ions is critical to water treatment and purification.[20] In many cases, the presence of metal or metal ions on the surface of the semiconductor photocatalysts significantly influences their photocatalytic reactivities. In addition to photooxidation, photoreduction reactions have also been demonstrated to generate hydrogen peroxide (H_2O_2) for several colloidal semiconductors including CdS,[120] α-Fe_2O_3,[121] TiO_2[96,121] and ZnO.[121] It has been noted that ZnO produces H_2O_2 more efficiently than TiO_2.[97]

7.4.2. Large Organic Molecules

Photocatalytic oxidation of organic and biological molecules is of great interest for environmental applications, especially in the destruction of hazardous wastes. The ideal outcome is complete mineralization of the organic or biological compounds, including aliphatic and aromatic chlorinated hydrocarbons, into small inorganic, non- or less- hazardous molecules, such as CO_2, H_2O, HCl, HBr,

SO_4^{2-}, NO_3^-, etc. Photocatalysts include various metal oxide semiconductors, such as TiO_2, in both bulk and particulate forms. Compounds that have been degraded by semiconductor photocatalysis include alkanes, haloalkanes, aliphatic alcohols, carboxylic acids, alkenes, aromatics, haloaromatics, polymers, surfatctsntas, herbicides, pesticides, and dyes, as summarized in an excellent review article by Hoffmann.[92] It has been found in many cases that the colloidal particles show new or improved photocatalytic reactivities over their bulk counterparts. For instance, photolysis of pentachlorophenyl (C_6Cl_5OH, PCP), a widely used pesticide and wood preservative, in homogeneous solution has been shown to produce toxic byproducts such as tetrachlorodioxins. However, in the presence of TiO_2 suspensions, the intermediate dioxines have been found to be destroyed effectively and the photooxidation of PCP was found to proceed with the following stoichiometry:[122]

$$2HOC_6Cl_5 + 7O_2 \xrightarrow{h\nu, TiO_2} HCO_2H + 8CO_2 + 10HCl \quad (7.4)$$

It has been found that glyoxylate can be oxidized to formate on the surface of illuminated ZnO colloidal particles with the formate serving as an electron donor for the reduction of dioxygen.[123] Similarly, photocatalytic oxidation of acetate ($CH_3CO_2^-$) on ZnO colloidal nanoparticles has been demonstrated with final products of CO_2 and H_2O.[123] Acetate can also be photocatalytically oxidized on TiO_2 particles, although the reaction mechanisms have been proposed to be different between TiO_2 and ZnO.[92] Likewise, photocatalytic reduction of CCl_4 and oxidation of $HCCl_3$ have been carried out in the presence of TiO_2 colloids.[124,125] In the case of CCl_4, the pH of the TiO_2 colloids were found to influence the rate of CCl_4 reduction either by altering the electrostatic interaction of electron donors on the TiO_2 surface or by changing the reduction potential of the conduction band electron.[125] In the case of photooxidation of 4-chlorophenol (ClC_6H_4OH) with TiO_2 colloids, several concurrent reaction pathways have been conjectured due to the formation of different surface structures by the adsorbate.[20,126] The different pathways are not equally efficient with respect to photooxidation.

7.4.3. Water Pollution Control

One of the most important areas of application of photocatalytic reactions is removal or destruction of contaminates in water treatment or purification.[20,127,128] Major pollutants in waste waters are organic compounds. Small quantities of toxic and precious metal ions or complexes are usually also present. As discussed above, semiconductor nanoparticles, most often TiO_2, afford an attractive system for degrading both organic and inorganic pollutants in water. Water treatment based on photocatalysis provides an important alternative to other advanced oxidation technologies such as UV-H_2O_2 and UV-O_3 designed for environmental remediation by oxidative mineralization.

The photocataytic mineralization of organic compounds in aqueous media typically proceeds through the formation of a series of intermediates of progressively higher oxygen to carbon ratios. For example, photodegradation of phenols yields hydroquinone, catechol, and benzoquinone as the major intermediates that are eventually oxidized quantitatively to carbon dioxide and water.[129]

Two mechanisms have traditionally been proposed in the photooxidation of organics in aqueous TiO_2 suspensions. The first one proposes that the photogenerated valence band holes react primarily with physisorbed H_2O and surface-bound HO^- on TiO_2 particles to produce surface-bound ·OH radicals ($E^0_{redox} = +1.5$ V vs. NHE).[130] The ·OH radicals then react with presorbed or photoadsorbed organic molecules on the TiO_2 particles. The second mechanism postulates a direct reaction between the photogenerated holes ($E^0_{redox} = +3.0$ V vs. NHE) and the organic molecules. These two mechanisms cannot be distinguished based on product analysis alone since both lead to the same primary radicals.[128,131]

Examples of organic molecules that have been studied extensively in terms of their photocatalytic reactivities and mechanisms include 4-chlorophenol,[131–139] 3,4-dimethylphenol,[140–143] and atrazine.[144–147] In all these cases, the ·OH radical bound on the TiO_2 particle surface is believed to be primarily responsible for the initial generation of organic radicals from the organic substrates adsorbed on the particle surface. In these reactions, the semiconductor nanoparticles, for example, TiO_2, not only function as a photocatalyst in generating conduction band electrons and valence band holes, but also serve to provide a surface for the organic molecule to adsorb onto. The second function is as important as the first one since close proximity between the photogenerated holes or electrons in the nanoparticles and the molecules involved in the reaction is critical for the reaction to be efficient. Figure 7.3 shows several events that occur in and on the surface

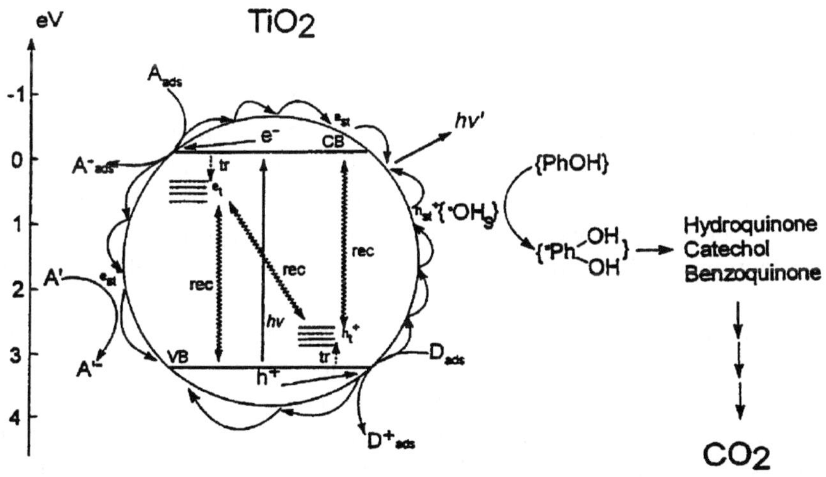

FIGURE 7.3. Schmetic illustration of several events in and on the TiO_2 nanoparticle and subsequent events occurring on the surface towards the ultimate oxidation of a phenolic substrate. Reproduced from Figure 6 of Ref. 128 with permission from Elsevier Science.

of TiO_2 nanoparticles following photoexcitation and subsequent events that lead to the final oxidation of a phenolic substrate.[128] One critical step in the overall process is the reaction between the ·OH radical with phenol to generate the dihydroxylcyclohexadienyl radical and subsequently hydroxylated intermediate products.

7.4.4. Air Pollution Control

As discussed in the last section, most efforts in applying photocatalytic oxidation processes for the destruction of contaminants have focused on purification of water that involves liquid–solid heterogeneous reactions. Gas–solid heterogeneous photocatalytic oxidations of vapor or gas phase contaminants have received considerable attention only recently.[127,148–152] These reactions are important for applications in air purification. It has been demonstrated that the photocatalytic reaction rates of some compounds, for example, trichloroethylene, are orders of magnitude faster in the gas phase than in aqueous solution.[153–155] These high reaction rates have stimulated strong interests in using such reactions for air or other gas or vapor purification.

Gas–solid photocatalytic oxidation for remediation of contaminants in gas streams has been successfully applied to treating a large variety of compounds, including alkenes and alkanes, aromatics, olefinsm ketones, aldehydes, alcohols, aliphatic carboxylic acids, and halogenated hydrocarbons.[156] Many semiconductors such as TiO_2, ZnO, Fe_2O_3, CdS, and ZnS have been found to show desired photocatalytic reactivities for these applications. Studies have shown that there are major differences in the photocatalytic reactivities between water and the gas phase. In general, the reaction rates in gas–solid photoreactors are much higher than those reported for liquid–solid photoreactors; photoefficiency higher than 100% has been found for some gas phase photooxidations.[153–155] Such a high efficiency has not been observed in similar liquid phase reactions. For example, trichloroethylene (TCE) in the gas–solid reactors has been found to be photooxidized with TiO_2 at a rate at least one order of magnitude higher than that in liquid–solid reactors.[157] Similar observations have been made for acetonitrile, methanol, and methylene chloride, perchloroethylene (PCE), ethanol, acetone, and methyl *t*-butyl ether.[153–155] There have been indications in the case study of TCE using TiO_2 as a photocatalyst that the reaction mechanism are different between the gas–solid and liquid–solid photocatalytic reactions.[157] A chain reaction mechanism has been proposed for the gas phase photodegradation.[154,158]

The photocatalytic reactivities in such gas–solid heterogeneous systems have been found to be influenced by presence of water vapor and reaction temperature.[148,149,151,153,159–162] Higher temperature usually leads to a higher reaction rate.[163,164] The reactivities can also be improved by modifying the photocatalysts. For example, titania-based binary oxide systems and platinization of TiO_2 have been found to improve the performance of the photocatalysts.[165–171] Increase in porosity by controlling preparation conditions for the photocatalysts has generally been found to improve the performance of the photocatalysts.[150]

7.5. MOLECULAR RECOGNITION AND SURFACE SPECIFIC INTERACTION

The surface of nanoparticles can be functionalized with molecular specificity so that they interact selectively with other molecules. Systems with such molecular recognition features are important for a number of applications including sensors, imaging, and labeling. For example, the surface of particles can be functionalized with known biological molecules that recognizes specifically other receptor molecules. Similarly, small molecules with such specificity can be used as a way to detect specific target molecules.

By chemically controlling the surface of the nanoparticles, they can be "programmed" to recognize and selectively bind molecules, another nanoparticle, or a suitably patterned substrate surface.[172] One approach used is similar to ideas developed in supramolecular chemistry. Supramolecular chemistry is concerned with the assembly of molecular components through covalent or non-covalent interactions where the intrinsic properties of the resulting supermolecules are not simple superpositions of those of the individual molecular components.[173] The supermolecules have functionalities that the molecular components do not have. Similarly, assembly of condensed phase, for example, nanoparticles, and molecular components can yield heterosupermolecules that afford new functionalities over the constituting nanoparticle or molecular components. For example, heterosupermolecules consisting of assembled condensed phase TiO_2 nanoparticles and a covalently or non-covalently linked molecular component, for example, viologen, have been demonstrated, as schematically illustrated in Fig. 7.4.[172] Both systems show light-induced vectorial electron transfer, which is a function only associated with the supermolecules.[174,175]

As another example, Au nanoparticles with a narrow size distribution and stabilized by a chemisorbed monolayer of a dodacane thiol covalently linked to dibenzo-24-crown-8, denoted as Au-I, have been demonstrated to recognize and selectively bind in solution a dibenzylamonium cation to form a pseudorotxane assembly, donated Au-(I + II), as shown in Fig. 7.5.[176] When a similar scheme was used for size monodisperse Ag nanoparticles with bis-dibenzylammonium diation, instead of a dibenzylamonium cation, aggregation of the silver nanoparticles has been established.[177] Interestingly, it was also found that addition of excess amount of the dibenzylammonium dication or dibelzo-24-crown-8-compound inhibits aggregation. This has been rationalized based on competition between interaction of the molecular components involved and interaction of the nanoparticles with the molecules. Likewise, gold nanoparticles with cyclodextrin receptors attached on the surface have been shown to form stable inclusion complexes with specific organic substrates such as ferrocene derivatives.[178] These studies demonstrate that nanoparticles can be capped with suitably functionalized molecular receptors (hosts) with well-defined binding sites for interaction with complementary substrates (guests).

Similarly, the surface of nanoparticles can be chemically controlled so they can recognize and selectively bind to other nanoparticles. For instance, size monodisperse silver nanoparticles have been stabilized by chemisorption of long-chain

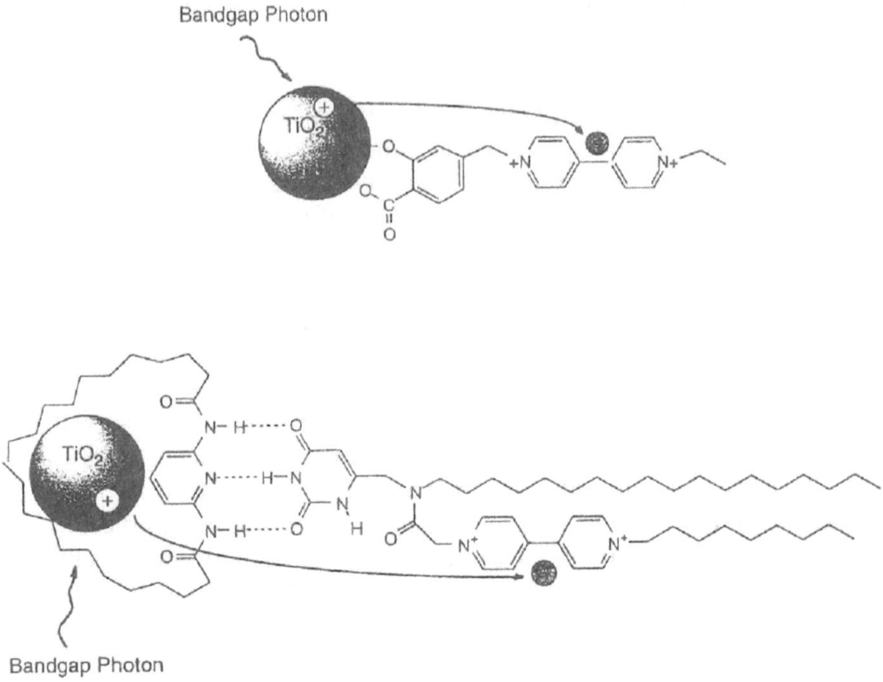

FIGURE 7.4. Two examples of heterosupermolecules consisting of a covalently and non-covalently assembled condensed phase (TiO_2 nanoparticle, and electron donor) and molecular (viologen, and electron acceptor) component. Reproduced with permission from Scheme I of Ref. 172 with permission from Elsevier Science.

alkane thiols incorporating a receptor site.[179] When dispersed in a suitable solvent, these nanoparticles recognize and selectively bind a long-chain alkane incorporating two complementary substrate sites, resulting in non-covalent linking between them. Interaction between the receptor and substrate sites leads to nanoparticle aggregation. Charge-stabilized Au nanoparticles with a narrow size distribution and modified by chemisorption of a disulfide biotin analog have been demonstrated to recognize and selectively bind to each other in solution with the addition of streptavidin.[180,181] The idea is schematically illustrated in Fig. 7.6. The biotin/streptavidin system has been considered as an ideal model link system since it has one of the largest free energies of association yet observed for non-covalent binding of a ligand by a protein in aqueous solution ($K_a > 10^{14}$ mol^{-1} dm^3). This type of strategy can be potentially useful in the development of biological sensors based on nanoparticles.[182]

The same approach can be used to modify the nanoparticle surface so the particles can selectively bind to specific sites of a solid surface. As an example, TiO_2 nanoparticles stabilized by chemisorbed decanoic acid incorporating a terminal thiol moiety are expected to recognize and selectively bind to the Au regions of a Au patterned silicon wafer when the wafer is immersed in an ethanolic dispersion of such TiO_2 nanoparticles. Experimental results based on XPS, TEM, and infrared absorption spectroscopy indicate that the expectation

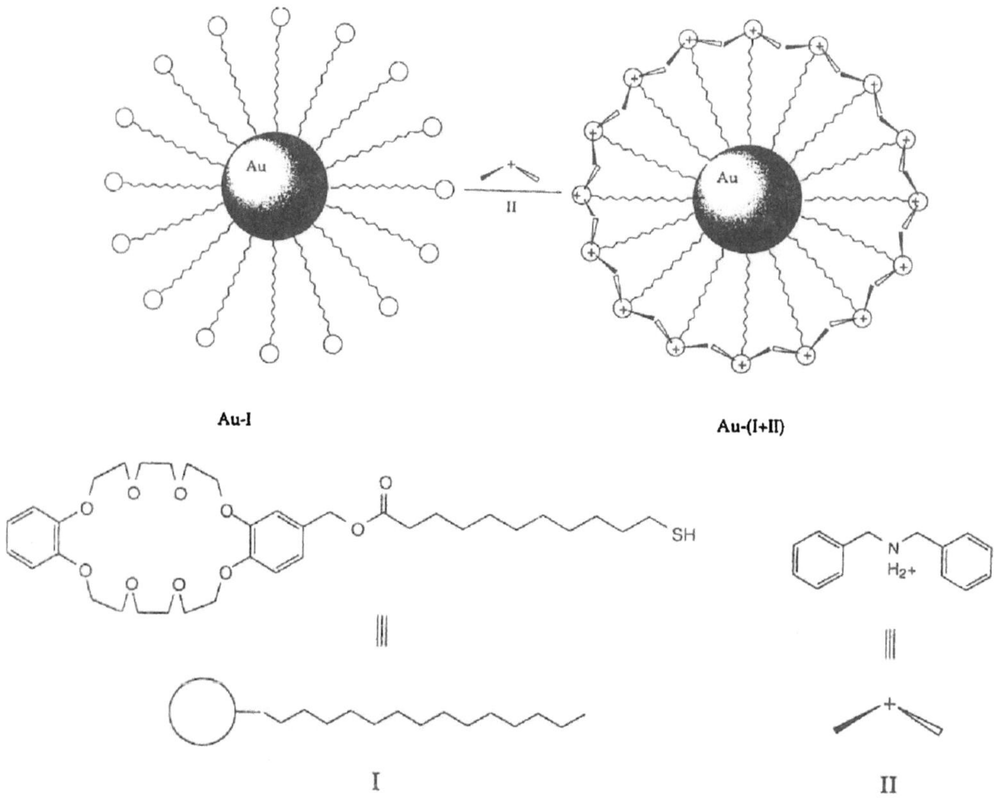

FIGURE 7.5. Schematic of Au nanoparticles stabilized by a monolayer of dodecane thiol covalently linked to dibenzo-24-crown-8 (Au-I) and their selective binding with a dibenzylammonium cation (Au-II). Reprinted from Scheme II of Ref. 172 with permission from Elsevier Science.

was justified.[183] This approach may be useful for self-assembling complex nanoparticle assemblies or superlattice structures from solution onto solid substrates, which is important for many solid device applications based on nanoparticles.[172,184]

The properties of heterosupramolecular assemblies based on nanoparticles with covalently or non-covalently linked molecular components on the particle surface are often modified or improved compared to the properties of bare nanoparticles. The improved properties can be useful for many applications. For example, heterosupramolecular assemblies based on nanostructured TiO_2 films modified by chemisorption of covalently linked ruthenium and viologen complex components have been found to exhibit efficient (95%) electron transfer from the ruthenium complex to the TiO_2 nanoparticle component upon visible light excitation.[185] Direct excitation of TiO_2 with visible light would otherwise not be possible. Such films have been considered for applications in optical storage devices[185,186] Similar TiO_2 films modified by chemisorption of a monolayer of redox chromophore bis(2-phosphonoethyl)-4,4′-bipyridinium dichloride has been found to be potentially useful for applications in electrochromic windows.[187]

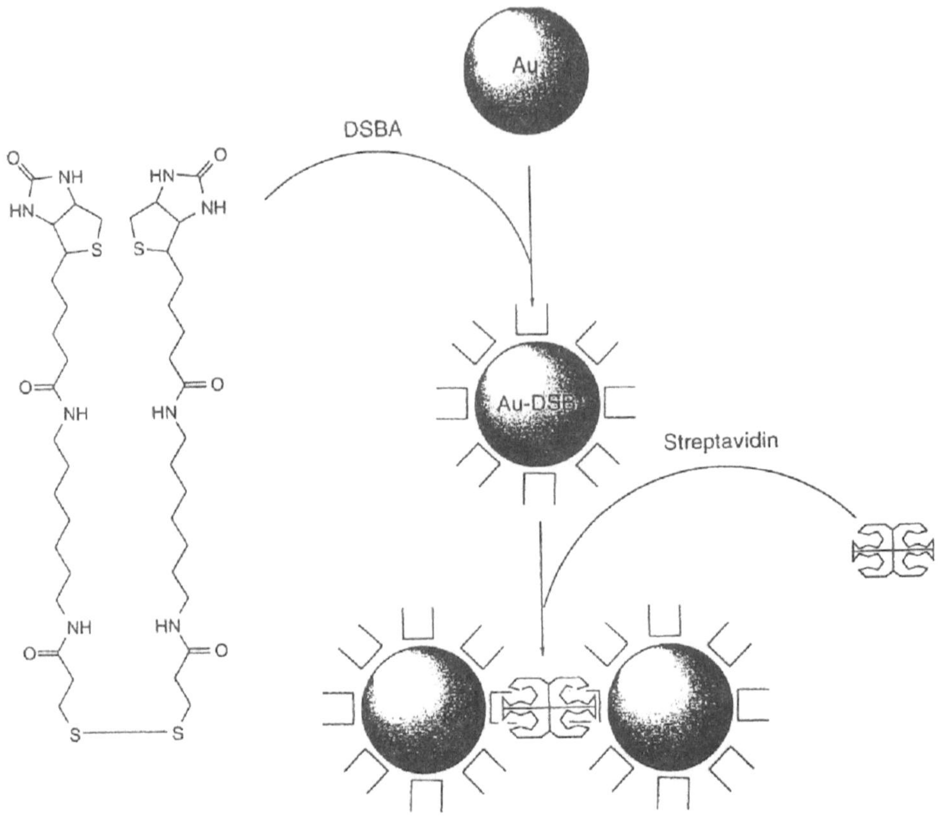

FIGURE 7.6. Addition of streptavidin is expected to enable the biotin (DSBA, disulfide biotin analogue) modified Au nanoparticles, denoted Au-DSBA, to recognize and selectively bind each other in solution. Reprinted from Scheme III of Ref. 172 with permission from Elsevier Science.

REFERENCES

1. B. A. Gregg and A. J. Nozik, Existence of a light intensity threshold for photoconversion processes, *J. Phys. Chem.* **97**, 13441–13443 (1993).
2. J. M. Nedeljkovic, M. T. Nenadovic, O. I. Micic, and A. J. Nozik, Enhanced photoredox chemistry in quatized semiconductor colloids, *J. Phys. Chem.* 12–13 (1986).
3. M. Anpo, T. Shima, S. Kodama, and Y. Kubokawa, Photocatalytic hydrogenation of CH_3CCH with H_2O on small-particle TiO_2: Size quantization effects and reaction intermediates, *J. Phys. Chem.* **91**, 4305–4310 (1987).
4. D. W. Bahnemann, C. Kormann, and M. R. Hoffmann, Preparation and characterization of quantum size zinc oxide: A detailed spectroscopic study, *J. Phys. Chem.* **91**, 3789–3798 (1987).
5. C. Kormann, D. Bahnemann, and M. R. Hoffmann, Preparation and characterization of quantum-size titanium dioxide, *J. Phys. Chem.* **92**, 5196–5201 (1988).
6. O. I. Micic, T. Rajh, J. M. Nedeljkovic, and M. I. Comor, Enhanced redox chemistry in quantized semiconductor colloids, *Isr. J. Chem.* **33**, 59–65 (1993).
7. Z. Y. Xu and C. L. Tang, Picosecond relaxation of hot carriers in highly photoexcited bulk GaAs and GaAs–AlGaAs multiple quantum wells, *Appl. Phys. Lett.* **44**, 692–694 (1984).

8. D. C. Edelstein, C. L. Tang, and A. J. Nozik, Picosecond relaxation of hot-carrier distributions in GaAs/GaAsP strained-layer superlattices, *Appl. Phys. Lett.* **51**, 48–50 (1987).
9. A. J. Nozik, J. A. Turner, and M. W. Peterson, Kinetics of electron transfer from photoexcited superlattice electrodes, *J. Phys. Chem.* **92**, 2493–2501 (1988).
10. A. J. Nozik, C. A. Parsons, D. J. Dunlavy, B. M. Keyes, and R. K. Ahrenkiel, Dependence of hot carrier luminescence on barrier thickness in GaAs/AlGaAs superlattices and multiple quantum wells, *Solid State Commun.* **75**, 297–301 (1990).
11. W. S. Pelouch, R. J. Ellingson, P. E. Powers, C. L. Tang, D. M. Szmyd, and A. J. Nozik, Comparison of hot-carrier relaxation in quantum wells and bulk GaAs at high carrier densities, *Phys. Rev. B: Condens. Matter* **45**, 1450–1453 (1992).
12. Y. Rosenwaks, M. C. Hanna, D. H. Levi, D. M. Szmyd, R. K. Ahrenkiel, and A. J. Nozik, Hot-carrier cooling in GaAs—quantum wells versus bulk, *Phys. Rev. B: Condens. Matter* **48**, 14675–14678 (1993).
13. D. S. Boudreaux, F. Williams, and A. J. Nozik, Hot carrier injection at semiconductor–electrolyte junctions, *J. of Appl. Phys.* **51**, 2158–2163 (1980).
14. R. T. Ross and A. J. Nozik, Efficiency of hot-carrier solar energy converters, *J. Appl. Phys.* **53**, 3813–3818 (1982).
15. J. Z. Zhang, Interfacial charge carrier dynamics of colloidal semiconductor nanoparticles, *J. Phys. Chem. B* **104**, 7239–7253 (2000).
16. M. A. Fox and M. T. Dulay, Heterogeneous photocatalysis, *Chemi. Rev.* **93**, 341–357 (1993).
17. C. Joyce-Pruden, J. K. Pross, and Y. Z. Li, Photoinduced reduction of aldehydes on titanium dioxide, *J. Org. Chem.* **57**, 5087–5091 (1992).
18. F. Mahdavi, T. C. Bruton, and Y. Z. Li, Photoinduced reduction of nitro compounds on semiconductor particles, *J. Org. Chem.* **58**, 744–746 (1993).
19. Y. Li and L. Wang, in: *Semiconductor Nanoclusters—Physical, Chemical, and Catalytic Aspects*, edited by P. V. Kamat and D. Meisel (Elsevier, New York, 1997), p. 391.
20. A. Mills, R. H. Davies, and D. Worsley, Water purification by semiconductor photocatalysis, *Chem. Soc. Rev.* **22**, 417–425 (1993).
21. M. A. Fox and T. Pettit, Use of organic molecules as mechanistic probes for semiconductor-mediated photoelectrochemical oxidations: Bromide oxidation, *J. Org. Chem.* **50**, 5013–5015 (1985).
22. D. M. Blake, J. Webb, C. Turchi, and K. Magrini, Kinetic and mechanistic overview of TiO_2-photocatalyzed oxidation reactions in aqueous solution, *Sol. Energy Mater.* **24**, 584–593 (1991).
23. F. Hussein, H. G. Pattenden, R. Rudham, and J. J. Russell, Photooxidation of alcohols catalysed by platinised titanium dioxide, *Tetrahedron Lett.* **25**, 3363–3364 (1984).
24. P. Pichat, M.-N. Mozzanega, and H. Courbon, Investigation of the mechanism of photocatalytic alcohol dehydrogenation over Pt/TiO_2 using poisons and labeled ethanol, *J. Chem. Soc., Faraday Trans. I* **83**, 697–704 (1987).
25. M. A. Fox and A. A. Abdelwahab, Photocatalytic oxidation of multifunctional organic molecules – the effect of an intramolecular aryl thioether group on the semiconductor-mediated oxidation dehydrogenation of a primary aliphatic alcohol, *J. Catal.* **126**, 693–696 (1990).
26. I. M. Fraser and J. R. MacCallum, Photocatalytic dehydrogenation of liquid propan-2-ol by TiO_2, *J. Chem. Soc., Faraday Trans. I* **82**, 2747–2754 (1986).
27. M. Kawai, T. Kawai, S. Naito, and K. Tamaru, The mechanism of photocatalytic reaction over Pt/TiO_2; production of H_2 and aldehyde from gaseous alcohol and water, *Chem. Phys. Lett.* **110**, 58–62 (1984).
28. P. Zuman and Z. Fijalek, Contribution to the understanding of the reduction mechanism of nitrobenzene, *J. Electroanal. Chem. and Interfacial Electrochem.* **296**, 583–588 (1990).
29. L. F. Lin and R. R. Kuntz, Photocatalytic hydrogenation of acetylene by molybdenum–sulfur complexes supported on TiO_2, *Langmuir* **8**, 870–875 (1992).
30. S. Kodama and S. Yagi, Reaction mechanisms for the photocatalytic isomerization and hydrogenation of cis-2-butane over TiO_2, *J. Phys. Chem.* **93**, 4556–4561 (1989).
31. M. Anpo, M. Sunamoto, and M. Che, Preparation of highly dispersed anchored vanadium oxides by photochemical vapor deposition method and their photocatalytic activity for isomerization of *trans*-2-butane, *J. Phys. Chem.* **93**, 1187–1189 (1989).

32. H. Reiche and A. J. Bard, Heterogeneous photosynthetic production of amino acids from methane–ammonia–water at Pt/TiO_2. Implications in chemical evolution, *J. Am. Chem. Soc.* **101**, 3127–3128 (1979).
33. W. W. Dunn, Y. Aikawa, and A.J. Bardm, Heterogeneous photocatalytic production of amino acids at Pt/TiO_2 suspensions by near ultraviolet light, *J. Am. Chem. Soc.*, 6893–6897 (1981).
34. H. Harada, T. Ueda, and T. Sakata, Semiconductor effect on the selective photocatalytic reaction of alfa-hydroxycarboxylic acids, *J. Phys. Chem.* **93**, 1542–1548 (1989).
35. J. Onoe and T. Kawai, Photochemical nucleic acid base formation with particulate semiconductor under irradiation, *J. Chem. Soc. Commun.* 681–683 (1988).
36. R. A. Barber, P. de Mayo, and K. Okada, Surface photochemistry: Semiconductor-mediated [4+4] photocycloreversion and [2+2] photocycloadition, *J. Chem. Soc., Chem. Commun.* 1073–1074 (1982).
37. A. M. Draper, M. Ilyas, P. Mayo, and V. Ramamurthy, Surface photochemistry: Semiconductor photoinduced dimerization of phenyl vinyl ether, *J. Am. Chem. Soc.* **106**, 6222–6230 (1984).
38. M. Ilyas and P. de Mayo, Surface photochemicstry: CdS-mediated dimerization of phenyl vinyl ether. The dark reaction, *J. Am. Chem. Soc.* **107**, 5093–5099 (1985).
39. H. Tada, M. Hyodo, and H. Kawahara, Photoinduced polymerization of 1,3,5,7-tetramethylcyclotetrasiloxane by TiO_2 particles, *J. Phys. Chem.* **95**, 10185–10188 (1991).
40. K. Kawai, N. Mihara, S. Kuwabata, and H. Yoneyama, Electrochemical synthesis of polypyrrole films containing TiO_2 powder particles, *J. Electrochem. Soc.* **137**, 1793–1796 (1990).
41. B. Kraeutler and A. J. Bard, Heterogeneous photocatalytic decomposition of saturated carboxylic acids on TiO_2 power. Decarboxylative route to alkanes, *J. Am. Chem. Soc.* **100**, 5985–5992 (1978).
42. I. Izumi, F. F. Fan, and A. Bard, Heterogeneous photocatalysis decomposition of benoic acid and adipic acid on platinized TiO_2 powder. The photo-Kolbe decarboxylative route to the breakdown of the benzene ring and to the production of butane, *J. Phys. Chem.* **85**, 218–223 (1981).
43. H. Yoneyama, Y. Takao, H. Tamura, and A. J. Bard, Factors influencing product distribution in photocatalytic decompostion of aqueous acetic acid on platinized TiO_2, *J. Phys. Chem.* **87**, 1417–1422 (1983).
44. M. A. Fox and M.-J. Chen, Photocatalytic formylation of primary and secondary amines on irradiated semiconductor powers, *J. Am. Chem. Soc.* **105**, 4497–4499 (1983).
45. M. A. Fox and A. A. Abdel-Wahab, Selectivity in the TiO_2-mediated photocatalytic oxidation of thioethers, *Tetrahedron Lett.* **31**, 4533–4536 (1990).
46. S. Yanagida, K. Mizumoto, and C. Pac, Semiconductor photocatalysis. *Cis–trans* photoisomerization of simple alkenes induced by trapped holes at surface states, *J. Am. Chem. Soc.* **108**, 647–654 (1986).
47. H. Al-Ekabi and P. de Mayo, Surface photochemistry: CdS photoinduced *cis–trans* isomerization of olefins, *J. Phys. Chem.* **89**, 5815–5821 (1985).
48. P. A. Carson and P. de Mayo, Surface photochemistry: semiconductor mediated reactions of some 1,2-diarylcyclopropanes, *Can. J. Chem.* **65**, 976–979 (1987).
49. I. G. Popovic, L. Katsikas, and H. Weller, The photopolymerisation of methacrylic acid by colloidal semiconductors, *Polym. Bull.* **32**, 597–603 (1994).
50. H. Y. Wang, R. E. Partch, and Y. Z. Li, Synthesis of 2-alkylbenzimidazoles via TiO_2-mediated photocatalysis, *J. Org. Chem.* **62**, 5222–5225 (1997).
51. K. Kubo, Y. Inada, Y. Kohara, Y. Sugiura, M. Ojima, K. Itoh, Y. Furukawa, K. Nishikawa, and T. Naka, Nonpeptide angiotensin–II receptor antagonists–synthesis and biological activity of benzimidazoles, *J. Med. Chem.* **36**, 1772–1784 (1993).
52. E. Alcalde, I. Dinares, L. Perezgarcia, and T. Roca, An advantageous synthesis of 2-substituted benzimidazoles using polyphosphoric acid-2-pyridyl)-1H-benzimidazoles, 1-alkyl-(1H-benzimidazol-2-Yl)pyridinium salts, their homologues and vinylogues, *Synthesis-Stuttgart* **4**, 395–398 (1992).
53. S. N. Raicheva, B. V. Aleksiev, and E. I. Sokolova, The effect of the chemical structure of some nitrogen-containing and sulphur-containing organic compounds on their corrosion inhibiting action, *Corros. Sci.* **34**, 343–350 (1993).
54. G. Schmid, Large clusters and colloids – metals in the embryonic state, *Chem. Rev.* **92**, 1709–1727 (1992).

55. P. V. Kamat and D. Meisel, *Semiconductor Nanoclusters – Physical, Chemical, and Catalytic Aspects* (Elsevier, New York, 1997).
56. J. J. Schneider, N. Czap, J. Hagen, J. Engstler, J. Ensling, P. Gutlich, U. Reinoehl, H. Bertagnolli, F. Luis, L. J. de Jongh, M. Wark, G. Grubert, G. L. Hornyak, and R. Zanoni, Metallorganic routes to nanoscale iron and titanium oxide particles encapsulated in mesoporous alumina: Formation, physical properties, and chemical reactivity, *Chem. Eur. J.* **6**, 4305–4321 (2000).
57. M. Q. Zhao and R. M. Crooks, Dendrimer-encapsulated Pt nanoparticles: Synthesis, characterization, and applications to catalysis, *Adv. Mater.* **11**, 217–220 (1999).
58. M. Q. Zhao and R. M. Crooks, Homogeneous hydrogenation catalysis with monodisperse, dendrimer-encapsulated Pd and Pt nanoparticles, *Angew. Chem. Int. Ed.* **38**, 364–366 (1999).
59. R. A. van Santen and H. P. C. E. Kuipers, The mechanism of ethylene epoxidation, *Adv. Catal.* **35**, 265–321 (1987).
60. M. Shelef and G. W. Graham, Why rhodium in automotive 3-way catalysts, *Catal. Rev. – Sci. Eng.* **36**, 433–457 (1994).
61. S. R. Seyedmonir, J. K. Plischke, M. A. Vannice, and H. W. Young, Ethylene oxidation over small silver crystallites, *J. Catal.* **123**, 534–549 (1990).
62. K. Luo, T. P. St Clair, X. Lai, and D. W. Goodman, Silver growth on TiO_2(110)(1×1) and (1×2), *J. Phys. Chem. B* **104**, 3050–3057 (2000).
63. C. L. Carnes and K. J. Klabunde, Synthesis, isolation, and chemical reactivity studies of nanocrystalline zinc oxide, *Langmuir* **16**, 3764–3772 (2000).
64. O. Koper, Y. X. Li, and K. J. Klabunde, Destructive adsorption of chlorinated hydrocarbons on ultrafine (Nanoscale) particles of calcium oxide, *Chem. Mater.* **5**, 500–505 (1993).
65. K. J. Klabunde, J. Stark, O. Koper, C. Mohs, D. G. Park, S. Decker, Y. Jiang, I. Lagadic, and D. J. Zhang, Nanocrystals as stoichiometric reagents with unique surface chemistry, *J. Phys. Chem.* **100**, 12142–12153 (1996).
66. H. Itoh, S. Utamapanya, J. V. Stark, K. J. Klabunde, and J. R. Schlup, Nanoscale metal oxide particles as chemical reagents – intrinsic effects of particle size on hydroxyl content and on reactivity and acid base properties of ultrafine magnesium oxide, *Chem. Mater.* **5**, 71–77 (1993).
67. J. V. Stark, D. G. Park, I. Lagadic, and K. J. Klabunde, Nanoscale metal oxide particles/clusters as chemical reagents – unique surface chemistry on magnesium oxide as shown by enhanced adsorption of acid gases (sulfur dioxide and carbon dioxide) and pressure dependence, *Chem. Mater.* **8**, 1904–1912 (1996).
68. J. V. Stark and K. J. Klabunde, Nanoscale metal oxide particles/clusters as chemical reagents – adsorption of hydrogen halides, nitric oxide, and sulfur trioxide on magnesium oxide nanocrystals and compared with microcrystals, *Chem. Mater.* **8**, 1913–1918 (1996).
69. I. Willner, F. Patolsky, and J. Wasserman, Photoelectrochemistry with controlled DNA-cross-linked CdS nanoparticle arrays, *Angew. Chem. Int. Ed.* **40**, 1861–1864 (2001).
70. C. Nasr, P. V. Kamat, and S. Hotchandani, Photoelectrochemical behavior of coupled SnO_2 vertical bar CdSe nanocrystalline semiconductor films, *J. Electroanal. Chem.* **420**, 201–207 (1997).
71. S. G. Hickey, D. J. Riley, and E. J. Tull, Photoelectrochemical studies of CdS nanoparticle modified electrodes: Absorption and photocurrent investigations, *J. Phys. Chem. B* **104**, 7623–7626 (2000).
72. G. Hodes, Size-quantized nanocrystalline semiconductor films, *Isr. J. Chem.* **33**, 95–106 (1993).
73. K. Keis, L. Vayssieres, H. Rensmo, S. E. Lindquist, and A. Hagfeldt, Photoelectrochemical properties of nano- to microstructured ZnO electrodes, *J. Electrochem. Soc.* **148**, A149–A155 (2001).
74. L. Zhang, M. Z. Yang, E. Q. Gao, X. B. Qiao, Y. Z. Hao, Y. Q. Wang, S. M. Cai, F. S. Meng, and H. Tian, Photoelectrochemical studies on the TiO_2 nanostructured porous film sensitized by pentamethylcyanine dye, *Chem. J. Chinese Universities – Chinese* **21**, 1543–1546 (2000).
75. Y. Ren, Z. Zhang, E. Gao, S. Fang, and S. Cai, A dye-sensitized nanoporous TiO_2 photoelectrochemical cell with novel gel network polymer electrolyte, *J. Appl. Electrochem.* **31**, 445–447 (2001).
76. X. M. Qian, D. Q. Qin, Q. Song, Y. B. Bai, T. J. Li, X. Y. Tang, E. K. Wang, and S. J. Dong, Surface photovoltage spectra and photoelectrochemical properties of semiconductor-sensitized nanostructured TiO_2 electrodes, *Thin Solid Films* **385**, 152–161 (2001).

77. Z. S. Wang, C. H. Huang, F. Y. Li, S. F. Weng, K. Ibrahim, and F. Q. Liu, Alternative self-assembled films of metal-ion-bridged 3,4,9,10-perylenetetracarboxylic acid on nanostructured TiO_2 electrodes and their photoelectrochemical properties, *J. Phys. Chem. B* **105**, 4230–4234 (2001).
78. M. S. Liu, M. Z. Yang, Y. Z. Hao, S. M. Cai, and Y. F. Li, Photoelectrochemical studies on nanoporous TiO_2 conducting polymer film electrode, *Acta Chim. Sinica* **59**, 377–382 (2001).
79. T. Yoshida, K. Terada, D. Schlettwein, T. Oekermann, T. Sugiura, and H. Minoura, Electrochemical self-assembly of nanoporous ZnO eosin Y thin films and their sensitized photoelectrochemical performance, *Adv. Mater.* **12**, 1214–1217 (2000).
80. J. F. Suyver, R. Bakker, A. Meijerink, and J. J. Kelly, Photoelectrochemical characterization of nanocrystalline ZnS : Mn^{2+} layers, *Phys. Status Solidi B: Basic Res.* **224**, 307–312 (2001).
81. C. Santato, M. Ulmann, and J. Augustynski, Photoelectrochemical properties of nanostructured tungsten trioxide films, *J. Phys. Chem. B* **105**, 936–940 (2001).
82. K. Vinodgopal, I. Bedja, and P. V. Kamat, Nanostructured semiconductor films for photocatalysis – photoelectrochemical behavior of SnO_2 TiO_2 composite systems and its role in photocatalytic degradation of a textile Azo Dye, *Chem. Mater.* **8**, 2180–2187 (1996).
83. S. K. Poznyak, D. V. Talapin, and A. I. Kulak, Structural, optical, and photoelectrochemical properties of nanocrystalline TiO_2–In_2O_3 composite solids and films prepared by sol–gel method, *J. Phys. Chem. B* **105**, 4816–4823 (2001).
84. N. J. Cherepy, G. P. Smestad, M. Gratzel, and J. Z. Zhang, Ultrafast electron injection: Implications for a photoelectrochemical cell utilizing an anthocyanin dye-sensitized TiO_2 nanocrystalline electrode, *J. Phys. Chem. B* **101**, 9342–9351 (1997).
85. S. Gottesfeld and T. F. Fuller, in: *Proceedings of the Second International Symposium on Proton Conducting Membrane Fuel Cells II* (Electrochemical Society Inc., Pennington, New Jersey, 1999).
86. M. S. Loffler, B. Gross, H. Natter, R. Hempelmann, T. Krajewski, and J. Divisek, New preparation technique and characterisation of nanostructured catalysts for polymer membrane fuel cells, *Scripta Mater.* **44**, 2253–2257 (2001).
87. R. Giorgi, P. Ascarelli, S. Turtu, and V. Contini, Nanosized metal catalysts in electrodes for solid polymeric electrolyte fuel cells: An XPS and XRD study, *Appl. Surf. Sci.* **178**, 149–155 (2001).
88. M. Watanabe, H. Uchida, and M. Emori, Polymer electrolyte membranes incorporated with nanometer-size particles of Pt and/or metal-oxides: Experimental analysis of the self-humidification and suppression of gas-crossover in fuel cells, *J. Phys. Chem. B* **102**, 3129–3137 (1998).
89. G. Lalande, D. Guay, J. P. Dodelet, S. A. Majetich, and M. E. McHenry, Electroreduction of oxygen in polymer electrolyte fuel cells by activated carbon coated cobalt nanocrystallites produced by electric arc discharge, *Chem. Mater.* **9**, 784–790 (1997).
90. W. Y. Choi, A. Termin, and M. R. Hoffmann, The role of metal ion dopants in quantum-sized TiO_2 – correlation between photoreactivity and charge carrier recombination dynamics, *J. Phys. Chem.* **98**, 13669–13679 (1994).
91. W. Y. Choi, A. Termin, and M. R. Hoffmann, Effects of metal-ion dopants on the photocatalytic reactivity of quantum-sized TiO_2 particles, *Angew. Chem., Int. Ed. Engl.* **33**, 1091–1092 (1994).
92. M. R. Hoffmann, S. T. Martin, W. Y. Choi, and D. W. Bahnemann, Environmental applications of semiconductor photocatalysis, *Chem. Rev.* **95**, 69–96 (1995).
93. C. H. Pollema, E. B. Milosavljevic, J. L. Hendrix, L. Solujic, and J. H. Nelson, Photocatalytic oxidation of aqueous ammonia (ammonium ion) to nitrite or nitrate at TiO_2 particles, *Monatsh. Chem.* **123**, 333–339 (1992).
94. A. Bravo, J. Garcia, X. Domenech, and J. Peral, Some aspects of the photocatalytic oxidation of ammonium ion by titanium dioxide, *J. Chem. Res.* S376–377 (1993).
95. J. Soria, J. C. Conesa, V. Augugliaro, L. Palmisano, M. Schiavello, and A. Sclafani, Dinitrogen photoreduction to ammonia over titanium dioxide powders doped with ferric ions, *J. Phys. Chem.* **95**, 274–282 (1991).
96. A. P. Hong, D. W. Bahnemann, and M. R. Hoffmann, Cobalt(II) tetrasulfophthalocyanine on titanium dioxide. II. Kinetics and mechanisms of the photocatalytic oxidation of aqueous sulfur dioxide, *J. Phys. Chem.* **91**, 6245–6251 (1987).
97. A. J. Hoffman, E. R. Carraway, and M. R. Hoffmann, Photocatalytic production of H_2O_2 and organic peroxides on quantum-sized semiconductor colloids, *Environ. Sci. Technol.* **28**, 776–785 (1994).

98. B. Ohtani, S. W. Zhang, T. Ogita, S. Nishimoto, and T. Kagiya, Photoactivation of silver loaded on titanium(IV) oxide for room-temperature decomposition of ozone, *J. Photochem. Photobiol. A: Chem.* **71**, 195–198 (1993).
99. N. W. Cant and J. R. Cole, Photocatalysis of the reaction between ammonia and nitric oxide on TiO_2 Surfaces, *J. Catal.* **134**, 317–330 (1992).
100. A. Kudo and T. Sakata, Photocatalytic decomposition of N_2O at room temperature, *Chem. Lett.* 2381–2384 (1992).
101. T. Ibusuki and K. Takeuchi, Removal of low concentration nitrogen oxides through photoassisted heterogeneous catalysis, *J. Mol. Catal.* **88**, 93–102 (1994).
102. A. Zafra, J. Garcia, A. Milis, and X. Domenech, Kinetics of the catalytic oxidation of nitrite over illuminated aqueous suspensions of TiO_2, *J. Mol. Catal.* **70**, 343–349 (1991).
103. A. Milis and X. Domenech, Photoassisted oxidation of nitrite to nitrate over different semiconducting oxides, *J. Photochem. Photobiol. A: Chem.* **72**, 55–59 (1993).
104. A. Milis, J. Peral, X. Domenech, and J. A. Navio, Heterogeneous photocatalytic oxidation of nitrite over iron-doped TiO_2 smples, *J. Mol. Catal.* **87** 67–74 (1994).
105. C. H. Pollema, J. L. Hendrix, E. B. Milosavljevic, L. Solujic, and J. H. Nelson, Photocatalytic oxidation of cyanide to nitrate At TiO_2 Particles, *J. Photochem. Photobiol. A: Chem.* **66**, 235–244 (1992).
106. B. V. Mihaylov, J. L. Hendrix, and J. H. Nelson, Comparative catalytic activity of selected metal oxides and sulfides for the photo-oxidation of cyanide, *J. Photochem. Photobiol. A: Chem.* **72**, 173–177 (1993).
107. W. S. Rader, L. Solujic, E. B. Milosavljevic, J. L. Hendrix, and J. H. Nelson, Sunlight-induced photochemistry of aqueous solutions of hexacyanoferrate(II) and hexacyanoferrate(III) ions, *Environ. Sci. Technol.* **27**, 1875–1879 (1993).
108. M. R. Prairie, L. R. Evans, B. M. Stange, and S. L. Martinez, An investigation of TiO_2 photocatalysis for the treatment of water contaminated with metals and organic chemicals, *Environ. Sci. Technol.* **27**, 1776–1782 (1993).
109. M. L. G. Gonzalez, A. M. Chaparro, and P. Salvador, Photoelectrochemical study of the TiO_2–Cr System – observation of strong(001) rutile photoetching in the presence of Cr(VI), *J. Photochem. Photobiol. A: Chem.* **73**, 221–231 (1993).
110. E. C. Butler and A. P. Davis, Photocatalytic oxidation in aqueous titanium dioxide suspensions – the influence of dissolved transition metals, *J. Photochem. Photobiol. A: Chem.* **70**, 273–283 (1993).
111. N. S. Foster, R. D. Noble, and C. A. Koval, Reversible photoreductive deposition and oxidative dissolution of copper ions in titanium dioxide aqueous suspensions, *Environ. Sci. Technol.* **27**, 350–356 (1993).
112. A. Wold, Photocatalytic properties of TiO_2, *Chem. Mater.* **5**, 280–283 (1993).
113. G. R. Bamwenda, S. Tsubota, T. Kobayashi, and M. Haruta, Photoinduced hydrogen production from an aqueous solution of ethylene glycol over ultrafine gold supported on TiO_2, *J. Photochem. Photobiol. A: Chem.* **77**, 59–67 (1994).
114. B. Ohtani and S. Nishimoto, Effect of surface adsorptions of aliphatic alcohols and silver ion on the photocatalytic activity of TiO_2 suspended in aqueous solutions, *J. Phys. Chem.* **97**, 920–926 (1993).
115. K. Tennakone, C. T. K. Thaminimulle, S. Senadeera, and A. R. Kumarasinghe, TiO_2-catalysed oxidative photodegradation of mercurochrome—an example of an organo-mercury compound, *J. Photochem. Photobiol. A: Chem.* **70**, 193–195 (1993).
116. S. O. Pehkonen, R. Siefert, Y. Erel, S. Webb, and M. R. Hoffmann, Photoreduction of iron oxyhydroxides in the presence of important atmospheric organic compounds, *Environ. Sci. Technol.* **27**, 2056–2062 (1993).
117. R. B. Draper and M. A. Fox, Titanium dioxide photosensitized reactions studied by diffuse reflectance flash photolysis in aqueous suspensions of TiO_2 Powder, *Langmuir* **6**, 1396–1402 (1990).
118. D. J. Fitzmaurice and H. Frei, Transient near-Infrared Spectroscopy of visible light sensitized oxidation of I – at colloidal TiO_2, *Langmuir* **7**, 1129–1137 (1991).
119. O. I. Micic, Y. N. Zhang, K. R. Cromack, A. D. Trifunac, and M. C. Thurnauer, Trapped holes on TiO_2 colloids studied by electron paramagnetic resonance, *J. Phys. Chem.* **97**, 7277–7283 (1993).

120. J. R. Harbour and M. L. Hair, Superoxide generation in the photolysis of aqueous cadmium sulfide dispersions. Detection by spin trapping, *J. Phys. Chem.* **81**, 1791–1793 (1977).
121. C. Kormann, D. W. Bahnemann, and M. R. Hoffmann, Environmental photochemistry – is iron oxide (hematite) an active photocatalyst—a Comparative Study—Alpha-Fe_2O_3, ZnO, TiO_2, *J. Photochem. Photobiol. A: Chem.* **48**, 161–169 (1989).
122. G. Mills and M. R. Hoffmann, Photocatalytic degradation of pentachlorophenol on TiO_2 particles – identification of intermediates and mechanism of reaction, *Environ. Sci. Technol.* **27**, 1681–1689 (1993).
123. E. R. Carraway, A. J. Hoffman, and M. R. Hoffmann, Photocatalytic oxidation of organic acids on quantum-sized semiconductor colloids, *Environ. Sci. Technol.* **28**, 786–793 (1994).
124. C. Kormann, D. W. Bahnemann, and M. R. Hoffmann, Photolysis of chloroform and other organic molecules in qqueous TiO_2 suspensions, *Environ. Sci. Technol.* **25**, 494–500 (1991).
125. W. Y. Choi and M. R. Hoffmann, Photoreductive mechanism of CCl_4 degradation on TiO_2 particles and effects of electron donors, *Environ. Sci. Technol.* **29**, 1646–1654 (1995).
126. A. Mills and S. Morris, Photomineralization of 4-chlorophenol sensitized by titanium dioxide—a study of the initial kinetics of carbon dioxide photogeneration, *J. Photochem. Photobiol. A: Chem.* **71**, 75–83 (1993).
127. D. F. Ollis and H. Al-Ekabi, Photocatalytic purification and treatment of water and air, in: *Proceedings of the 1st International Conference on TiO* Photocatalytic Purification and Treatment of Water and Air, London, Ontario, Canada, 8–13 November 1992* (Elsevier, Amsterdam, 1993).
128. N. Serpone and R. F. Khairutdinov, in: *Semiconductor Nanoclusters-Physical, Chemical, and Catalytic Aspects*, edited by P. V. Kamat and D. Meisel (Elsevier, New York, 1997), pp. 417–444.
129. K. Okamoto, Y. Yamamoto, H. Tanaka, M. Tanaka, and A. Itaya, Heterogeneous photocatalytic decomposition of phenol over TiO_2 powder, *Bull. Chem. Soc. Jpn.* **58**, 2015–2022 (1985).
130. D. Lawless, N. Serpone, and D. Meisel, Role of OH radicals and trapped holes in photocatalysis—a pulse radiolysis study, *J. Phys. Chem.* **95**, 5166–5170 (1991).
131. U. Stafford, K. A. Gray, and P. V. Kamat, Radiolytic and TiO_2-assisted photocatalytic degradation of 4-chlorophenol—a comparative study, *J. Phys. Chem.* **98**, 6343–6351 (1994).
132. R. Terzian, N. Serpone, C. Minero, E. Pelizzetti, and H. Hidaka, Kinetic studies in heterogeneous photocatalysis. 4. The photomineralization of a hydroquinone and a catechol, *J. Photochem. Photobiol. A: Chem.* **55**, 243–249 (1990).
133. R. Terzian, N. Serpone, C. Minero, and E. Pelizzetti, Kinetic Studies in heterogeneous photocatalysis. 5. Photocatalyzed mineralization of cresols in aqueous media with irradiated titania, *J. Catal.* **128**, 352–365 (1991).
134. J. C. Doliveira, G. Al-Sayyed, and P. Pichat, Photodegradation of 2-chlorophenol and 3-chlorophenol in TiO_2 aqueous suspensions, *Environ. Sci. Technol.* **24**, 990–996 (1990).
135. G. Al-Sayyed, J. C. D'Oliveira, and P. Pichat, Semiconductor-sensitized photodegradation of 4-chlorophenol in water, *J. Photochem. Photobiol. A: Chem.* **58**, 99–114 (1991).
136. A. Mills, S. Morris, and R. Davies, Photomineralisation of 4-chlorophenol sensitised by titanium dioxide – a study of the intermediates, *J. Photochem. Photobiol. A: Chem.* **70**, 183–191 (1993).
137. A. Mills and S. Morris, Photomineralisation of 4-chlorophenol sensitised by titanium dioxide—a study of the effect of annealing the photocatalyst at different temperatures, *J. Photochem. Photobiol. A: Chem.* **71**, 285–289 (1993).
138. J. Kochany and J. R. Bolton, Mechanism of photodegradation of aqueous organic pollutants. 1. EPR spin-trapping technique for the determination of ·OH radical rate constants in the photooxidation of chlorophenols following the photolysis of H_2O_2, *J. Phys. Chem.* **95**, 5116–5120 (1991).
139. K. Vinodgopal, U. Stafford, K. A. Gray, and P. V. Kamat, Electrochemically assisted photocatalysis. 2. The role of oxygen and reaction intermediates in the degradation of 4-chlorophenol on immobilized TiO_2 particulate films, *J. Phys. Chem.* **98**, 6797–6803 (1994).
140. R. Terzian and N. Serpone, Heterogeneous photocatalyzed oxidation of creosote components—mineralization of xylenols by illuminated TiO_2 in oxygenated aqueous media, *J. Photochem. Photobiol. A: Chem.* **89**, 163–175 (1995).

141. R. Terzian, N. Serpone, R. B. Draper, M. A. Fox, and E. Pelizzetti, Pulse radiolytic studies of the reaction of pentahalophenols with OH radicals—formation of pentahalophenoxyl, dihydroxypentahalocyclohexadienyl, and semiquinone radicals, *Langmuir* **7**, 3081–3089 (1991).
142. R. B. Draper, M. A. Fox, E. Pelizzetti, and N. Serpone, Pulse radiolysis of 2,4,5-trichlorophenyl: Formation, kinetics, and properties of hydroxytrichlorocyclohexadienyl, trichlorophenoxyl, and dihydroxytrichlorocyclohexadienyl radicals, *J. Phys. Chem.* **93**, 1938–1944 (1989).
143. R. Terzian, N. Serpone, and M. A. Fox, Primary radicals in the photo-oxidation of qromatics—reactions of xylenols with (OH)-O-center-dot, N-3(center-dot) and H-center-dot radicals and formation and characterization of dimethylphenoxyl, dihydroxydimethylcyclohexadienyl and hydroxydimethylcyclohexadienyl radicals by pulse radiolysis, *J. Photochem. Photobiol. A: Chem.* **90**, 125–135 (1995).
144. E. Pelizzetti, V. Maurino, C. Minero, V. Carlin, E. Pramauro, O. Zerbinati, and M. L. Tosato, Photocatalytic degradation of atrazine and other S-triazine herbicides, *Environ. Sci. Technol.* **24**, 1559–1565 (1990).
145. E. Pelizzetti, V. Carlin, V. Maurino, C. Minero, M. Dolci, and A. Marchesini, Degradation of atrazine in soil through induced photocatalytic processes, *Soil Sci.* **150**, 523–526 (1990).
146. V. Carlin, C. Minero, and E. Pelizzetti, Effect of chlorine on photocatalytic degradation of organic contaminants, *Environ. Technol.* **11**, 919–926 (1990).
147. E. Pelizzetti, V. Carlin, C. Minero, and M. Gratzel, Enhancement of the rate of photocatalytic degradation on TiO_2 of 2-chlorophenol, 2,7-dichlorodibenzodioxin and atrazine by inorganic oxidizing species, *New J. Chem.—Nouveau J. De Chimie* **15**, 351–359 (1991).
148. T. N. Obee and R. T. Brown, TiO_2 photocatalysis for indoor air applications—effects of humidity and trace contaminant levels on the oxidation rates of formaldehyde, toluene, and 1,3-butadiene, *Environ. Sci. Technol.* **29**, 1223–1231 (1995).
149. S. Sampath, H. Uchida, and H. Yoneyama, Photocatalytic degradation of gaseous pyridine over zeolite-supported titanium dioxide, *J. Catal.* **149**, 189–194 (1994).
150. S. Yamazaki-Nishida, K. J. Nagano, L. A. Phillips, S. Cerveramarch, and M. A. Anderson, Photocatalytic degradation of trichloroethylene in the gas phase using titanium dioxide pellets, *J. Photochem. Photobiol. A: Chem.* **70**, 95–99 (1993).
151. L. A. Dibble and G. B. Raupp, Kinetics of the gas–solid heterogeneous photocatalytic oxidation of trichloroethylene by near UV illuminated titanium dioxide, *Catal. Lett.* **4**, 345–354 (1990).
152. J. Peral and D. F. Ollis, Heterogeneous photocatalytic oxidation of gas-phase organics for air purification—acetone, 1-butanol, butyraldehyde, formaldehyde, and meta-xylene oxidation, *J. Catal.* **136**, 554–565 (1992).
153. N. N. Lichtin and M. Avudaithai, TiO_2-photocatalyzed oxidative degradation of CH_3CN, CH_3OH, C_2HCl_3, and CH_2Cl_2 supplied as vapors and in aqueous solution under similar conditions, *Environ. Sci. Technol.* **30**, 2014–2020 (1996).
154. N. N. Lichtin, M. Avudaithai, E. Berman, and J. Dong, Photocatalytic oxidative degradation of vapors of some organic compounds over TiO_2, *Res. Chem. Intermed.* **20**, 755–781 (1994).
155. G. B. Raupp and C. T. Junio, Photocatalytic oxidation of oxygenated air toxics, *Appl. Surf. Sci.* **72**, 321–327 (1993).
156. X. Fu, W. A. Zeltner, and M. A. Anderson, in: *Semiconductor Nanoclusters—Physical, Chemical, and Catalytic Aspects*, edited by P. V. Kamat and D. Meisel (Elseveir, New York, 1997), pp. 445–461.
157. S. A. Larson and J. L. Falconer, in: *Photocatalytic Purification and Treatment of Water and Air*, edited by D. F. Ollis and H. Al-Ekabi (Elsevier, New York, 1993), pp. 473–479.
158. M. R. Nimlos, W. A. Jacoby, D. M. Blake, and T. A. Milne, Direct mass spectrometric studies of the destruction of hazardous wastes. 2. Gas-phase photocatalytic oxidation of trichloroethylene over TiO_2—products and mechanisms, *Environ. Sci. Technol.* **27**, 732–740 (1993).
159. X. Z. Fu, W. A. Zeltner, and M. A. Anderson, The gas-phase photocatalytic mineralization of benzene on porous titania-based catalysts, *Appl. Catal. B: Environ.* **6**, 209–224 (1995).
160. L. A. Dibble and G. B. Raupp, Fluidized-bed photocatalytic oxidation of trichloroethylene in contaminated airstreams, *Environ. Sci. Technol.* **26**, 492–495 (1992).
161. W. A. Jacoby, M. R. Nimlos, D. M. Blake, R. D. Noble, and C. A. Koval, Products, intermediates, mass balances, and reaction pathways for the oxidation of trichloroethylene in air via heterogeneous photocatalysis, *Environ. Sci. Technol.* **28**, 1661–1668 (1994).

162. X. Z. Fu, L. A. Clark, W. A. Zeltner, and M. A. Anderson, Effects of reaction temperature and water vapor content on the heterogeneous photocatalytic oxidation of ethylene, *J. Photochem. Photobiol. A: Chem.* **97**, 181–186 (1996).
163. R. W. Matthews, Photooxidation of organic impurities in water using thin films of titanium dioxide, *J. Phys. Chem.* **91**, 3328–3333 (1987).
164. J.-M. Herrmann, M.-N. Mozzanega, and P. Pichat, Oxiation of oxalic acid in aqueous suspensions of semiconductor illuminated with UV or visible light, *J. Photochem. Photobiol. A: Chem.* **22**, 333–343 (1983).
165. J. Fung and I. Wang, Dehydrocyclization of C6–C8 normal-paraffins to aromatics over TiO_2–ZrO_2 catalysts, *J. Catal.* **130**, 577–587 (1991).
166. M. A. Cauqui, J. J. Calvino, G. Cifredo, L. Esquivias, and J. M. Rodriguezizquierdo, Preparation of rhodium catalysts dispersed on TiO_2–SiO_2 aerogels, *J. Non-Cryst. Solids* **147**, 758–763 (1992).
167. J. G. Weissman, E. I. Ko, and S. Kaytal, Titania zirconia mixed oxide aerogels as supports for hydrotreating catalysts, *Appl. Catal. A: General* **94**, 45–59 (1993).
168. F. H. Hussein and R. Rudham, Photocatalytic dehydrogenation of liquid alcohols by platinized anatase, *J. Chem. Soc., Faraday Trans. I* **83**, 1631–1639 (1987).
169. M. A. Aguado and M. A. Anderson, Degradation of formic acid over semiconducting membranes supported on glass—effects of structure and electronic doping, *Sol. Energy Mater. Sol. Cells* **28**, 345–361 (1993).
170. J. Papp, S. Soled, K. Dwight, and A. Wold, Surface acidity and photocatalytic activity of TiO_2, WO_3/TiO_2, and MoO_3/TiO_2 photocatalysts, *Chem. Mater.* **6**, 496–500 (1994).
171. X. Z. Fu, L. A. Clark, Q. Yang, and M. A. Anderson, Enhanced photocatalytic performance of titania-based binary metal oxides—TiO_2/SiO_2 and TiO_2/ZrO_2, *Environ. Sci. Technol.* **30**, 647–653 (1996).
172. S. Connolly, S. N. Rao, R. Rizza, N. Zaccheroni, and D. Fitzmaurice, Heterosupramolecular chemistry: Toward the factory of the future, *Coord. Chem. Rev.* **186**, 277–295 (1999).
173. P. D. Beer, P. A. Gale, and D. K. Smith, *Supramolecular Chem.* (Oxford University Press, Oxford, 1999).
174. X. Marguerettaz, R. Oneill, and D. Fitzmaurice, Heterodyads – electron transfer at a semiconductor electrode liquid electrolyte interface modified by an adsorbed spacer acceptor complex, *J. Am. Chem. Soc.* **116**, 2629–2630 (1994).
175. L. Cusack, X. Marguerettaz, S. N. Rao, J. Wenger, and D. Fitzmaurice, Heterosupramolecular chemistry: Self-assembly of an electron donor (TiO_2 nanocrystallite)-acceptor (viologen) complex, *Chem. Mater.* **9**, 1765–1772 (1997).
176. D. Fitzmaurice, S. N. Rao, J. A. Preece, J. F. Stoddart, S. Wenger, and N. Zaccheroni, Heterosupramolecular chemistry: Programmed pseudorotaxane assembly at the surface of a nanocrystal, *Angew. Chem. Int. Ed.* **38**, 1147–1150 (1999).
177. D. Ryan, S. N. Rao, H. Rensmo, D. Fitzmaurice, J. A. Preece, S. Wenger, J. F. Stoddart, and N. Zaccheroni, Heterosupramolecular chemistry: Recognition initiated and inhibited silver nanocrystal aggregation by pseudorotaxane assembly, *J. Am. Chem. Soc.* **122**, 6252–6257 (2000).
178. J. Liu, J. Alvarez, and A. E. Kaifer, Metal nanoparticles with a knack for molecular recognition, *Adv. Mater.* **12**, 1381–1383 (2000).
179. S. Fullam, S. N. Rao, and D. Fitzmaurice, Noncovalent self-assembly of silver nanocrystal aggregates in solution, *J. Phys. Chem. B* **104**, 6164–6173 (2000).
180. S. Connolly and D. Fitzmaurice, Programmed assembly of gold nanocrystals in aqueous solution, *Adv. Mater.* **11**, 1202–1205 (1999).
181. S. Connolly, S. Cobbe, and D. Fitzmaurice, Effects of ligand-receptor geometry and stoichiometry on protein-induced aggregation of biotin-modified colloidal gold, *J. Phys. Chem. B* **105**, 2222–2226 (2001).
182. S. Connolly, S. N. Rao, and D. Fitzmaurice, Characterization of protein aggregated gold nanocrystals, *J. Phys. Chem. B* **104**, 4765–4776 (2000).
183. R. Rizza, D. Fitzmaurice, S. Hearne, G. Hughes, G. Spoto, E. Ciliberto, H. Kerp, and R. Schropp, Self-assembly of monolayers of semiconductor nanocrystallites, *Chem. Mater.* **9**, 2969–2982 (1997).

184. J. H. Fendler, Nanoparticles and nanostructured films: Preparation, characterization and applications (Wiley-VCH, Weinheim, 1998), p. 468.
185. A. Merrins, C. Kleverlaan, G. Will, S. N. Rao, F. Scandola, and D. Fitzmaurice, Time-resolved optical spectroscopy of heterosupramolecular assemblies based on nanostructured TiO_2 films modified by chemisorption of covalently linked ruthenium and viologen complex components, *J. Phys. Chem. B* **105**, 2998–3004 (2001).
186. G. Will, J. Rao, and D. Fitzmaurice, Heterosupramolecular optical write-read-erase device, *J. Mater. Chem.* **9**, 2297–2299 (1999).
187. D. Cummins, G. Boschloo, M. Ryan, D. Corr, S. N. Rao, and D. Fitzmaurice, Ultrafast electrochromic windows based on redox-chromophore modified nanostructured semiconducting and conducting films, *J. Phys. Chem. B* **104**, 11449–11459 (2000).

8

Optical, Electronic, and Dynamic Properties of Semiconductor Nanomaterials

8.1. ENERGY LEVELS AND DENSITY OF STATES IN REDUCED DIMENSION SYSTEMS

8.1.1. Energy Levels

The electronic energy levels and density of states (DOS) determine the properties of materials, including optical and electronic properties as well as their functionalities. For nanoscale materials, the energy levels and DOS vary as a function of size, resulting in dramatic changes in the material's properties. The energy level spacing increases with decreasing dimension and this is known as the quantum size confinement effect. It can be understood using the classic particle-in-a-box model. With decreasing particle size, the energy level spacing increases quadratically as the length of the box decreases. The quantum confinement effect can be qualitatively explained using the effective mass approximation.[1–5] For a spherical particle with radius R, the effective band gap, $E_{g,eff}(R)$, is given by:

$$E_{g,eff}(R) = E_g(\infty) + \frac{\hbar^2\pi^2}{2R^2}\left(\frac{1}{m_e} + \frac{1}{m_h}\right) - \frac{1.8e^2}{\varepsilon R} \tag{8.1}$$

where $E_g(\infty)$ is the bulk band gap, m_e and m_h are the effective masses of the electron and hole, and ε is the bulk optical dielectric constant or relative permittivity. The second term on the right-hand side shows that the effective band gap is inversely proportional to R^2 and increases as size decreases. On the other hand, the third term shows that the band gap energy decreases with decreasing R due to increased Coulombic interaction. However, since the second term becomes dominant with small R, the effective band gap is expected to increase with decreasing R, especially when R is small. The quantum size confinement effect becomes significant particularly when the particle size becomes comparable to or smaller than the Bohr exciton radius, a_B, which is given by

$$\alpha_B = \varepsilon_0 \varepsilon h^2 / \pi \mu e^2 \tag{8.2}$$

where ε_0 and ε are the permittivity of vacuum and relative permittivity of the semiconductor, μ is the reduced mass of the electron and hole, $m_e m_h/(m_e + m_h)$, and e the electron charge. For instance, the Bohr radius of CdS is around 2.4 nm[6] and particles with radius smaller or comparable to 2.4 nm show strong quantum confinement effects, as indicated by a significant blueshift of their optical absorption relative to that of bulk.[7-9] Likewise, the absorption spectra of CdSe nanoparticles (NPs) show a dramatic blueshift with decreasing particle size, as shown in Fig. 8.1. The emission spectra usually show a similar blueshift with decreasing size.

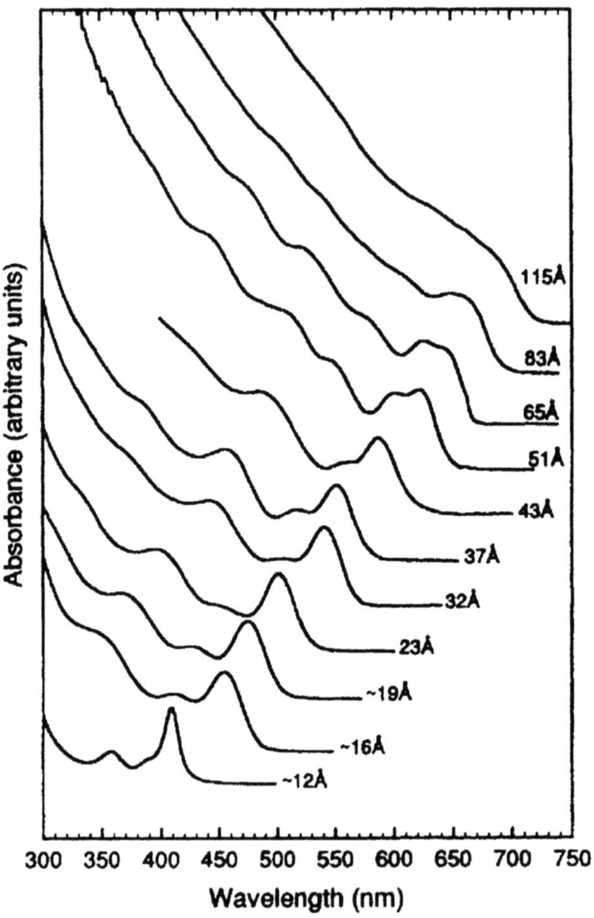

FIGURE 8.1. Room-temperature electronic absorption spectra of CdSe nanocrystals dispersed in hexane as a function of particle size from 1.2 to 11.5 nm. Reproduced with permission from Fig. 3 of Ref. 89.

The quantum size confinement effect can also be qualitatively understood from Fig. 8.2. Just like a particle-in-a-box, the energy level spacing decreases with increasing box size. Therefore, as the number of atoms increases from a single atom or molecule to a cluster or particle of atoms or molecules and eventually to a bulk solid, the spatial size of the system increases and the energy level spacing decreases. The band gap of a semiconductor also becomes smaller with decreasing size. There are some interesting differences between semiconductors and metals in terms of quantum confinement. A major difference is in their electronic band structure. The band gap that plays a central role in semiconductors is much less pronounced in metals, since electrons are free to move even at fairly low temperature in their half-filled conduction band (CB). As the particle size increases, the center of the band develops first and the edge last.[10] In metals with

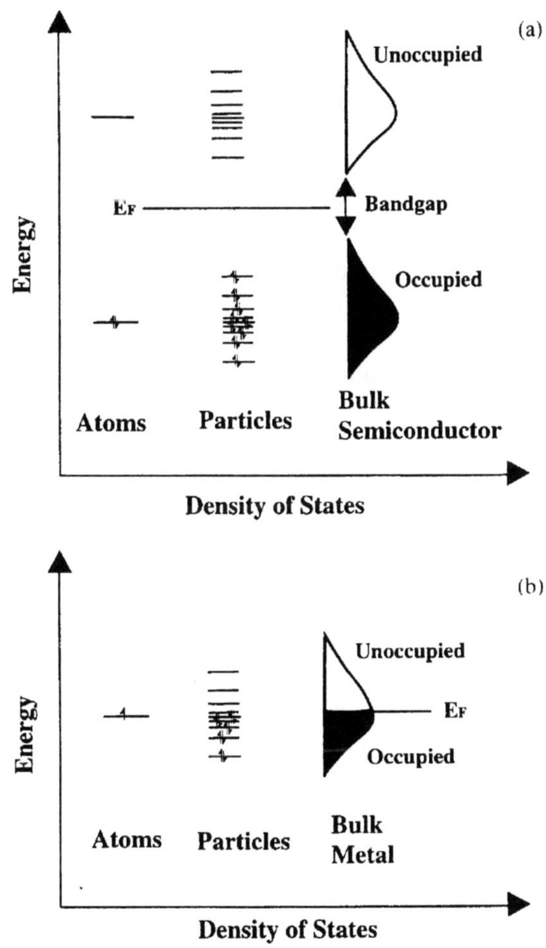

FIGURE 8.2. Schematic illustration of the energy levels as a function of DOS for different sized semiconductor (top panel) and metal (bottom panel) systems. The small arrows indicate occupation of states by electrons. Adapted from Fig. 1 of Ref. 314 with permission.

Fermi level lying in the band center, the relevant energy level spacing is very small even in relatively small particles, whose properties resemble those of the bulk.[11] For semiconductors, in contrast, the Fermi level lies between two bands and the edges of the bands dominate the low energy optical and electronic properties. Band gap optical excitation depends strongly on particle size. As a consequence, the quantum size confinement effect is expected to occur at much smaller sizes for metal than for semiconductor nanoparticles. This difference has fundamental consequences on their optical, electronic, and dynamic properties. It is interesting to note that for very small metal nanoparticles, a significant band gap can be observed,[12] as discussed in more detail in Chapter 10.

8.1.2. Density of States

The DOS of a reduced dimension system also changes significantly with decreasing size. For example, for a 3D bulk material, the DOS is proportional to the square root of energy, E. For a system with confinement in one dimension (a 2D material or quantum well), the DOS is a step function. For systems with confinement in two dimensions (a 1D material or quantum wire), the DOS has a peculiarity. For systems with confinement in three dimensions (zero-dimensional (0D) material or the so-called QDs), the DOS has the shape of a δ-function.[13]

The DOS strongly influences the electronic and optical properties of semiconductors as well as metals. The changes in DOS, along with changes in electronic and vibrational energy levels, alter the properties of the nanomaterials compared to bulk. This results in changes in both the dynamic and equilibrium properties of the materials. For example, changes in the DOS of the electrons and phonons affect the electron–phonon coupling and thereby the electronic relaxation due to electron–phonon interaction. A direct measure of this effect is usually difficult since the surface often plays a dominant role in electronic relaxation through trapping and the effect of electron–phonon interaction on relaxation cannot be determined unambiguously.

8.2. ELECTRONIC STRUCTURE AND ELECTRONIC PROPERTIES

8.2.1. Electronic Structure of Nanomaterials

The electronic band structure of semiconductors and metals determines their optical and electronic properties. The electronic band structure is a description of the energy levels of electronic states as a function of lattice phonon wave vector **k**. If the band maximum for the valence band (VB) matches the band minimum of the CB for the same **k** value, electronic transition from the VB to the CB is electrical dipole allowed and this is called an allowed transition, as shown in Fig. 8.3(a). Otherwise, the transition is electrical dipole forbidden. In this case, transition can be made possible with phonons involved. These transitions are called phonon-assisted transitions and are usually much weaker than electric dipole allowed transitions, as shown in Fig. 8.3(b).

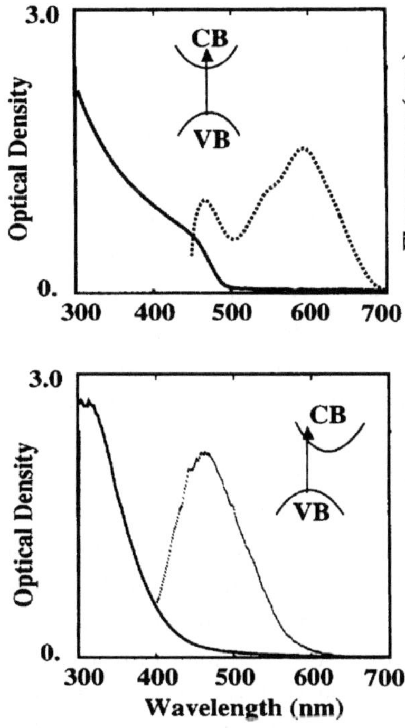

FIGURE 8.3. Comparison of electronic aborption (left) and emission (right) spectra of NPs of CdS (top), a direct band gap, and Si (bottom), an indirect band gap, semiconductor. The fluorescence was collected with 390 nm excitation. The emission spectrum of CdS NPs with an average diameter of 3 nm shows both band edge and trap state emission. Adapted from Fig. 3 of Ref. 23 with permission.

According to their electronic band structures, semiconductors are generally divided into two classes: direct and indirect band gap. Optical absorption and emission as well as electronic transport properties of these two types of semiconductors can be very different. For direct band gap semiconductors with electrical dipole allowed transitions from the VB to the CB, their electronic absorption as well as emission is usually strong with a distinct excitonic feature in the absorption spectrum. For indirect band gap semiconductors with electrical dipole forbidden transitions from the VB to the CB, the transition is phonon-assisted, that is, both energy and momentum of the electron–hole pair are changed in the transition.[14] Both their absorption and emission are weaker compared to those of direct band gap semiconductors.

For nanoscale materials, the electronic band structure could be altered as compared to bulk. As a result, the properties of the materials can change with reducing dimension of the system. Reducing the system dimension could result in changes in the electronic band structure. Unfortunately, the band structure of nanomaterials is not as well characterized as with bulk materials. One example is silicon nanoparticles for which there have been intensive debates over if the strong luminescence observed was due to changes in electronic band structure.[15–21] It is most likely that the luminescence from nanostructured silicon was partly due to surface effect, especially the red emission, and partly due to quantum confinement (electronic structure change), especially the bluer emission. This is similar to the red trap state emission due to surface states vs. bandedge emission determined by the band structure observed typically for CdS.[22] In principle, the band structure could be altered in nanostructured materials due to confinement. However, clear demonstration of this effect is rare due to complication of surface effect and requires further theoretical and experimental research.

8.2.2. *Electron–Phonon Interaction*

Electron–phonon interaction plays a critical role in the materials properties and functionalities of bulk as well as nanoscale materials. With reduced dimensions in nanomaterials, the electron–phonon interaction is expected to change. From the standpoint of DOS, one may expect that the interaction becomes weaker with decreasing size, since the DOS of both the electrons and phonons decreases with decreasing size and their spectral overlap should decrease. However, other factors, such as surface, that could play a competing role by introducing electronic trap states should be considered in practice. The change in electron–phonon interaction and electron–surface interaction with particle size will in turn affect the charger carrier lifetime, as to be discussed in more detail later.

Compared to bulk materials, another unique aspect of nanostructured materials, in terms of electron–phonon interaction, is the involvement of surface phonons or vibrations. These include phonons due to surface atoms of the nanocrystals, which tend to have a lower frequency than bulk phonons, and vibrations of molecules or atoms adsorbed on the surface of the nanocrystal that can have frequencies higher or lower compared to bulk phonon frequencies. The combination of changes of DOS and surface phonon frequencies could significantly affect the electron–phonon interaction in nanostructured materials.[23]

8.3. OPTICAL PROPERTIES OF SEMICONDUCTOR NANOMATERIALS

8.3.1. *Absorption: Direct and Indirect Transitions*

Optical absorption of semiconductors is determined by their electronic structure and is often used as a way to probe their optical and electronic properties. Optical absorption is a result of interaction between the material and light. When the frequency of light is on resonance with the energy difference between states and the transition is allowed or partly allowed by selection rules, a photon is absorbed by the material. This is reflected as a decrease of transmission or an increase in absorbance of the light passing through the sample. By measuring the transmission or absorbance of samples as a function of the frequency of light, one obtains an absorption spectrum of the sample. The spectrum is characteristic of a given material. Figure 8.3 shows a comparison of the electronic absorption and emission spectra of nanoparticles of CdS, a direct band gap semiconductor, and Si, an indirect band gap material. The absorption peak near 430 nm for CdS NPs is known as the exciton peak [Fig. 8.3(a)]. Such excitonic features are absent in indirect band gap semiconductor materials such as Si [Fig. 8.3(b)].

As mentioned earlier, one of the most striking features of nanoparticles compared to bulk crystalline materials is the blueshift of the absorption spectrum due to quantum size confinement. This has been observed for many semiconductor nanoparticles with CdSe receiving the most attention. As shown in Fig. 8.1, the electronic absorption spectra of CdSe nanoparticles show a significant blueshift with decreasing size. Many studies have been conducted on high quality samples of CdSe to measure and assign their size-dependent optical spectrum.[24–26] The measurements of the absorption and emission spectra were often carried out at low temperature (~10 K) to reduce inhomogeneous spectral broadening due to thermal effect. These studies have shown that the size dependence of up to 10 excited states in the absorption spectra of CdSe nanocrystals can be successfully described by uncoupled multiband effective mass (MBEM) theory that includes VB degeneracy but not coupling between the conduction and VBs. The assignment of the states provides a foundation for discussion of electronic structure of semiconductor nanoparticles.[27] The MBEM theory has recently been extended to include CB–VB coupling and applied to describe the size-dependent electronic structure of InAs nanocrystals.[28] Similar to CdSe, the absorption properties of CdS have also been extensively studied and similar blueshift in the spectra has been observed.[7,9] The absorption spectra of CdS nanoparticles are blueshifted compared to those of CdSe mainly due to the lighter mass and lower electron density of S than Se.

Other metal sulfide nanoparticles such as PbS, Cu_xS, and Ag_2S have also been studied with respect to their optical properties. PbS NPs are interesting in that their particle shapes can be readily varied by controlling synthetic conditions.[29–33] Also, since the Bohr radius of PbS is relatively large, 18 nm, and its bulk band gap is small, 0.41 eV,[34] it is easy to prepare particles with size smaller than the Bohr radius that show strong quantum confinement effects and still absorb in the visible part of the spectrum. Attempts have been made to study the surface and shape dependence of electronic relaxation in different shaped PbS NPs. It was observed that the ground state electronic absorption spectrum changes significantly when particle shapes are changed from mostly spherical to needle and cube shaped by changing the surface capping polymers.[30]

Copper sulfides (Cu_xS, $x = 1$–2) are interesting due to their ability to form with various stoichiometries. The copper–sulfur system ranges between the chalcocite (Cu_2S) and covellite (CuS) phases with several stable and metastable phases of varying stoichiometry inbetween. Their complex structures and valence states result in some unique properties.[35-45] In a study of CuS nanoparticles by Drummond *et al.* a near-IR (NIR) band was found to disappear following reduction to Cu_2S via viologen and was, therefore, attributed to the presence of Cu(II).[35,36] The IR band was assigned to a state in the band gap due to surface oxidation, which lies 1 eV below the CB. It was believed that this new middle-gap state was occupied by electrons and thus has electron donor character. However, the model appeared to have some inconsistencies and a recent study has suggested instead that the state has electron acceptor character.[46]

Ag_2S is potentially useful for photoimaging and photodetection in the IR region due to its strong absorption in the IR.[47] Ag_2S nanoparticles are usually difficult to synthesize due to their tendency to aggregate into bulk. One approach was to synthesize them in reverse micelles.[48] A new synthetic method for preparing Ag_2S nanoparticles has recently been developed using cysteine (Cys) and glutathione (GSH) as capping molecules.[49] The ground state electronic absorption spectra of the Ag_2S nanoparticles show a simple continuous increase in absorption cross section towards shorter wavelengths starting from the red (600–800 nm). There is no apparent excitonic feature in the visible and UV region of the spectrum. The absorption spectrum of Au_2S nanoparticles also shows featureless absorption that increases with decreasing wavelength,[50] similar to that of Ag_2S.

Silver halide nanoparticles, for example, AgI and AgBr, play an important role in photography[51,52] and their synthesis is relatively simple.[53,54] The optical absorption spectrum of AgI features a sharp excitonic peak at 416 nm, while the absorption spectrum of AgBr nanoparticles lacks such as excitonic feature and the spectrum is blue shifted with respect to that of AgI.[55]

GaN nanoparticles have received considerable attention as a potential blue emitting laser or light emitting diole (LED) material.[56] For 3 nm particles, the electronic absorption spectrum features an excitonic band around 330 nm and the emission around 350–550 nm was broad and featureless.[57]

Another important class of nanoparticles is metal oxide nanoparticles, for example, TiO_2, SnO_2, Fe_2O_3, and ZnO. Metal oxides play an important role in catalysis and photocatalysis and as paint pigments. Many metal oxides are considered as insulators because of their large band gap. Therefore, they usually appear as white in color and have little absorption in the visible region of the spectrum. Their nanoparticles can usually be prepared by hydrolysis. Among the different metal oxide nanoparticles, TiO_2 has received the most attention because of its stability, easy availability, and promise for applications, for example, solar energy conversion.[58] In contrast to most metal oxide NPs that have no color or visible absorption, Fe_2O_3 NPs show a red color due to absorption in the blue and green region. Fe_2O_3 can exist in the γ-phase (maghemite) or α-phase (hematite). In its α-phase, it can be used as a photocatalyst and in its γ-phase it can be used as a component in magnetic recording media because of its magnetic properties.[59–63] The optical properties of the two forms of Fe_2O_3 NPs are very similar despite their very different magnetic properties.[64]

Silicon NPs have attracted considerable attention recently because of their photoluminescence (PL) properties. Since bulk silicon is an indirect band gap semiconductor with a band gap of 1.1 eV, it is very weakly luminescent. For opto-electronics applications, it is highly desirable to develop luminescent materials that are compatible with the current existing silicon technology developed and matured for the electronics industry. The weak luminescence of bulk silicon presents a major obstacle to its use for the fast-growing opto-electronics industry. The discovery that porous and nanocrystalline Si emit visible light with high quantum yield in 1990[65] has raised hopes for new photonic devices based on silicon and stimulated strong research interest in porous silicon and Si nanoparticles.[15–21] Various methods have been used to make Si nanoparticles, including slow combustion of silane,[19] reduction of $SiCl_4$ by Na,[66] separation from porous Si following HF acid electrochemical etching,[67–69] microwave discharge,[70] laser vaporization/controlled condensation,[15] high pressure aerosol reaction,[16] laser-induced chemical vapor deposition,[18] and chemical vapor deposition.[17] Si NPs are difficult to make using wet colloidal chemistry techniques. The absorption spectrum of Si nanoparticles shows no excitonic features in the near IR to near UV region and the absorption cross section increases with decreasing wavelength [Fig. 8.3(b)]. Interestingly, Si nanowires synthesized with a supercritical fluid solution approach was found to show sharp discrete absorption features in the electronic absorption spectrum and relatively strong "bandedge" PL.[71] These optical properties have been suggested to be due to quantum confinement effects, even though surface effects cannot be ruled out.

Nanoparticles of Ge, an element with similar properties to those of Si, have also been synthesized and studied spectroscopically.[72] The absorption spectrum starting in the NIR was dominated by direct band gap transitions for the largest dots while was dominated by indirect transitions for the smallest dots.

Layered semiconductors such as PbI_2, Bi_2I_3, and MoS_2 form an interesting class of semiconductors with some unique properties.[73] Nanoparticles of layered semiconductors can be prepared using techniques similar to those used for other semiconductors. Some can also be made by simply dissolving bulk crystals in suitable solvents. For PbI_2 nanoparticles, there have been some controversies over the nature of the optical absorption spectrum (Fig. 8.4), whether it is from the nanoparticles or from some kind of iodine complexes, and if the three major absorption peaks are due to different sized "magic" numbered particles.[73-75] These two questions have been addressed in detail in a recent study by Sengupta *et al.*[76] which found no evidence for "magic" numbered particles of different size correlating with the three absorption peaks and the optical absorption seemed to be dominated by PbI_2 NPs. Using a particle-in-a-rectangular box model, the peak positions and the observed blueshift of the peaks with simultaneous decrease in particle size upon aging under light has been satisfactorily explained.[77] TEM images and optical studies led to the proposal that in the photodecomposition process the initially formed large, single-layered particles break down into smaller, multi-layered particles, resulting in significant increase in the optical

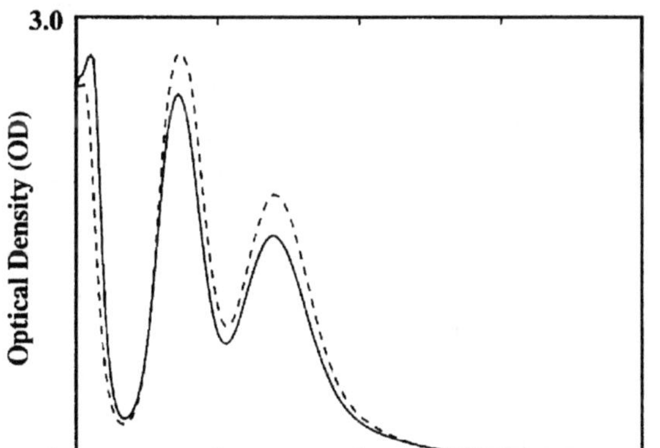

FIGURE 8.4. Electronic absorption spectra of PbI_2 nanoparticles in propanol aged under light for a few days (solid line) and a few weeks (dashed line). There is little shift in the absorption peaks during the aging process. However, the particle size became smaller based on TEM data. Adapted from Fig. 1 of Ref. 76 with permission.

absorption intensity in the visible region and slight blueshift of the absorption peaks.[77]

Similar studies conducted on two related layered semiconductors, bismuth iodide, BiI_3, and bismuth sulfide, Bi_2S_3, support the model developed for PbI_2.[77,78] BiI_3 is promising for non-silver based and thermally controlled photographic applications.[79] The band gap of bulk BiI_3 has been reported as 2.1 eV.[80] Colloidal BiI_3 nanoparticles and BiI_3 clusters in zeolite LTA have been synthesized.[81–83] Similar to PbI_2, the peak positions and the blueshift of the peaks with simultaneous decrease in particle size in BiI_3 NPs can be explianed using the particle-in-a-rectangular-box model.[77] In contrast to PbI_2 and BiI_3, Bi_2S_3 nanoparticles show no sharp peaks in their absorption spectrum and no evidence of photodegradation based on TEM measurements.[78]

Optical absorption of isolated nanostructures is relatively simple in comparison to that of assembled systems in which there is strong interparticle interaction. Assemblies of nanoparticles often exhibit properties modified from those of isolated particles.[10,84,85] Most samples studied exist in the form of a collection of a large number of particles with weak or no interaction between them. The optical absorption observed is an ensemble average of that of all particles. Since each particle has different size, shape, and surface properties, the spectrum observed is inhomogeneously broadened. It would be ideal to study the properties of a single particle to avoid the problem of inhomogeneous distribution. There have been recent attempts to use single particle spectroscopy to study single particles. Even when there are a large number of particles, the interaction between particles can usually be ignored if they are far apart, for example, in colloidal solutions. However, when the particles are close in distance such as in films or assembled systems, their interaction becomes important and the interaction can be reflected in changes in properties of the particles, for example, optical absorption and emission. The absorption spectrum often becomes broader and red shifted compared to that of isolated particles. For example, as shown in Fig. 8.5, the absorption spectra of close-packed CdSe particles demonstrated a distinct broadening and a redshift of the optical bands compared to those of isolated particles.[86] This has been considered to be a result of delocalization and formation of collective electronic states in the close-packed assembled particles. It was also found that the collective states can collapse and primary localization of electronic states can be restored under strong electric fields. This observed large electro-absorption response was thought to be potentially useful for the development of large area electro-optic devices.[86]

The attractive potential for the dispersion interaction, $V(D)$, between two spheres of finite volume as a function of distance between them can be explained using the theory derived by Hamaker:[87]

$$V(D) = \frac{-A_{\mathrm{H}}}{12}\left\{\frac{R_{12}}{D[1 + D/2(R_1 + R_2)]} + \frac{1}{1 + D/R_{12} + D^2/4R_1R_2} + 2\ln\left(\frac{D[1 + D/2(R_1 + R_2)]}{R_{12}[1 + D/R_{12} + D^2/4R_1R_2]}\right)\right\} \tag{8.3}$$

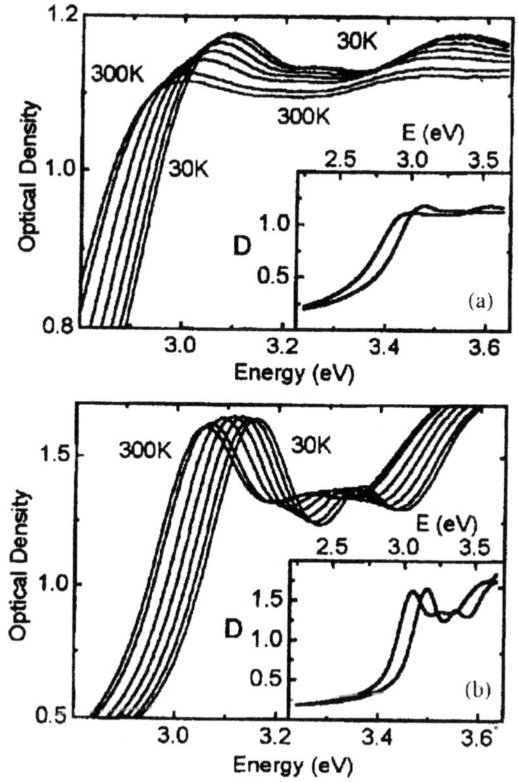

FIGURE 8.5. Electronic absorption spectra of thin films of close-packed (a) and isolated (b) CdSe nanocrystals at different temperatures (from right to left curves): 30, 80, 180, 280, and 300 K. The inserts show the full-range spectrum of optical density D for lowest and highest temperature. Reproduced from Fig. 2 of Ref. 86 with permission.

where A_H is the Hamaker constant, which is material dependent, R_1 and R_2 are the radii of two particles with the separation between them as D, and R_{12} is the reduced radius of particles 1 and 2. Equation (8.3) has very different behavior for two extremes. When $D \gg R_{12}$, $V(D)$ simply becomes the Van der Waals potential (D^{-6} dependence), and when $D \ll R_{12}$, $V(D)$ shows D^{-1} dependence. Another aspect of the theory is that $V(D)$ has to be comparable to kT at room temperature, otherwise the driving force towards ordering becomes negligible, even though the entropy term is still present.[88] If $V(D)$ is much greater than kT, the interparticle attraction is strong and the particles will form nonequlibrium aggregate structures.[85]

Potentially, superlattices formed from assembled nanocrystals can have enhanced charge transport properties compared to isolated particles. The modification of electrical conductivity of a solid will depend on various parameters including spatial arrangement of lattice sites with respect to each other, charging energy of the individual lattice site, and coupling between various sites in a unit cell.[85] The properties of nanocrystal superlattices are expected to depend on factors such as size, stoichiometry of nanoparticles, interparticle separation, and symmetry of the nanocrystals. Murray *et al.*[89] did extensive work on developing techniques to produce narrow size distribution of II–VI semiconductor nanoparticles and on understanding inter-QD interactions. They crystallized long-range ordered structure of CdSe QD with extremely narrow size distribution ($\pm 3\%$).[90] Long-range resonance energy transfer (LRRT) between CdSe nanoparticles in the superlattice structure has been observed and Forster theory has been used to determine the distance of LRRT from experimental data.[91,92] It was shown that the site charging energy, the coupling interaction between sites, which is mostly dipole–dipole in origin, and the lattice symmetry can be controlled in a semiconductor superlattice. However, due to the presence of surface states in semiconductor nanoparticles, the individual particle dipole may not be aligned in a crystallographic order with respect to each other. Therefore, true quantum mechanical wavefunction overlap may not be possible. For this reason, the role of the superlattice symmetry is not clear.[85] The coupling between nanoparticles in a superlattice has important implications in QD lasers, resonant tunneling devices and quantum cellular automata.[93,94] Close-packed nanoparticle superlattices have also been used to develop various photonics devices, such as LEDs[95,96] and photovoltaics.[97]

The self-assembly and self-organization of nanomaterials in solution are important in terms of unique chemistry and material applications. However, to date, progress in this area has been very limited. Covalent assembly of nanocrystals in solution to build superlatttices or heterosupermolecules is another area related to the assembly of nanoparticles to yield well-defined functions. The method to make covalent-linked nanoparticles has its advantages and disadvantages. On one hand, when they form irreversible cross-linkages, the superlattices are more stable than those formed based on noncovalent interparticle interaction. This has been tested to produce devices like single electron tunnel junctions and nanoelectrodes.[85] On the other hand, long-range order is hard to achieve by covalent linkages.

Most optical studies of nanoparticles have focused on the effects of size and surface. Another very important factor to consider is the shape dependence of the absorption as well as emission properties. A limited number of studies have been carried to address the issue of shape on absorption and luminescence properties of nanoparticles. The extreme case is nanowires that can be considered as a limiting case of nanoparticle in the general sense. For instance, the absorption spectra of InAs/InP self-assembled quantum wires show a dependence on polarization of the excitation light as well as a polarization anisotropy in the emission spectrum.[98] More work needs to be done to explore the dependence of optical properties on the shape of particles and this is expected to be an interesting area of research in the future.

8.3.2. Emission: Photoluminescence and Electroluminescence

Light emission from nanoparticles serves as a sensitive probe of their electronic properties. Emission is also the basis of applications in lasers, optical sensors, and LEDs. Light emission can be photoinduced or electrically induced, commonly referred to as PL and electroluminescence (EL). In PL, the emission is a result of photoexcitation of the material. The energy of the emitted photons is usually lower than the energy of the incident excitation photons and this emission is considered as the Stokes emission. In EL, the light emission is a result of electron–hole recombination following electrical injection of the electron and hole. This is schematically illustrated in Fig. 8.6. The application of EL in LEDs will be discussed in more detail in Section 8.4.1

Photoluminescence can be generally divided into bandedge emission, including excitonic emission, and trap state emission. Trap state emission is usually red shifted compared to bandedge emission. For instance, in Fig. 8.3(a), the emission spectrum clearly shows both trap state and bandedge emission with the latter at a shorter wavelength. The ratio between the two types of emission is determined by the density and distribution of trap states. A high trap state emission indicates a high density of trap states and efficient trapping. It is possible to prepare high quality samples that have mostly bandedge emission when the surface is well capped. For example, TOPO capped CdSe show mostly bandedge emission and weak trap state emission, which is an indication of high quality of the sample.[89,99,100] Luminescence can be enhanced by surface modification[22,101–105] or using core/shell structures.[106–109] Nanoparticles that have been found to show strong PL include CdSe, CdS, ZnS.[110] Other nanoparticles have generally been found to be weakly luminescence or non-luminescent at room temperature, for example, PbS,[30] PbI_2,[76] CuS,[46] and Ag_2S.[49] The low luminescence can be due to either indirect nature of the semiconductor or a high density of internal and/or surface trap states that quench the luminescence. The luminescence usually increases at lower temperature due to suppression of electron–phonon interaction

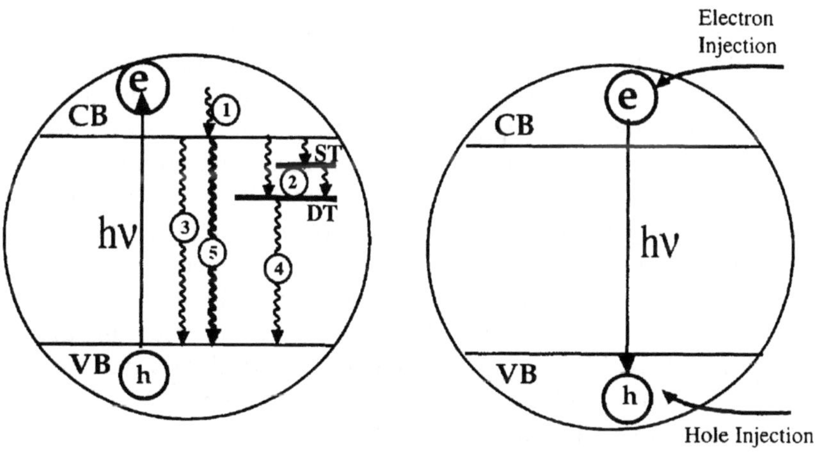

FIGURE 8.6. Schematic illustration of photoexcitation (upward arrow on left) and PL, left, and EL, right. In the left panel, the curve lines with downward arrows indicate different relaxation processes: (1) electronic relaxation within the conduction band (CB), (2) trapping into shallow trap (ST) and deep trap (DT) states as well as further trapping from ST to DT, (3) bandedge electron–hole recombination, (4) trapped electron–hole recombination, and (5) exciton–exciton annihilation. In the right panel, electrical electron and hole injections are indicated. EL emission resulting from the electron–hole recombination is indicated by the downward arrow.

and thereby lengthened excited electronic state lifetime. Controlling the surface by removing surface trap states can lead to significant enhancement of luminescence as well as of the ratio of bandedge over trap state emission.[22,101–105] The surface modification often involves capping of the particle surface with organic, inorganic or biological molecules or ions that can result in reduction of trap states that fall within the band gap and quench the luminescence. This scheme of surface states reduction and luminescence enhancement is important for many applications that requires high luminescence yield of nanoparticles, for example, laser, LEDs, fluorescence imaging, and optical sensing.

In contrast to Stokes emission for which the emission is red shifted compared to the excitation wavelength, the energy of emitted photons can be higher than that of the incident photons and this is called anti-Stokes emission, similar to anti-Stokes Raman scattering. Interestingly, in a number of nanoparticle systems, anti-Stokes emission or up-conversion, whose wavelength is blue shifted relative to the excitation wavelength, has been observed, including CdS,[111] InP,[112,113] CdSe,[112] InAs/GaAs,[114] Er^{3+}-doped $BaTiO_3$,[115] and Mn-doped ZnS.[116] This will be discussed in more detail in Section 8.3.3.

One interesting class of luminescent semiconductor nanoparticles is those doped with transition metal or rare earth metal ions. Doped bulk semiconductor materials play a critical role in various technologies, including the semiconductor industry.[117] Compared to undoped semiconductors, doped materials offer the possibility of using the dopant to tune their electronic, magnetic, and optical properties. Therefore, in addition to existing advantages nanomaterials offer in terms of controllable parameters such as size and surface, dopants offer the additional flexibility for designing new nanomaterials. Luminescent semiconductor nanoparticles are of strong interest for possible use in opto-electronics such as LEDs and lasers or as novel phosphors. One example is Mn^{2+}-doped ZnS nanoparticles, commonly denoted as ZnS : Mn, that has received considerable attention recently. Bulk counterparts of these materials have found wide technological applications as phosphors and in EL.[117] In these materials, a small amount of transition metal ions, such as Mn^{2+}, is incorporated into the nanocrystalline lattice of the host semiconductor, for example, ZnS or CdS. These doped particles have interesting magnetic[118–121] and electro-optical properties.[122–126] The host semiconductor usually absorbs light and transfers energy to the metal ion that emits photons with energies characteristic of the metal ion. Luminescence can also result from direct photoexcitation through transitions of the metal ions.[116]

The PL and EL of Mn^{2+}-, Cu^{2+}-, and Er^{3+}-doped ZnS nanoparticles as free colloids[125–127] and in polymer matrices and thin films[124,128–131] have been reported. As in bulk Mn^{2+}-doped ZnS, the Mn^{2+} ion acts as a luminescence color center, emitting near 585 nm as a result of $^4T_1$ to $^6A_1$ transition.[120,125,127,132,133] Mn^{2+}-doped ZnS NP was first reported by Becker and Bard[134] in 1983. It was found that the Mn^{2+} emission at 538 nm with a quantum yield of about 8% was sensitive to chemical species on the particle surface and that the emission can be enhanced with photoirradition in the presence of oxygen, which was attributed to photoinduced adsorption of O_2. In 1994, Bhargava *et al.*[122] made the claim that the luminescence yield is much higher and the emission lifetime is much shorter in

ZnS : Mn NPs than in bulk. There are currently still debates over if the emission quantum yield is higher in ZnS : Mn NPs relative to bulk. Several recent studies,[135-140] including a theoretical study,[141] made claims of enhancement that seem to support the original claim of enhanced luminescence by Bhargava *et al.*[122] However, most of the studies failed to provide a calibrated, quantitative measure of the luminescence quantum yield in comparison to bulk ZnS : Mn. Enhancement has been observed mostly with respect to different NPs, instead of a true calibrated measure against the quantum yield of the corresponding bulk. Such a measurement is critically needed to establish if there is a true enhancement relative to bulk, which is an issue to be resolved. However, several recent time-resolved studies, to be discussed later, have found that the emission lifetime in ZnS : Mn NPs is the same as in bulk.[142-144] These lifetime studies seem to suggest that the luminescence yield in NPs should not be higher than that of bulk.

Another interesting observation is photoenhanced luminescence in a number of nanoparticle systems, including CdSe,[145-147] porous Si,[148] as well as Mn^{2+}-doped ZnS.[134,136] Several tentative explanations have been provided, including decreasing trapping rates due to trap state filling,[149] surface transformation,[146] change in density of dangling bonds[148] for CdSe, and increasing energy transfer rate from ZnS to Mn^{2+},[150] photooxidation of the surface,[151] or photoinduced adsorption of oxygen[134] in the case of ZnS : Mn. All these studies show that the surface plays a critical role in the photoinduced enhancement. However, there still lacks a molecular level model for the observed photoinduced enhancement.

A special type of light emission is Raman scattering. Raman spectroscopy is a powerful technique for studying vibrational or phonon modes, electron–phonon coupling, as well as symmetries of excited electronic states of nanoparticles. Raman spectra of NPs have been studied in a number of cases, including CdS,[152-157] CdSe,[158-160] ZnS,[155] InP,[161] Si,[162-165] and Ge.[72,166-169] Resonance Raman spectra of GaAs[170] and CdZnSe/ZnSe[171] quantum wires have also been determined. For CdS nanocrystals, resonance Raman spectrum reveals that the lowest electronic excited state is coupled strongly to the lattice and the coupling decreases with decreasing nanocrystal size.[154] For 4.5 nm nanocrystals of CdSe, the coupling between the lowest electronic excited state and the LO phonons is found to be 20 times weaker than in the bulk solid.[158] For CdZnSe/ZnSe quantum wires, resonanace Raman spectrsocpy revealed that the ZnSe-like LO phonon position depends on the Cd content as well as excitation wavelength due to relative intensity changes of the peak contributions of the wire edges and of the wire center.[171]

Most optical emission studies have been conducted on isolated particles, that is, particles with weak or no interparticle interaction. A few studies have been done on nanoparticle assemblies. Similar to optical absorption, the emission properties can be altered for assembled particles relative to isolated particles. Interparticle interaction due to dipole–dipole or electrostatic couplings usually results in spectral line broadening and redshift.[172]

Similar to optical absorption, the emission properties of semiconductor nanoparticles can be strongly affected by the shape of nanoparticles. Only a limited number of studies have been carried to address the issue of shape on luminescence properties of nanoparticles. In the case of InAs/InP self-assembled quantum wires, a polarization anisotropy was observed in the emission spectrum due to wire and strain geometry.[98] It was also found that the nonradiative decay mechanism limiting the emission intensity at room temperature is related to thermal escape of carriers out of the wire.

8.3.3. Nonlinear Optical Properties

Nanoparticles have interesting nonlinear optical properties at high excitation intensities, including absorption saturation, shift of transient bleach, third and second harmonic generation, and up-conversion luminescence. The most commonly observed nonlinear effects in semiconductor nanoparticles are absorption saturation and transient bleach shift at high intensities.[22,102,173–181] Similar nonlinear absorption has been observed for quantum wires of GaAs[182,183] and porous Si.[184,185] These nonlinear optical properties have been considered potentially useful for optical limiting and switching applications.[186] The mechanism for nonlinear absorption will be discussed in more detail in Section 8.5.4 in relation to the dynamics of electron–hole recombination.

Another nonlinear optical phenomenon is harmonic generation, mostly based on the third-order nonlinear optical properties of semiconductor nanoparticles.[33,187–189] The third-order nonlinearity is also responsible for phenomena such as the Kerr effect and degenerate four wave mixing (DFWM).[190] For instance, the third-order nonlinear susceptibility, $\chi^{(3)}$ ($\sim 5.6 \times 10^{-12}$ esu) for PbS nanoparticles has been determined using time-resolved optical Kerr effect spectroscopy and it was found to be dependent on surface modification.[33] Third-order nonlinearity of porous silicon has been measured with the Z-scan technique and found to be significantly enhanced over crystalline silicon.[186] DFWM studies of thin films containing CdS nanoparticles found a large $\chi^{(3)}$ value, $\sim 10^{-7}$ esu, around the excitonic resonance at room temperature.[191]

Only a few studies have been carried out on second-order nonlinear optical properties since it is usually believed that the centrosymmetry or near centrosymmetry of the spherical nanoparticles reduces their first-order hyperpolarizability (β) to zero or near zero. Using hyper-Rayleigh scattering, second harmonic generation in CdSe nanocrystals has been observed.[192] The first hyperpolarizibility β per nanocrystal was found to be dependent on particle size, decreasing with size down to about 1.3 nm in radius and then increasing with further size reduction. The results are explained in terms of surface and bulk-like contributions. A similar technique has been used for CdS nanoparticles for which the β-value per particle (4 nm mean diameter) was found to be on the order of 10^{-27} esu, which is quite high for solution species.[193] Second harmonic generation has also been observed for magnetic cobalt ferrite ($CoFe_2O_4$) colloidal particles when oriented with a magnetic field.[194] The nonlinear optical properties of nanoparticles are found to be strongly influenced by the surface.

As discussed earlier, the optical properties of isolated nanoparticles can be very different from those of assembled nanoparticle films. This is true for both linear and nonlinear optical properties. Theoretical calculations on nonlinear optical properties of nanoparticle superlattice solids have shown that an ideal resonant state for a nonlinear optical process is the one that has large volume and narrow line width.[195–197] The calculations also showed that nonlinear optical responses could be enhanced greatly with a decrease in interparticle separation distance.

Anti-Stokes PL or PL up-conversion is another interesting nonlinear optical phenomenon. In contrast to Stokes emission, the photon energy of the luminescence output is higher than the excitation photon energy. This effect has been previously reported for both doped[198,199] and high purity bulk semiconductors.[200,201] For bulk semiconductors, the energy up-conversion is usually achieved by (i) an Auger recombination process, (ii) anti-Stokes Raman scattering mediated by thermally populated phonons, or (iii) two-photon absorption.[112,202] Luminescence up-conversion has recently been observed in semiconductor heterojunctions and quantum wells[202–218] and has been explained based on either Auger recombination[206,211,219] or two-photon absorption.[212] Long-lived intermediate states have been suggested to be essential for luminescence up-conversion in some heterostructures such as $GaAs/Al_xGa_{1-x}As$.[211] For semiconductor nanoparticles or QDs with confinement in three dimensions, luminescence up-conversion has only recently been reported for CdS,[111] InP,[112,113] CdSe,[112] InAs/GaAs[114] and Er^{3+}-doped $BaTiO_3$.[115] Surface states have been proposed to play an important role in the up-conversion in nanoparticles such as InP and CdSe.[112]

Very recently, luminescence up-conversion in ZnS:Mn nanoparticles and bulk has been observed.[116] When 767 nm excitation was used, Mn^{2+} emission near 620 nm was observed with intensity increasing almost quadratically with excitation intensity (Fig. 8.7). The shift of Mn^{2+} emission from the usually observed 580–620 nm could be due to difference in particle size. Comparison with 383.5 nm excitation showed similar luminescence spectrum and decay kinetics, indicating that the up-converted luminescence with 767 nm excitation is due to a two-photon excitation process. The observation of luminescence up-conversion in Mn^{2+}-doped ZnS opens up some new and interesting possibilities of applications in optoelectronics, for example, as infrared phosphors. There remain some unanswered questions, especially in terms of some intriguing temperature dependence of the up-converted luminescence.[220] It was found that the up-conversion luminescence of ZnS : Mn nanoparticles first decreases and then increases with increasing temperature. This is in contrast to bulk ZnS : Mn in which the luminescence intensity decreases monotonically with increasing temperature due to increasing electron–phonon interaction. The increase in luminescence intensity with increasing temperature for nanoparticles was attributed tentatively to involvement of surface trap states. With increasing temperature, surface trap states can be thermally activated, resulting in increased energy transfer to the excited state of Mn^{2+} and thereby increased luminescence. This factor apparently is significant enough to overcome the increased electron–phonon coupling with increasing temperature that usually results in decreased luminescence.[220]

8.3.4. Single Particle Spectroscopy

Most spectroscopy studies of nanoparticles have been carried out on ensembles of a large number of particles. The properties measured are thus

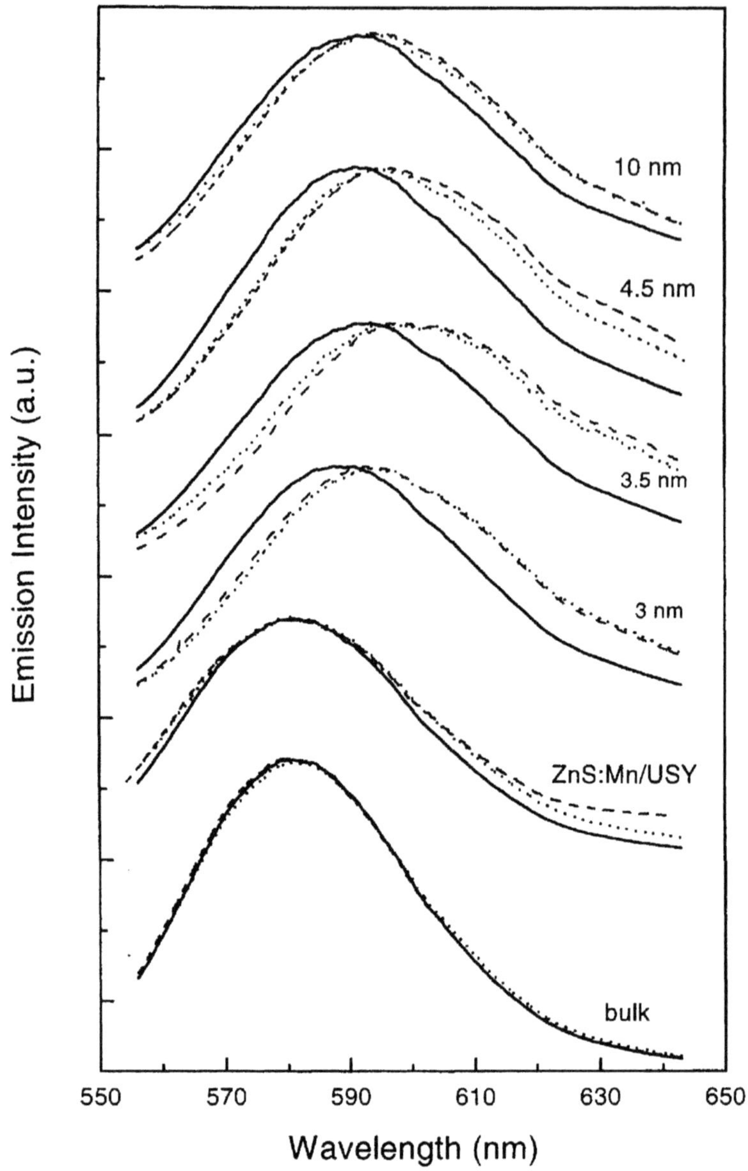

FIGURE 8.7. The Stokes luminescence spectra at 300 nm excitation (Solid), at 383.5 nm excitation (Dash), and the anti-Stokes luminescence spectra at 767 nm excitation (dot) of ZnS : Mn^{2+} bulk and nanoparticles. Reproduced from Fig. 1 of Ref. 116. Copyright of American Physical Society.

ensemble averages of the properties of individual particles. Due to heterogeneous distributions in size, shape, environment, and surface properties, the spectrum measured is thus inhomogeneously broadened. This results in loss of spectral information.[221] For instance, it has been predicted by theory that nanocrystallites should have a spectrum of discrete, atomic-like energy states.[1,2] However, transition line widths observed experimentally appear significantly broader than expected, even though the discrete nature of the excited states has been verified.[26,222] This is true even when size-selective optical techniques are used to extract homogeneous line widths.[26,221–226]

One way to solve the above problem is to make particles with truly uniform size, surface, environment, and shape. However, this is almost impossible or at least very difficult. Another, perhaps simpler, approach to remove the heterogeneity is to conduct the measurement on one single particle. This approach is similar to that used in the field of single-molecular spectroscopy.[227,228] A number of single nanoparticle studies have been reported on semiconductor nanoparticles, including CdS[229] and CdSe.[230–236] Compared to ensemble averaged samples, single particle spectroscopy studies of CdSe nanoparticles revealed several new features, including fluorescence blinking, ultranarrow transition line width, a range of phonon couplings between individual particles, and spectral diffusion of the emission spectrum over a wide range of energies.[236,237] Furthermore, electrical field studies showed both polar and polarizable character of the emitting state. The polar component has been attributed to an induced excited state dipole resulting from the presence of local electrical fields. The fields could be the result of trapped charge carriers on the particle surface. These fields seemed to change over time and result in spectral diffusion.[235,237] It was noted that the spectral line widths are strongly dependent on experimental conditions such as integration time and excitation intensity. Therefore, it has been concluded that the line widths are primarily the result of spectral shifting and not the intrinsic properties of the nanoparticle.[235]

Figure 8.8 shows the emission spectra of a single 5.65 nm overcoated nanocrystallite with different integration time. The data exhibit two interesting features. First, the line shape of a single particle contains information about changes in the surrounding environment and not the intrinsic physics of the nanocrystallite. Second, changes in line shape of a single nanocrystallite spectrum resulting from different experimental conditions are likely to be caused by changes in spectral diffusion. The line width was found to become broader with increasing excitation intensity.[237]

With low-temperature (20 K) confocal microscopy, fluorescence spectra of isolated single CdS nanoparticles have been recorded.[229] The narrowest measured line width of the main fluorescence band has been determined to be about 5 meV at the lowest excitation intensities. At higher intensities, the main fluorescence band broadens and two new peaks appeared on each side of the main band, one of which was attributed to the coupling to a longitudinal optical phonon. In a single CdS/HgS/CdS quantum well QD structure, the spectral and intensity fluctuations are strongly reduced, attributed to charge carrier localization in the HgS region of the nanocrystal.[238] Similarly, for CdSe nanoparticles, the elimination of spectral inhomogeneities reveals resolution limited spectral line widths of <120 meV at 10 K, more than 50 times narrower than expected from ensemble measurements.[231]

One interesting observation in measuring the emission of a single nanoparticle is an intermittent on–off behavior in emission intensity under CW light excitation, first discovered by Nirmal *et al.*[232] The intermittency was analyzed in terms of random telegraph nose and the on–off times were extracted. The off times were found to be power independent while the on times, over

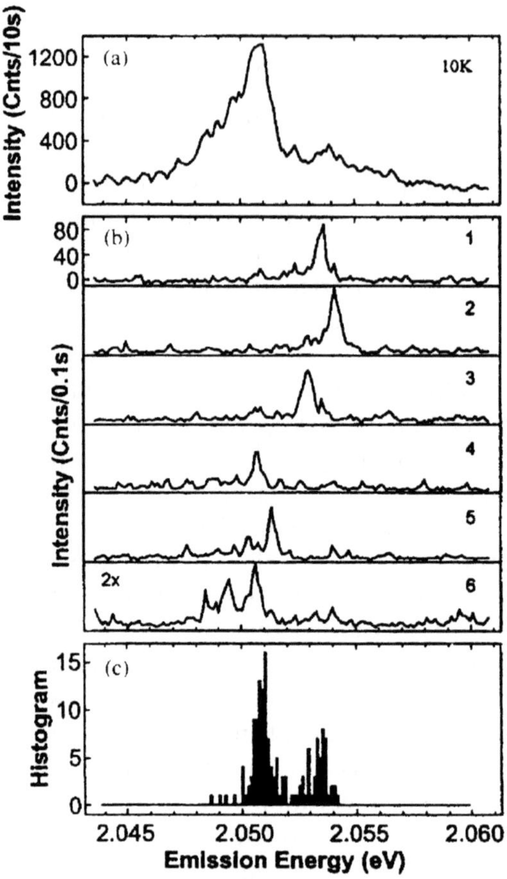

FIGURE 8.8. (a) Spectrum of a single overcoated CdSe quantum dot with a 10 s integration time at 10 K. (b) A representative sample of six spectra taken at ~10 frames/s, frames 1–6 indicate spectrum number 1, 16, 35, 59, 84, and 150, respectively. Each peak in frames 1–5 corresponds to ~20 collected photons. (c) Histogram of peak positions from each of the 150 spectra described in (b). The excitation intensity for all the spectra was 200 W cm^{-2}. Reproduced from Fig. 1 of Ref. 236 with permission.

a narrow intensity range, revealed a linear dependence on power. The intermittency observed was attributed to an Auger photoionization mechanism that leads to ejection of one charge (electron or hole) outside the particle. A "dart exciton" state was assigned to such ionized nanocrystal in which the emission is quenched because of the excess charge. The feasibility of this mechanism was further demonstrated in theoretical calculations which also consider the possibility of thermal ionization.[239] Similar studies on single CdSe/CdS core/shell nanocrystals as a function of temperature and excitation intensity and the observations are consistent with a darkening mechanism that is a combination of Auger photoionization and thermal trapping of charge.[240]

Time-resolved studies of single nanoparticles or QDs have also been carried out to probe the dynamic properties of single particles. For example, time-resolved emission from self-assembled InGaAs/GaAs QDs has been measured at low temperature by the time-correlated single photon counting method using near-field microscopy.[241] The decay time of the emission from discrete levels of a single dot was found to increase with the decrease in emission energy and with the increase in excitation intensity. The results are explained with a model that includes initial filling, of the states, cascade relaxation, state filling and carrier feeding from a wetting layer. Room temperature PL was also acquired and the homogeneous line widths were found to be around 9.8–14.5 meV.[242]

Interestingly, single particle spectroscopy also reveals nonlinear optical properties of single nanoparticles. For example, in the low temperature near-field absorption spectroscopy study of InGaAs single QDs, the absorption change was found to depend nonlinearly on the excitation intensity.[243] This nonlinearity was suggested to originate from state filling of the ground state.

8.4. APPLICATIONS OF OPTICAL PROPERTIES

8.4.1. Lasers and Light Emitting Diodes

One area of application of nanomaterials that has attracted considerable interest is lasers. It is in principle possible to built lasers with different wavelengths by simply changing the particle size. There are two practical problems with this idea. First, the spectrum of most nanoparticles is usually quite broad due to homogeneous and inhomogeneous broadening. Second, the high density of trap states leads to fast relaxation of the excited charge carrier, making it difficult to build up population inversion necessary for lasing. When the surface of the particles is clean and has little defects, the idea of lasing can indeed be realized. This has been demonstrated mostly for nanoparticles self-assembled in clean environments based on physical methods, for example, molecular beam epitaxy (MBE)[244–247] or MOCVD.[248] Examples of quantum dot lasers include InGaAs,[244] InAs,[246] AlInAs,[245,247] and InP.[249] Stimulated emission has also been observed in GaN QDs by optical pumping.[248] The lasing action or stimulated emission has been observed mostly at low temperature.[247,249] However, some room temperature lasing has also been achieved.[244,245]

Only very recently, lasing action has been observed in colloidal nanoparticles of CdSe based on wet chemistry synthesis and optical pumping.[250] It was found that, despite highly efficient intrinsic nonradiative Auger recombination, large optical gain can be developed at the wavelength of the emitting transition for close-packed solids of CdSe QDs. Narrow band stimulated emission with a pronounced gain threshold at wavelengths tunable with size of the nanocrystal was observed. This work demonstrates the feasibility of nanocrystal QD lasers based on wet chemistry synthesis. Whether real laser devices can be built based on this type of nanoparticles remains to be seen. Also, it is unclear if electrical pumping of such lasers can be realized. Likewise, nanoparticles can be potentially used for laser amplification and such an application has yet to be explored. Nanoparticles such as TiO_2 have also been used to enhance stimulated emission for conjugated polymers based on multiple reflection effect.[251]

Very recently, room-temperature ultraviolet lasing in ZnO nanowire arrays has been demonstrated.[252] The ZnO nanowires grown on sapphire substrates were synthesized with a simple vapor transport and condensation process. The nanowires form a natural laser cavity with diameters varying from 20 to 150 nm and lengths up to 10 μm. Under optical excitation at 266 nm, surface-emitting lasing action was observed at 385 nm with emission line width less than 0.3 nm. Such miniaturized lasers could have many interesting applications ranging from optical storage to integrated optical communication devices.

Nanoparticles have been used for LED application in two ways. First, they are used to enhance light emission of LED devices with other materials, for example, conjugated polymers, as the active media. The role of the nanoparticles, such as TiO_2, is not completely clear but thought to enhance either charge injection or transport.[253] In some cases, the presence of semiconductor nanocrystals in carrier-transporting polymers has been found to not only enhance the photoinduced charge generation efficiency but also extend the sensitivity range of the polymers, while the polymer matrix is responsible for charge transport.[254] This type of polymer/nanocrystal composite materials can have improved properties over the individual constituent components and may have interesting applications. Second, the nanoparticles are used as the active materials for light generation directly.[95,107,255,256] In this case, the electron and hole are injected directly into the CB and VB, respectively, of the NPs and the recombination of the electron and hole results in light emission (Fig. 8.6). Several studies have been reported with the goal to optimize injection and charge transport in such device structures using CdS[256] and CdSe nanoparticles.[95,255] Since the mobility of the charge carriers is usually much lower than in bulk single crystals, charge transport is one of the major limitations in efficient light generation in such devices. For example, photoconductivity and electric field induced PL quenching studies of close-packed CdSe quantum dot solids suggest that photoexcited, quantum confined excitons are ionized by the applied electric field with a rate dependent on both the size and surface passivation of the quantum dots.[257,258] Separation of electron–hole pairs confined to the core of the dot requires significantly more energy than separation of carriers trapped at the surface and occurs through tunneling processes. New nanostructures such nanowires,[20,252] nanorods,[259–262] and nanobelts[263] may provide some interesting alternatives with better transport properties than nanoparticles. Devices such as LEDs and solar cells based on such nanostructures are expected to be developed in the next few years.

8.4.2. Photovoltaic Solar Cells

There have been considerable efforts devoted towards developing solar cells for energy conversion in the last several decades. Solar energy conversion into electricity or chemical energy, such as hydrogen, is one of the most attractive alternatives to current energy resources based on fossil fuels or nuclear energy. Chapin *et al.*[264] at Bell Labs developed the first crystalline silicon solar cell in the 1950s with a solar power conversion efficiency of 6%.[264] Since then silicon-based solar cells have become a mature commercial technology. Modern solar power conversion efficiencies are between 15% and 20%.[265] However, the relatively high cost of manufacturing these silicon cells and use of toxic chemicals in their manufacturing have prevented them from widespread use. These aspects have prompted the search for environmentally friendly and low cost solar cell alternatives.

On example of such effort is solar cells built based on inexpensive nanocrystalline materials such as TiO_2 in conjunction with dye sensitization. In 1991, O'Regan and Gratzel[58] developed the first such as dye-sensitized nanocrystalline solar cell based on TiO_2 nanoparticle films sensitized with a Ru complex dye, Ru(4,4′-dicarboxyl-2,2′-bipyridine)$_2$ (NCS)$_2$, known as N3.[266] The nanocrystalline nature of the materials provides significantly increased surface area relative to volume as compared to bulk materials. In a nanocrystalline TiO_2 solar cell, dye molecules are adsorbed onto the surface of the nanoparticles. Photoexcitation of the dye molecules results in injection of electrons into the CB of TiO_2. The resulting oxidized dye is reduced by electron transfer from a redox couple such as I^-/I^{3-}, for the dye to be regenerated. Charge transport occurs through particles that are interconnected and collected by the electrodes. These types of cells have reached solar power conversion efficiencies of 7–10% under full sun conditions.[58,266]

However, there are some practical problems with sealing these cells, for example, leakage of the eleectrolyte at the working conditions. This has led to the search for replacing the liquid electrolyte with a solid hole conducting material using organic conjugated polymers or inorganic solid compounds.[267–270] Conjugated polymers that have been studied and show promise as a hole conductor include MEH-PPV (poly[2-methoxy-5(2′-ethylhexyloxy)-1,4-phenylene vinylene])[271–275] and polythiophene.[260,269,270,276–278] One interesting finding from these studies is that the conjugated polymer can function both as a hole transporter as well as a sensitizer.[269,270,273–275] It was also found that the relative size of the sensitizer molecule and the pores of the nanocrystalline films may be a critical factor to consider in designing photovoltaic devices such as solar cells based on nanoporous materials.[270]

It should be pointed out that the mobility of charge carriers in organic conjugated polymers is usually very low compared to inorganic crystalline materials, which limits their transport properties. For example, polythiophenes have a hole mobility on the order of 0.1 $cm^2\ V^{-1}\ s^{-1}$.[279,280] As mentioned earlier in Section 8.4.2, the transport properties of charge carriers in nanocrystalline materials are also limited compared to single crystalline materials. Therefore, charge transport may be the limiting factor in this type of solar cells, especially given that charge injection has generally been found to be fast and efficient. Other nanostructures such as nanowires, nanorods, and nanobelts may offer some advantages over nanoparticles in terms of charge transport properties. Applications of these new nanostructures for solar cells are yet to be explored and expected to happen in the near future.

8.4.3. Optical Filters: Photonic Band Gap Materials

One of the earliest applications of semiconductor nanoparticles is in glass filters.[281,282] The nanoparticles are usually produced at high temperature in glass and the size is controlled by the annealing temperature. The different sized particles generate different colored filters. This is one of the simplest and best uses of the quantum confinement effects of nanoparticles. Many glass filters or stained glass are based on this arguably oldest nanotechnology.

A special type of filter can be constructed from the so-called photonic band gap materials that are composed of nanoparticles with periodically packed structures.[283,284] Mesoscale nanoparticles assembled to form patterned surperlattice nanostructures over large areas can be used to function as tunable, narrow-band optical filters.[285,286] The position (λ_{max}) of the first-order diffraction peak is related to the spacing between planes (d), the refractive index of the crystalline assembly, and the angle (ϕ) between the incident light and the normal to the surface of the filter by the Bragg equation:[285,286]

$$\lambda_{max} = 2d(n^2 - \sin^2\phi)^{1/2} \tag{8.4}$$

The peak wavelength is thus strongly dependent on the incident angle and plane spacing. As shown in Fig. 8.9, the peak position can be tuned from ~574 to 538 nm when the angle (ϕ) was changed from 0° to 40°. These materials are also considered as diffractive components for sensors,[287,288] tunable gratings,[289] and photonic band gap structures.[290–292]

In some case, semiconductor nanoparticles are introduced in the voids of the photonic band gap materials to influence the gap and optical properties such as photoemission. It was found that the emission lifetime of Mn^{2+} in Mn-doped ZnS nanoparticles is dependent on the dielectric constant of the photonic material.[293]

8.4.4. Other Applications

Semiconductor nanoparticles have many other applications or potential applications. For example, detectors and sensors based on nanomaterials have been reported in a few cases. Mixed SnO_2 (15–20 mol%)-α-Fe_2O_3 particles were found to exhibit high methane sensitivity which makes it suitable as a gas-leak sensor.[294,295] Copper iodide nanocrystals have been suggested as a potentially economical material for UV detectors.[296] Nanocomposite thin films of lead titanate/vinylidene fluoride-trifluoroethylene have been used in the design of pyroelectric detectors.[297]

Semiconductor nanoparticles have also been explored for application in single electron transistor[298] and other nanoscale quantum devices for information and telecommunication technologies.[299] This is an area of growing interest and requires much further investigation before their implementation in electronic device technology.

In principle, nanomaterials can be used for any bulk material-based devices. The key question is whether the nanomaterials offer any advantages or improved performance or functionality over bulk materials. The answer varies from application to application. In many cases, nanomaterials can offer improved properties and performances over bulk materials, while in some other cases bulk materials are

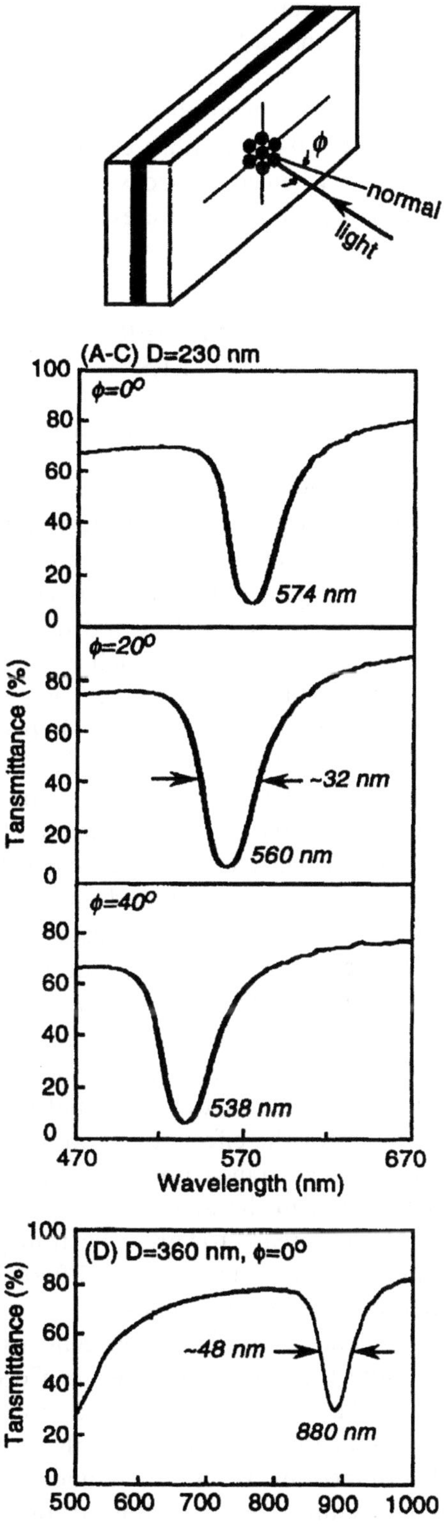

FIGURE 8.9. (A–C) UV–visible transmission spectra of a crystal of polystyrene beads ($D = 230$ nm) assembled in a 12 μm thick cell with the normal to the surface of the crystal tilted with different angles (ϕ) relative to the incident light beam. This assembly consists of ~50 layers. (D) UV–visible transmission spectrum of a crystal of polystyrene beads ($D = 360$ nm) assembled in a 12 μm thick cell with $\phi = 0°$. This assembly consists of ~30 layers. Reproduced with permission from Fig. 10 of Ref. 283.

more advantageous. It is thus important to recognize in which situation nanomaterials, instead of bulk materials, should be used and vice versa. For example, the large surface area afforded by nanomaterials can itself be useful or harmful, depending on the application of interest. One of the major advantages offered by nanomaterials over bulk crystalline materials is processibility over large areas, especially when combined with a matrix material such as polymers. Such composite materials have many interesting new properties and potential applications.

8.5. CHARGE CARRIER DYNAMICS IN SEMICONDUCTOR NANOPARTICLES

8.5.1. Spectral Line Width and Electronic Dephasing

The homogeneous spectral line width in electronic absorption spectrum is usually determined by the fastest event in dynamics, which is electronic depahsing. Pure dephasing, commonly refereed as the T_2 process, is considered as a dynamic process in which the energy is conserved while momentum is changed.[221] The homogeneous line widths and time for pure dephasing can be determined using experimental techniques such as hole burning[109,300,301] or photon echo.[223,302,303] The hole burning technique has been successfully applied to measure homogeneous line widths and line shapes of semiconductor nanoparticles, including CdSe, CdS, CuCl, CuBr,[300,301,304,305] and CdSe/ZnS core/shell.[109] The hole burning studies, mostly carried out at low temperature, for example, 2 K, also found that quantum confinement of carriers and the resulting strong Coulomb interaction between confined carriers and trapped carriers are essential for the energy change as reflected in the observed persistent spectral hole burning phenomenon.[300] Exciton localization and photoionization of the nanocrystals have been suggested to play an important role in the observed hole burning and hole filling processes. Homogeneous line width as narrow as 32 meV was obtained at low temperature (10 K) for CdSe/ZnS core/shell structures.[109] At low temperature, the temperature dependence of the homogeneous line width was found to deviate from the usual linear dependence, which was considered as a reflection of the effects of phonon quantization. Time-resolved hole burning has been applied to measure the dephasing time in CdSe nanocrystals and the energy dependence in the gain region was found to be rather constant for nanocrystals while to increase towards the transparency point for bulk-like samples.[304] The difference was attributed to different gain mechanisms in the strong and weak quantum confined regimes.

Time-resolved photon echo experiments have been conducted to directly measure the dephasing times in CdSe[223,302] and InP[303] nanocrystals. For CdSe, the dephasing times were found to vary from 85 fs for 2 nm diameter nanocrystals to 270 fs for 4 nm nanocystals at low temperature. The dephasing times are determined by several dynamical processes that are dependent sensitively on nanocrystal size, including trapping of the electronic excitation to surface states, which increases with increasing size, and coupling of the excitation to low-frequency vibrational modes, which peaks at intermediate size.[223] Contributions from acoustic phonons were found to dominate the homogeneous line width at room temperature.[302] Dephasing time in CuCl nanocrystals has been measured using femtosecond transient DFWM and for $R < 5$ nm a reservoir correlation time of 4.4–8.5 ps, which increases with increasing R, was found based on the stochastic model.[305] The results are explained based on an increase of exciton-acoustic phonon coupling strength and a change in the acoustic phonon DOS due to quantum confinement of acoustic phonons.

8.5.2. Charge Carrier Relaxation

Above band gap photoexcitation of a semiconductor creates an electron in the CB and a hole in the VB. This is a non-equilibrium situation. To go back to equilibrium, the photogenerated carriers have a number of pathways. The simplest is radiative electron–hole recombination, releasing a photon. Usually, the electron and hole lose some of their initially acquired energy before they recombine. As a result, the emitted photon has a lower energy, or red-shifted, than the incident photon. For single crystal semiconductors, radiative electron–hole recombination is the dominant mechanism of relaxation and the material is thus highly luminescent, with quantum yield > 50%.

If the crystal has impurities or defects, it tends to have electronic states within the band gap that can trap the carriers. Trapping is a major non-radiative pathway for semiconductors with a high density of trap states.

Ignoring surface effect for a moment, in the first order of the perturbation the rate of carrier relaxation through the emission of phonons is given by the Fermi golden rule:[13]

$$\frac{1}{\tau} = \sum_{\mathbf{q}} |<f|W|i>|^2 \delta(Ef - Ei + \hbar\omega_{\mathbf{q}})(n_B(\hbar\omega_{\mathbf{q}}, T) + 1) \tag{8.5}$$

where $|i>$ and $|f>$ stand for the initial and final states of the exciton, respectively, $\mathbf{q}$ is the wave vector of the emitted phonons, $\hbar\omega$ is the phonon energy, and n_B represents the Bose–Einstein distribution of phonons in the crystal lattice at temperature T ($n_B = 0$ when $T = 0$). The electron–phonon coupling operator W is characteristic of the semiconductor and depends on the relevant phonon frequencies and electronic band structure. Therefore, changes in phonon frequencies and distribution or electron–phonon coupling with size will affect the carrier relaxation time, according to Eq. (8.5). Theoretical investigations of III–V semiconductor QDs indicate that the electron–phonon scattering rates decrease strongly with increasing spatial quantization.[306]

One important factor that influences the relaxation is the surface. This is much more difficult to quantify at the first principle level. Qualitatively, the surface will have two effects. First, the surface phonon frequencies and distribution are expected to change relative to bulk. This will affect the overall electron–phonon interaction and thereby the relaxation time. Second, the surface introduces a high density of surface trap (electronic) states within the band gap that will act to trap the charge carriers. This usually significantly shortens the carrier lifetime and is usually the dominant pathway of relaxation. Trapped carriers can further relax radiatively or non-radiatively. The following sections will provide a number of examples of semiconductor nanoparticle systems that have been studied to illustrate the different timescales for various relaxation mechanisms.

8.5.3. Charge Carrier Trapping

One of the most important characteristics of nanoparticles is their extremely large surface relative to volume. This creates a high density of surface states due to surface defects and dangling bonds. These surface states may fall energetically within the band gap of a semiconductor and serve to trap charge carriers. Charge carrier trapping thus plays a critical role in the electronic relaxation process of photoinduced carriers. The trapped electrons and holes significantly influence the optical properties, for example, emission, and chemical reactivities of the nanoparticles. One measure of the presence of trap states is trap state emission, which is usually substantially red shifted relative to band edge emission, as a result of relaxation (trapping) of the electron and/or hole. The timescale for trapping is generally very short, on the order of a few hundreds of fs to tens of ps, depending on the nature of the nanoparticles. A higher DOS leads to faster trapping. The trapping time also depends on the energy difference between the trap states and the location of the bandedge (bottom of the CB for the electron and top of the VB of the hole). The smaller the energy difference, the faster the trapping is expected, provided that other factors are similar or the same.

For CdS NPs, an electron trapping time constant of about 100 fs has been suggested.[9,307] A longer trapping time (0.5–8 ps) was deduced for CdSe NPs based on time-resolved photon echo experiment.[223] An even longer trapping time of 30 ps has been reported for CdS NPs based on measurement of trap state emission.[308,309] A similar 30 ps electron trapping time for CdS NPs has been reported based on study of the effects of adsorption of electron acceptors such as viologen derivatives on the particle surface.[310] A hole trapping time of 1 ps has been reported for CdS based on time-resolved PL measurements.[309] The difference in trapping times reported could be either due to a difference in the samples used or different interpretations of the data obtained. It can be concluded, however, that the trapping time is on the order of a few hundred femtosecond to tens of picosecond, depending on the nature of the NPs and quality of the sample. The trapping time may also be different for shallow traps and deep traps or further trapping from shallow traps to deep traps.

8.5.4. Electron–Hole Recombination

Electron–hole recombination can occur before or after the electron and hole are trapped. In nanoparticles, since trapping is very fast due to a high density of trap states, most of the recombination takes place after one or both of the charge carriers is trapped. The recombination can be radiative or non-radiative.

Non-radiative recombination mediated by trap states is dominant for nanoparticles. A number of examples will be discussed next to illustrate the similarities and differences between different nanoparticle systems in terms of their electronic relaxation dynamics.

CdS is among the most extensively studied semiconductor nanoparticle systems. Earlier dynamics studies on the picosecond timescale identified a strong transient bleach feature near the excitonic absorption region of the spectrum.[102,173–178] It was noticed that the peak of the bleach feature shifts with time and one explanation proposed was increased screening by charge carriers for the particles.[179] It was also observed that there is a redshift of the transient absorption features in CdSe, which was explained by some as a result of formation of biexcitons.[180,181] Later, femtosecond measurements were carried out and a power dependence of the bleach recovery time was found for CdS.[311] The bleach recovery follows a double exponential rise with the fast component increasing with power faster than the slower component.

Recent work based on transient absorption measurements found a similar power dependence of the electronic relaxation dynamics featuring a double exponential decay behavior with a fast (2–3 ps) and slow (50 ps) component.[9,22,312] As shown in Fig. 8.10, the amplitude of the fast decay component increases with excitation intensity faster than that of the slow component. It grows nonlinearly, slightly subquadratic, with excitation intensity. This nonlinear fast decay was first attributed to nongeminate electron–hole recombination at high excitation intensities.[9] Subsequent studies using femtosecond transient absorption in conjunction with nanosecond time-resolved fluorescence found that

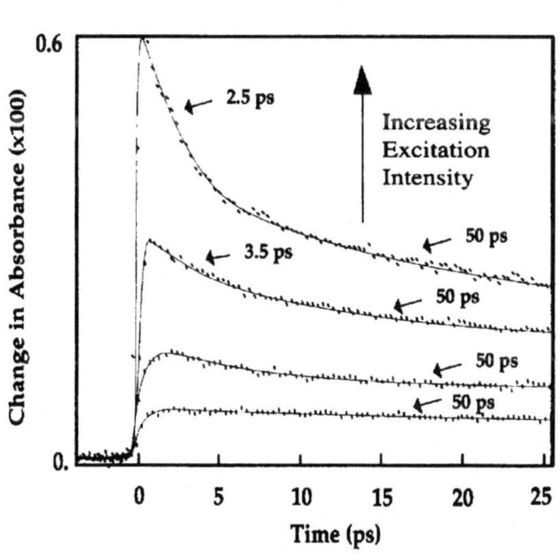

FIGURE 8.10. Excitation intensity dependence of photoinduced electronic relaxation dynamics of CdS NPs probed at 780 nm following excitation at 390 nm. The fast decay component with a time constant of 2.5–3.5 ps increases nonlinearly with excitation intensity and is absent at low intensity. Adapted with permission from Fig. 5 of Ref. 9.

the bandedge fluorescence was also power dependent.[22] These results led to the proposal that the fast decay is due to exciton–exciton annihilation upon trap state saturation, as suggested previously,[313] and the slow decay is due to trapped charge carrier recombination.[22] Therefore, the transient absorption signal observed has contributions from both bandedge electrons (excitons) and trapped electrons. At early times, the bandedge electrons contribute significantly to the signal, especially when trap states are saturated at high excitation intensities, while as time progresses the contribution from trapped electrons becomes more dominant. On long timescales (hundreds of picosecond to nanasecond), the signal is essentially all from trapped charge carriers. It was believed that this is true to many other colloidal semiconductor nanoparticles.[30,55,64,76,314] A more recent study of the emission lifetime on the nanosecond time scale of surface passivated and unpassivated CdS nanoparticles showed that the passivated sample with a lower density of surface trap states has a lower excitation threshold for observing exciton–exciton annihilation compared to the unpassivated sample.[315] This supports the model of exciton–exciton annihilation upon trap state saturation at high excitation intensities.[22]

The charge carrier dynamics in CdSe nanocrystals ranging in size from 2.7 to 7 nm have recently been determined using femtosecond fluorescence upconversion spectroscopy.[147] It has been found that both the rise time and decay of the bandedge emission show a direct correlation to the particle size and the rise time depends on excitation intensity. The long lifetime of the bandedge emission was suggested to originate from a triplet state. The deep trap emission that appears within 2 ps was attributed to relaxation of a surface selenium dangling bond electron to the VB where it recombines radiatively with the initial photogenerated hole. This is also believed to be responsible for the large amplitude, fast (2–6 ps) decay of the bandedge emission. This work seems to indicate that the hole trapping is slower compared to electron trapping.

To date, no systematic studies have been reported on the temperature dependence of charge carrier dynamics in semiconductor nanoparticles on the ultrafast timescales. At low temperature, non-radiative relaxation pathways should be suppressed and radiative fluorescence quantum yield is usually enhanced. Since the non-radiative relaxation processes are typically faster than radiative processes, suppression of non-radiative pathways at low temperature is expected to result in longer lifetime of the charge carriers or slower overall relaxation. This has been clearly demonstrated in the temperature-dependent emission lifetime observed for CdSe nanocrystals measured on the nanosecond timescale.[316] Direct study of the temperature dependence of charge carrier dynamics on the ultrafast timescale will help to gain further insight into the electronic relaxation mechanisms in semiconductor nanoparticles, for example, the rates of non-radiative trapping and trap state-mediated recombination.

Other metal sulfide nanoparticles such as PbS, Cu_xS, and Ag_2S have also been studied in terms of their charge carrier dynamics. For PbS nanoparticles, attempts have been made to study the surface and shape dependence of electronic relaxation in different shaped PbS NPs. When particle shapes are changed from mostly spherical to needle and cube shaped by changing the surface capping polymers, the electronic relaxation dynamics remain about the same for the apparently different shaped particles.[30] For all cases studied, the electronic relaxation was found to feature a double exponential decay with time constants of 1.2 and 45 ps that are independent of probe wavelength and excitation intensity. The shape independence was attributed to the dominance of the surface effects on the electronic relaxation. While the shapes are different, the different samples may have similar surface properties. Therefore, if the dynamics are dominated by the surface, change in shape may not affect the electronic relaxation dynamics substantially.

In the dynamics studies of CuS and Cu_2S,[46] a dominant fast decay component with a slower decay component observed for two types of Cu_2S nanoparticles was found to be independent of excitation intensity, probe wavelength, and capping agent. The fast decay (1.1 ps) was assigned to charge carrier trapping at shallow trap sites, while the long decay (80 ps) component was assigned primarily to deep trapping. The dynamics of crystalline CuS showed interesting power dependence. At low excitation intensities a bleach, due to ground state absorption of the probe, with a fast recovery (430 fs) followed by a long-time offset was observed. The fast recovery is due to carrier trapping, leading to an increase in transient absorption. The long-time offset is assigned to relaxation from shallow to deep traps and further relaxation of charge carriers from deep traps. At high excitation intensities, a transient absorption signal with a 1.1 ps decay and a slow rise with a lifetime > 1 ns was seen in crystalline CuS. The power dependence of the crystalline CuS could be attributed to trap state saturation.[46] Based on a kinetic model developed to simulate the data for both Cu_2S and CuS, an alternative explanation was proposed for the middle-gap state, this state is an electron-acceptor state or unoccupied and the IR absorption band corresponds to a transition from the valance band to this state.[46]

Femtosecond transient absorption studies have been recently conducted for Ag_2S nanoparticles capped with cystine and glutathione.[49] The dynamics of photoinduced electrons feature a pulse-width limited (< 150 fs) rise followed by a fast decay (750 fs) and a slower rise (4.5 ps). The signal has contributions from both transient absorption and transient bleach. An interesting excitation intensity dependence was observed for all the samples: the transient absorption contribution becomes more dominant over bleach with increasing excitation intensity. A kinetic model developed to account for the main features of the dynamics suggests that the difference in dynamics observed between the different samples is due to different absorption cross sections of deep trap states. The observed excitation intensity dependence of the dynamics is attributed to shallow trap state saturation at high intensities.[49]

The ultrafast charge carrier dynamics in AgI and core/shell structured AgI/Ag_2S and ArBr/Ag_2S NPs have recently been studied.[55] The electronic relaxation of AgI was found to follow a double exponential decay with time constants of 2.5 ps and > 0.5 ns, which are independent of excitation intensity at 390 nm. The fast decay was attributed to trapping and non-radiative electron–hole recombination dominated by a high density of trap states. The slow decay was assigned to reaction of deep trapped electrons with silver cations to form silver (Ag) atom, which is the basis for latent image formation in photography. The slow decay agrees with early nanosecond studies.[317] When the two core/shell systems, AgI/Ag_2S and AgBr/Ag_2S were compared, a new 4 ps rise component with AgBr/Ag_2S was observed which was taken as an indication of electron transfer from Ag_2S to AgBr.[55] However, subsequent experiments conducted on pure Ag_2S nanoparticles capped with cysteine and glutathione found similar features to that of the core/shell structured AgBr/Ag_2S.[49] Therefore, the 4 ps rise feature is more likely due to transient bleach, since there is noticeable ground state absorption at the probe wavelength.[49]

Electronic relaxation dynamics have been studied for several metal oxide nanoparticle systems, for example, TiO_2, ZnO, and Fe_2O_3. For TiO_2 NPs with excitation at 310 nm, the photoinduced electrons were found to decay following second-order kinetics with a second-order recombination rate constant of 1.8×10^{-10} cm^3 s^{-1}.[318] The electron trapping was suggested to occur on the timescale of 180 fs.[307] Similar studies have been done on ZnO. For ZnO NPs with 310 nm excitation, the photoinduced electron decay was found to follow second-order kinetics,[319] similar to TiO_2.[318] It was also found that the initial electron trapping and subsequent recombination dynamics were size dependent. The trapping rate was found to increase with increasing particle size, which was explained with a trap-to-trap hopping mechanism. The electron–hole recombination is faster and occurs to a greater extent in larger particles because there are two different types of trap states. A different explanation, based on exciton–exciton annihilation upon trap state saturation, has been recently proposed for similar excitation intensity dependent and size dependent relaxation observed in CdS and CdSe NPs.[22] This explanation would also seem to be consistent with the results observed for TiO_2[318] and ZnO.[319]

Cherepy *et al.*[64] have recently carried out femtosecond dynamic studies of electronic relaxation in both γ-and α-phased Fe_2O_3 nanoparticles with 390 nm excitation. The relaxation dynamics were found to be very similar between the two types of nanoparticles, despite their difference in magnetic properties and particle shape: γ being mostly spherical and paramagnetic, while α being mostly spindle-shaped and diamagnetic.[64] The relaxation featured a multiexponential decay with time constants of 0.36, 4.2, and 67 ps. The overall fast relaxation, in conjunction with very weak fluorescence, indicates extremely efficient nonradiative decay processes, possibly related to the intrinsic dense band structure or a high density of trap states. The fast relaxation of the photoinduced electrons is consistent with the typically low photocurrent efficiency of Fe_2O_3 electrodes, since the short lifetime due to fast electron–hole recombination does not favor charge transport that is necessary for photocurrent generation.[64]

A recent dynamics study on ion-implanted Si nanocrystals using femtosecond transient absorption identified two photoinduced absorption features, attributed to charge carriers in nanocrystal quantized states with higher energy and faster relaxation and Si/SiO_2 interface states with lower energy and slower relaxation.[320] Red emission observed in this sample was shown to be from surface trap states and not from quantized states. The faster relaxation of the blue emission relative to that of the red emission is similar to that observed for CdS NPs.[22]

A few studies have been carried out on the dynamic properties of layered semiconductor nanoparticles. A picosecond transient absorption study of charge carrier relaxation in MoS_2 NPs has been reported[321] and the relaxation was found to be dominated by trap states. The relaxation from shallow traps to deep traps is fast (40 ps) at room temperature and slows down to 200 ps at 20 K. Sengupta *et al.*[76] have very recently conducted a femtosecond study of charge carrier relaxation dynamics in PbI_2 nanoparticles and found that the relaxation was dominated by surface properties and independent of particle size in the size range (3–100 nm) studied. The early time dynamics were found to show some signs of oscillation with a period varying with solvent but not with size (6 ps in acetonitrile and 1.6 ps in alcohol solvents). The origin of the oscillation is not completely clear and such features are rarely observed for colloidal semiconductor NPs. The relaxation in aqueous PVA solution was found to be much faster, which is possibly due to aggregate formation of the NPs as evidenced by a new absorption band in the electronic absorption spectrum and by TEM data.[76]

Similar findings have been made for BiI_3 nanoparticles in different solvents.[77] The electronic relaxation dynamics were found to be sensitive to solvent, insensitive to particle size, and independent of excitation intensity. There also appear to be oscillations at early times with a period changing with solvent but not with particle size, similar to that found for PbI_2 nanoparticles. For BiI_3 the oscillation periods were slightly shorter and overall relaxation was somewhat faster than that in PbI_2. The decay was much faster in aqueous PVA (9 ps) and in inverse micelles (1.2 and 33 ps) with no oscillations observed. The results suggest that the surface play a major role in the electronic relaxation process of BiI_3 nanoparticles, just like in PbI_2 nanoparticles. The independence on particle size could be because the relaxation is dominated by surface characteristics that do not vary significantly with size and/or the size is much larger than the exciton Bohr radius (0.61 nm for bulk BiI_3)[83] and thereby spatial confinement is not significant in affecting the relaxation process.

To examine if the observed oscillation in PbI_2 and BiI_3 nanoparticles is unique to iodide semiconductor nanoparticles, a similar study was conducted on another layered semiconductor, Bi_2S_3.[78] In this case, while most of the dynamic features observed are similar to those observed for PbI_2 and BiI_3, for example, solvent dependence and size independence, no oscillations were observed. This is either because the overall electronic relaxation is faster in Bi_2S_3, which makes the observation of oscillation difficult, or because the oscillation is simply absent for Bi_2S_3. If the latter is correct, the results indicate that the oscillations observed for PbI_2 and BiI_3 are unique to layered, iodide semiconductor nanoparticles. Further study is clearly needed for a more complete explanation.

The dynamic properties of doped semiconductor nanoparticles are complex and intriguing. The most extensively studied example is Mn-doped ZnS. In 1994, Bhargava *et al.* reported that the emission lifetime of Mn^{2+} was significantly shorter than that in bulk and the luminescence efficiency was greater in the nanocrystalline system compared with bulk.[122,125,133] They observed a double exponential decay with time constants of 3 and 20 ns, which were five orders of magnitude faster than what has been observed in bulk (1.8 ms).[322] Since there was no significant offset or indication of slower decays on the nanosecond or longer timescales, no attempt was made to look for or show if there is any slow decay with lifetime on the millisecond timescale. To explain the fast nanosecond decay, Bhargava *et al.* had proposed that, due to quantum confinement, there is a rehybridization between the s–p CB of the ZnS host and the 3d states of the Mn^{2+}. Given the strong coupling between donor and acceptor, there is a rapid energy transfer and consequently fast radiative decay. There was a similar explanation for rapid decay kinetics observed in Mn^{2+} doped 2D quantum well structures.[323] This was considered as significant because sensors, display devices and lasers utilizing these nanoparticles could have much improved performance.[125]

However, more recent work has thrown some doubt as to the true timescale for radiative energy relaxation and suggested that the Mn^{2+} emission lifetime in nanoparticles is essentially the same as that in bulk (1.8 ms).[142] First, Bol and Meijerink[142] observed nanosecond decay rates for the blue ZnS emission but for the orange Mn^{2+} emission a normal 1.9 ms decay time along with a small amplitude nanosecond decay was seen. Furthermore, in their system the blue 420 nm emission band was observed to have a tail that extended into the orange Mn^{2+} region and could be observed with a 2 μs gate.[142] Given this they stated that the fast nanosecond decay observed by Bhargava *et al.* was due to ZnS trap state emission and not the Mn^{2+} emission. Unfortunately, the apparatus used in their experiment had limited time resolution (a 2 μs gate was used). A subsequent study by Murase *et al.*[324] also suggested that the Mn^{2+} luminescence lifetime in ZnS : Mn nanoparticles was similar to that of bulk.

Recently, Smith *et al.*[143] have successfully synthesized Mn^{2+}-doped ZnS nanoparticles using a new method based on reverse micelles and studied carefully the emission kinetics on the picasecond, nanosecond, microsecond to millisecond timescales with adequate time resolution for each timescale region. The samples have shown Mn^{2+} emission in addition to some weak trap state emission from ZnS in the 585 nm region and that the Mn^{2+} emission lifetime in these ZnS nanoparticles is ~1.8 ms, very similar to that of bulk.[143] Faster nanosecond and microsecond decays are observed for both doped and undoped particles and attributed to ZnS trap state emission. The main difference between doped and undoped samples is on the millisecond timescales, as shown in Fig. 8.11. The conclusions are consistent with those of Bol and Meijerink.[142] Similar conclusions have been reached in a more recent report by Chung *et al.*,[144] who have further suggested that surface bound and lattice bound Mn^{2+} have different emission lifetimes, 0.18 and 2 ms, respectively. These recent studies have consistently shown that the Mn^{2+} emission lifetime is essentially the same in nanoparticles as in bulk ZnS : Mn.

Charge carrier dynamics in semiconductor quantum wires have also been investigated in a few cases, including notably GaAs.[182,325–327] High luminescence efficiency was found in some cases and the luminescence was found to be completely dominated by radiative electron–hole recombination.[325] A more direct, systematic comparison of electronic relaxation among 2D, 1D, and 0D quantum confined systems should be interesting but is made difficult by practical issues such as sample quality, for example, surface characteristics, and

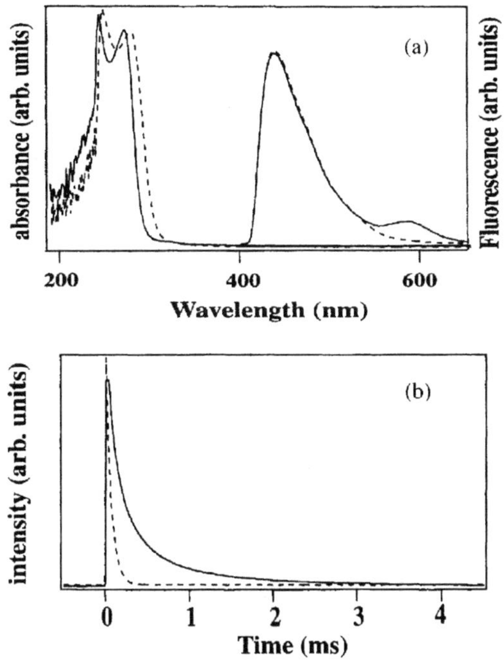

FIGURE 8.11. (a) Normalized absorption (left) and emission (right) spectra of Mn^{2+}-doped (solid lines) and undoped (dahsed lines) ZnS NPs. (b) 600 nm emission intensity vs. time collected using 285 nm excitation on the millisecond timescale. The decay for doped NPs can be fit with a 10 ms fast component and a slower 1.8 μs component, while for undoped NPs lacks the 1.8 ms component. Reproduced with permission from Fig. 8 of Ref. 23.

experimental conditions, for example, excitation wavelength and intensities, that can significantly influence the measurements. Such studies should be carried out in the future, both experimentally and theoretically.

8.5.5. *Charge Transfer*

Photoinduced charge transfer is one of the most important fundamental processes involved in liquid–semiconductor or liquid–metal interfaces. It plays a critical role in photocatalysis, photodegradation of wastes, photoelectrochemistry, and solar energy conversion.[328–333] Charge transfer competes with relaxation processes such as trapping and recombination. In many applications, charge transfer is the desired process. For example, charge transfer has been extensively studied for dye-sensitized metal oxide nanoparticles for the purpose of solar cell application. Dye sensitization of TiO_2 is the most notable example, primarily because of its potential use for solar energy conversion[58] and photocatalysis.[334] Since TiO_2 alone does not absorb visible light, dye sensitization helps to extend the absorption into the visible region.

In dye sensitization, the electron is injected from a dye molecule on the TiO_2 nanoparticle surface, as shown in the top of Fig. 8.12. There are several

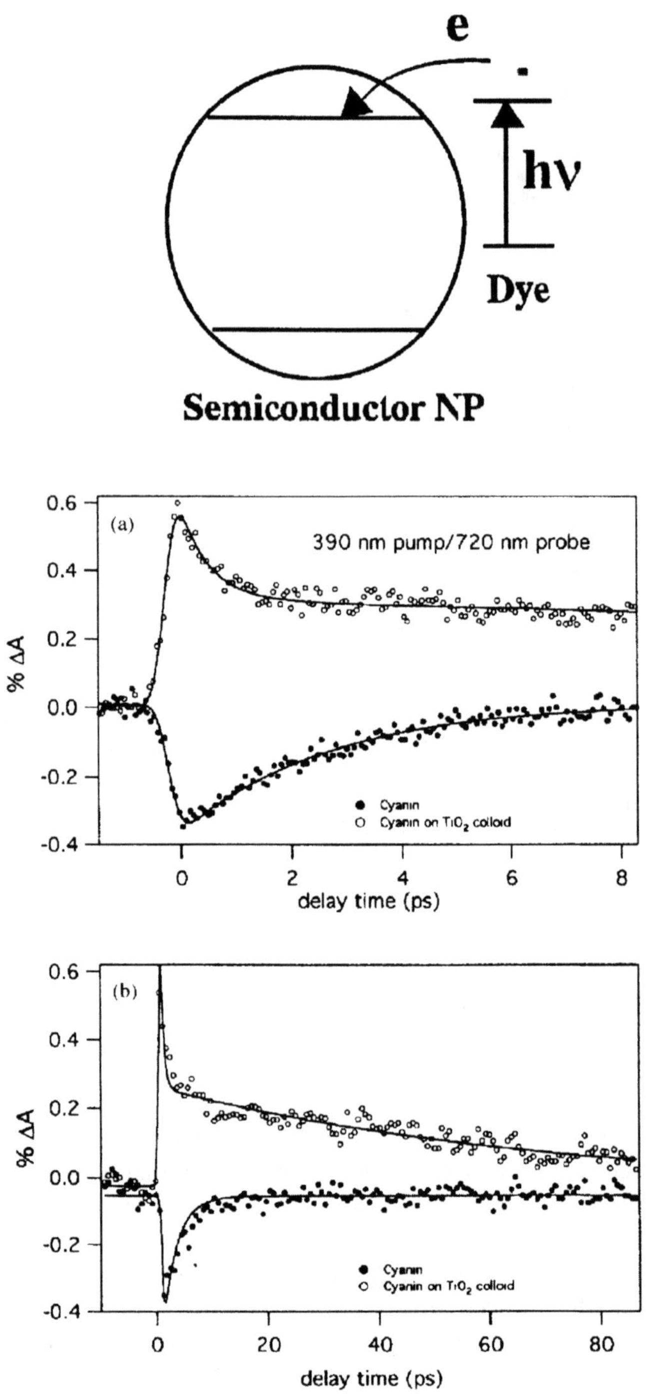

FIGURE 8.12. Top: schematic illustration of electron injection in dye sensitization of a semiconductor NP. Middle and bottom: transient decay profiles of photoinduced electron transfer from an anthocyanin dye molecule to TiO_2 NPs on a short (middle) and long (bottom) time scale probed at 780 nm following excitation at 390 nm. Reproduced from Fig. 9 of Ref. 23 with permission.

requirements for this to work effectively. First, the excited state of the dye molecule needs to lie above the bottom of the CB of the TiO_2 nanoparticle. Second, strong binding of the dye onto the TiO_2 nanoparticle surface is desired for fast and efficient injection. Third, back electron transfer to the dye cation following injection should be minimal. Fourth, the dye molecule must have strong absorption in the visible region of spectrum for solar energy conversion. A number of dye molecules have been studied and tested for solar energy conversion applications over the years.[335] To date, the dye molecule that shows most promise for applications is the N3 dye, which showed the highest reported light-to-electricity conversion efficiency of 10%.[58,266] This work has stimulated strong interest in understanding the mechanism of charge injection and recombination in such dye-sensitized nanocrystalline systems.

The rates of electron injection and subsequent recombination or back electron transfer in dye sensitization are expected to depend on the nature of the dye molecule and the NPs, especially the surface characteristics of the NPs. The interaction between the dye and the NP surface will determine the rates and yields of forward as well as reverse electron transfer.[6,328] The shape (facets) and size of the particles could also be important and tend to vary from sample to sample depending on the preparation methods used. Figure 8.13 shows schematically the potential energy as a function of nuclear corrdinate for electron transfer from an adsorbate excited state to a semiconductor.

To describe the ET process from an adsorbate molecule to the nanoparticle, one can adopt an approach similar to that of Marcus and coworkers[336–338] for ET between a discrete electron donor and acceptor level in solution. The total ET rate

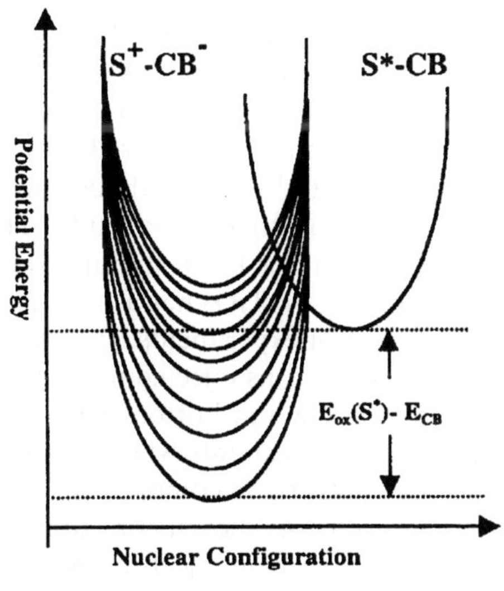

FIGURE 8.13. Schematic diagram of potential energy as a function of nuclear configuration for electron injection from an adsorbate excited state (S*) to a semiconductor CB. The reactant state (S*-CB) connects to a continuum of product states (S^+-CB^-), corresponding to the electron in different energy levels of the CB. Reproduced with permission from Fig. 3 of Ref. 339.

from adsorbate with discrete states to semiconductor with a nearly continuum of product states can be expressed as[339]

$$K_{ET} = \frac{2\pi}{\hbar}\int_{0}^{\infty} dE\rho(E)\left|\bar{H}(E)\right|^2 \frac{1}{\sqrt{4\pi\lambda k_B T}} \exp\left[-\frac{(\lambda + \Delta G_0 - E)^2}{4\pi\lambda k_B T}\right] \tag{8.6}$$

where $\Delta G_0 = E_{CB} - E_{ox}$ is the energy difference between CB edge and the redox potential of adsorbate excited state, $\rho(E)$ is the DOS at energy E from the CB edge, which can contain both the bulk states and surface states, $\bar{H}(E)$ is the average electronic coupling between the adsorbate excited state and different k states in the semiconductor at the same energy E, and λ is the total reorganization energy. Furthermore, a distribution of adsorbate/semiconductor interactions exists and gives rise to a distribution of electronic coupling matrix elements, H, and thus injection rate.[339] Therefore, the electron transfer rate depends on the detailed energetics of both the semiconductor and the adsorbate. In general, a strong electronic coupling and large energy difference ($E_{ox} - E_{CB}$) between the adsorbate excited state and the CB lead to faster electron transfer. It is challenging to independently examine the effects of different factors experimentally, since they are often interrelated. They can be evaluated separately under certain approximations or under special circumstances.

Experimental studies of the rate of electron injection in dye sensitization of semiconductor metal oxide nanoparticles have found that electron injection (forward electron transfer) is generally extremely fast, ~100 fs, in many systems. For instance, for the coumarin 343 dye on TiO_2, the electron injection rate was found to be around 200 fs.[340] For the N3 dye on TiO_2, the first direct femtosecond measurement reported a hot electron injection time of <25 fs.[341] However, there has been some debate over possible degradation of the dye sample used.[342,343] A picosecond infrared study showed an upper limit of 20 ps for the electron injection time.[344] More recent work by Ellingson *et al.* on N3 on TiO_2 reported an injection time of <50 fs based on transient infrared measurements.[339,345]

Cherepy *et al.* have recently studied an anthocyanin dye adsorbed on TiO_2 NPs and found an electron injection time of <100 fs. The assignment is made simple in this system since the dye alone has a stimulated emission signal (similar to a transient bleach), while the dye on TiO_2 has a transient absorption signal. By carefully performing several control experiments, the transient absorption signal was unambiguously assigned to electrons injected into TiO_2, which has a rise time of <100 fs. As shown in the middle and bottom panels of Fig. 8.12, a positive transient absorption signal was observed only when the dye is adsorbed onto TiO_2 NPs while the dye alone or on ZrO_2 or complexing with Al^{3+} ions showed a negative signal that is a result of bleach or stimulated emission.[64] This unequivocally establishes that the electron injection from the excited dye molecule to the TiO_2 NP takes place in <100 fs.

While forward electron transfer has generally been found to be very fast, back electron transfer was found to occur on a range of timescales, from about 10 ps to μs, depending on the nature of the dye and the nanoparticle.[346–348] For example, both forward and reverse electron transfer have been studied in the case of anthracenecarboxylic acids adsorbed on different types of TiO_2 NPs and were found to be dependent on the dye molecular structure and the method used to synthesize the TiO_2 particles.[348]

In a series of time-resolved studies using femtosecond NIR spectroscopy, Lian *et al.* systematically studied the dependence of the charge injection rate on several factors including the nature of the dye and the semiconductor as well as the coupling between them.[339,345,349–354] It was found out that fast electron injection (< 100 fs) from two Ru dye molecules into TiO_2 can compete with vibrational energy relaxation within the electronic excited state of the adsorbed dye molecule and the injection yield is thus dependent on the excited state redox potential. The injection rate from the N3 dye to different semiconductors was found to follow the trend of $TiO_2 > SnO_2 > ZnO$, indicating an interesting dependence on the nature of the semiconductor. Several factors, including the electronic coupling, DOS, and driving force, that control the interfacial electron transfer rate, were examined separately. It was found that the ET rate decreased with decreasing electronic coupling strength, while the ET rate increased with the excited state redox potential of the adsorbates (increasing driving force). The results agreed qualitatively with the theoretical prediction for a nonadiabatic interfacial electron transfer process.[339]

Similar dye-sensitization studies have been conducted on SnO_2 NPs.[355–357] For instance, femtosecond transient absorption and bleach studies have been performed on cresyl violet H-aggregate dimers adsorbed on SnO_2 colloidal particles.[357] It was found that the electron injection from the higher energy state, resulting from exciton splitting, of the dimer to SnO_2 NPs occurs in < 100 fs and back electron transfer occurs with a 12 ps time constant.

Electron transfer can also take place from the semiconductor nanoparticle to an electron acceptor near or on the surface of the NP. This is the basis of photoreduction reaction on the particle surface. This is less studied compared to electron transfer from an adsorbate to the NP. For instance, electron transfer dynamics from CdS and CdSe NPs to electron acceptors, for example viologen derivatives, adsorbed on the particle surface have been studied using transient absorption, transient bleach, and time-resolved fluorescence.[310,358] Electron transfer was found to take place on the timescale of 200–300 fs and competes effectively with trapping and electron–hole recombination. These results are important to understanding interfacial charge transfer involved in photocatalysis and photoelectrochemistry applications.

REFERENCES

1. A. L. Efros and A. L. Efros, Interband absorption of light in a semiconductor sphere, *Fizika i Tekhnika Poluprovodnikov* **16**, 1209–1214 (1982).
2. L. E. Brus, Electron–electron and electron–hole interactions in small semiconductor crystallites: The size dependence of the lowest excited electronic state, *J. Chem. Phys.* **80**, 4403–4409 (1984).
3. A. I. Ekimov, A. L. Efros, M. G. Ivanov, A. A. Onushchenko, and S. K. Shumilov, Quantum size effect in semiconductor microcrystals, *Solid State Commun.* **56**, 921–924 (1985).

4. L. Brus, Electronic wave functions in semiconductor clusters: Experiment and theory, *J. Phys. Chem.* **90**, 2555–2560 (1986).
5. M. G. Bawendi, W. L. Wilson, L. Rothberg, P. J. Carroll, T. M. Jedju, M. L. Steigerwald, and L. E. Brus, Electronic structure and photoexcited-carrier dynamics in nanometer-size CdSe clusters, *Phys. Rev. Lett.* **65**, 1623–1626 (1990).
6. M. Gratzel, *Heterogeneous Photochemical Electron Transfer* (CRC Press, Boca Raton, 1989).
7. V. L. Colvin, A. N. Goldstein, and A. P. Alivisatos, Semiconductor nanocrystals covalently bound to metal surfaces with self-assembled monolayers, *J. Am. Chem. Soc.* **114**, 5221–5230 (1992).
8. D. Duonghong, J. J. Ramsden, and M. Gratzel, Dynamics of interfacial electron-transfer processes in colloidal semiconductor systems, *J. Am. Chem. Soc.* **104**, 2977–2985 (1982).
9. J. Z. Zhang, R. H. O'Neil, and T. W. Roberti, Femtosecond studies of photoinduced electron dynamics at the liquid–solid interface of aqueous CdS colloids, *J. Phys. Chem.* **98**, 3859–3864 (1994).
10. A. P. Alivisatos, Perspectives on the physical chemistry of semiconductor nanocrystals, *J. Phys. Chem.* **100**, 13226–13239 (1996).
11. W. A. de Heer, The physics of simple metal clusters – experimental aspects and simple models, *Rev. Modern Phys.* **65**, 611–676 (1993).
12. S. W. Chen, R. S. Ingram, M. J. Hostetler, J. J. Pietron, R. W. Murray, T. G. Schaaff, J. T. Khoury, M. M. Alvarez, and R. L. Whetten, Gold nanoelectrodes of varied size: Transition to molecule-like charging, *Science* **280**, 2098–2101 (1998).
13. L. Jacak, A. Wójs, and P. Hawrylak, *Quantum Dots* (Springer, Berlin 1998).
14. J. I. Pankove, *Optical Properties in Semiconductors* (Dover Publications Inc., New York, 1971).
15. S. T. Li, S. J. Silvers, and M. S. ElShall, Surface oxidation and luminescence properties of weblike agglomeration of silicon nanocrystals produced by a laser vaporization-controlled condensation technique, *J. Phys. Chem. B* **101**, 1794–1802 (1997).
16. K. A. Littau, P. J. Szajowski, A. J. Muller, A. R. Kortan, and L. E. Brus, A luminescent silicon nanocrystal colloid via a high-temperature aerosol reaction, *J. Phys. Chem.* **97**, 1224–1230 (1993).
17. W. Q. Cao and A. J. Hunt, Photoluminescence of chemically vapor deposited si on silica aerogels, *Appl. Phys. Lett.* **64**, 2376–2378 (1994).
18. W. X. Wang, S. H. Liu, Y. Zhang, Y. B. Mei, and K. X. Chen, Influence of preparation parameters on the particle size of nanosized silicon, *Physica B* **225**, 137–141 (1996).
19. A. Fojtik and A. Henglein, Luminescent colloidal silicon particles, *Chem. Phys. Lett.* **221**, 363–367 (1994).
20. L. Brus, Luminescence of silicon materials—chains, sheets, nanocrystals, nanowires, microcrystals, and porous silicon, *J. Phys. Chem.* **98**, 3575–3581 (1994).
21. L. B. Zhang, J. L. Coffer, W. Xu, and T. W. Zerda, Luminescent Si nanoparticles in sol–gel matrices stabilized by amino acids, *Chem. Mater.* **9**, 2249–2251 (1997).
22. T. W. Roberti, N. J. Cherepy, and J. Z. Zhang, Nature of the power-dependent ultrafast relaxation process of photoexcited charge carriers in II–VI cemiconductor quantum dots: Effects of particle size, surface, and electronic structure, *J. Chem. Phys.* **108**, 2143–2151 (1998).
23. J. Z. Zhang, Interfacial charge carrier dynamics of colloidal semiconductor nanoparticles, *J. Phys. Chem. B* **104**, 7239–7253 (2000).
24. A. I. Ekimov, F. Hache, M. C. Schanneklein, D. Ricard, C. Flytzanis, I. A. Kudryavtsev, T. V. Yazeva, A. V. Rodina, and A. L. Efros, Absorption and intensity-dependent photoluminescence measurements on CdSe quantum dots – assignment of the 1st electronic transitions, *J. Opt. Soc. Am. B: Opt. Phys.* **10**, 100–107 (1993).
25. D. J. Norris, A. Sacra, C. B. Murray, and M. G. Bawendi, Measurement of the size dependent hole spectrum in CdSe quantum dots, *Phys. Rev. Lett.* **72**, 2612–2615 (1994).
26. D. J. Norris and M. G. Bawendi, Measurement and assignment of the size-dependent optical spectrum in CdSe quantum dots, *Phys. Rev. B: Condens. Matter* **53**, 16338–16346 (1996).
27. L. E. Brus, A. L. Efros, and T. Itoh, Special issue on spectroscopy of isolated and assembled semiconductor nanocrystals – introduction, *J. Lumin.* **70**, R7–R8 (1996).
28. U. Banin, C. J. Lee, A. A. Guzelian, A. V. Kadavanich, A. P. Alivisatos, W. Jaskolski, G. W. Bryant, A. L. Efros, and M. Rosen, Size-dependent electronic level structure of InAs

nanocrystal quantum dots: Test of multiband effective mass theory. *J. Chem. Phys.* **109**, 2306–2309 (1998).
29. M. T. Nenadovic, M. I. Comor, V. Vasic, and O. I. Micic, Transient bleaching of small PbS colloids – influence of surface properties, *J. Phys. Chem.* **94**, 6390–6396 (1990).
30. A. A. Patel, F. X. Wu, J. Z. Zhang, C. L. Torres-Martinez, R. K. Mehra, Y. Yang, and S. H. Risbud, Synthesis, optical spectroscopy and ultrafast electron dynamics of PbS nanoparticles with different surface capping, *J. Phys. Chem. B* **104**, 11598–11605 (2000).
31. M. Y. Gao, Y. Yang, B. Yang, J. C. Shen, and X. C. Ai, Effect of the surface chemical modification on the optical properties of polymer-stabilized Pbs nanoparticles, *J. Chem. Soci. Faraday Trans.* **91**, 4121–4125 (1995).
32. T. Schneider, M. Haase, A. Kornowski, S. Naused, H. Weller, S. Forster, and M. Antonietti, synthesis and characterization of PbS nanoparticles in block copolymer micelles, *Ber. Bunsen-Ges. Phys. Chem. Chem. Phys.* **101**, 1654–1656 (1997).
33. X. C. Ai, L. Guo, Y. H. Zou, Q. S. Li, and H. S. Zhu, The effect of surface modification on femtosecond optical Kerr effect of PbS nanoparticles, *Mater. Lett.* **38**, 131–135 (1999).
34. J. L. Machol, F. W. Wise, R. C. Patel, and D. B. Tanner, Vibronic quantum beats in Pbs microcrystallites, *Phys. Rev. B: Condens. Matter* **48**, 2819–2822 (1993).
35. E. J. Silvester, F. Grieser, B. A. Sexton, and T. W. Healy, Spectroscopic studies on copper sulfide sols, *Langmuir* **7**, 2917–2922 (1991).
36. K. M. Drummond, F. Grieser, T. W. Healy, E. J. Silvester, and M. Giersig, Steady-state radiolysis study of the reductive dissolution of ultrasmall colloidal CuS, *Langmuir* **15**, 6637–6642 (1999).
37. H. J. Gotsis, A. C. Barnes, and P. Strange, Experimental and theoretical investigation of the crystal structure of Cus, *J. Phys: Condens. Matter* **4**, 10461–10468 (1992).
38. H. Grijalva, M. Inoue, S. Boggavarapu, and P. Calvert, Amorphous and crystalline copper sulfides, CuS, *J. Mater. Chem.* **6**, 1157–1160 (1996).
39. H. Nozaki, K. Shibata, and N. Ohhashi, Metallic hole conduction in CuS, *J. Solid State Chem.* **91**, 306–311 (1991).
40. S. Saito, H. Kishi, K. Nie, H. Nakamaru, F. Wagatsuma, and T. Shinohara, Cu-63 NMR studies of copper sulfide, *Phys. Rev. B: Condens. Matter* **55**, 14527–14535 (1997).
41. C. Sugiura, H. Yamasaki, and T. Shoji, X-ray spectra and electronic structures of CuS and Cu_2S, *J. Phys. Soc. J.* **63**, 1172–1178 (1994).
42. K. V. Yumashev, P. V. Prokoshin, A. M. Malyarevich, V. P. Mikhailov, M. V. Artemyev, and V. S. Gurin, Optical transient bleaching and induced absorption of surface-modified copper sulfide nanocrystals, *Appl. Phys. B: Lasers and Optics* **64**, 73–78 (1997).
43. V. Klimov, P. H. Bolivar, H. Kurz, V. Karavanskii, V. Krasovskii, and Y. Korkishko, Linear and nonlinear transmission of Cu_xS quantum dots, *Appl. Phys. Lett.* **67**, 653–655 (1995).
44. I. Grozdanov and M. Najdoski, Optical and electrical properties of copper sulfide films of variable composition, *J. Solid State Chem.* **114**, 469–475 (1995).
45. M. V. Artemyev, V. S. Gurin, K. V. Yumashev, P. V. Prokoshin, and A. M. Maljarevich, Picosecond absorption spectroscopy of surface modified copper sulfide nanocrystals in polymeric film, *J. Appl. Phys.* **80**, 7028–7035 (1996).
46. M. C. Brelle, C. L. Torres-Martinez, J. C. McNulty, R. K. Mehra, and J. Z. Zhang, Synthisis and characterization of Cu_xS nanopartices: Nature of infrared band and charge carrier dynamics, *Pure Appl. Chem.* **72**, 101 (2000).
47. S. Kitova, J. Eneva, A. Panov, and H. Haefke, Infrared photography based on vapor-deposited silver sulfide thin films, *J. Imaging Sci. Technol.* **38**, 484–488 (1994).
48. L. Motte, F. Billoudet, and M. P. Pileni, Synthesis in situ of nanosize silver sulphide semiconductor particles in reverse micelles, *J. Mater. Sci.* **31**, 38–42 (1996).
49. M. C. Brelle, J. Z. Zhang, L. Nguyen, and R. K. Mehra, Synthesis and ultrafast study of cysteine- and glutathione-capped Ag_2S semiconductor colloidal nanoparticles, *J. Phys. Chem. A* **103**, 10194–10201 (1999).
50. T. Morris, H. Copeland, and G. Szulczewski, Synthesis and characterization of gold sulfide nanoparticles, *Langmuir* **18**, 535–539 (2002).
51. B. H. Carroll, G. C. Higgins, and T. H. James, *Introduction to Photographic Theory: The Silver Halide Process* (John Wiley, New York, 1980).

52. J. F. Hamilton, The silver halide photographic process, *Adv.Phys.* **37**, 359–441 (1988).
53. H. Saijo, M. Iwasaki, T. Tanaka, and T. Matsubara, Electron microscopic study of the growth of sub-microcrystals in nascent silver iodide and silver bromide hydrosols, *Photo. Sci. Eng.* **26**, 92–97 (1982).
54. T. Tanaka, H. Saijo, and T. Matsubara, Optical absorption studies of the growth of microcrystals in nascent suspensions. III. Absorption spectra of nascent silver iodide hydrosols, *J. Photo. Sci.* **27**, 60–65 (1979).
55. M. C. Brelle and J. Z. Zhang, Femtosecond sudy of photo-induced electron dynamics in AgI and core/shell structured AgI/Ag_2S and AgBr/Ag_2S colloidal nanoparticles, *J. Chem. Phys.* **108**, 3119–3126 (1998).
56. J. I. Pankove and T. D. Moustakas, *Gallium Nitride (GaN)* (Academic Press, San Diego, CA, 1998).
57. O. I. Micic, S. P. Ahrenkiel, D. Bertram, and A. J. Nozik, Synthesis, structure, and optical properties of colloidal GaN quantum dots, *Appl. Phys. Lett.* **75**, 478–480 (1999).
58. B. Oregan and M. Gratzel, A low-cost, high-efficiency solar cell based on dye-sensitized colloidal TiO_2 films, *Nature* **353**, 737–740 (1991).
59. Y. S. Kang, S. Risbud, J. F. Rabolt, and P. Stroeve, Synthesis and characterization of nanometer-size Fe_3O_4 and gamma-Fe_2O_3 particles, *Chem. Mater.* **8**, 2209+ (1996).
60. B. C. Faust, M. R. Hoffmann, and D. W. Bahnemann, Photocatalytic oxidation of sulfur dioxide in aqueous suspensions of alfa-Fe_2O_3, *J. Phys. Chem.* **93**, 6371–6381 (1989).
61. J. K. Leland and A. J. Bard, *J. Phys. Chem.* **91**, 5076–5083 (1987).
62. R. F. Ziolo, E. P. Giannelis, B. A. Weinstein, M. P. Ohoro, B. N. Ganguly, V. Mehrotra, M. W. Russell, and D. R. Huffman, Matrix-mediated synthesis of nanocrystalline gamma-Fe_2O_3 – a new optically transparent magnetic material, *Science* **257**, 219–223 (1992).
63. R. M. Cornell and U. Schwertmann, *The Iron Oxides* (VCH, New York, 1996).
64. N. J. Cherepy, D. B. Liston, J. A. Lovejoy, H. M. Deng, and J. Z. Zhang, Ultrafast studies of photoexcited electron dynamics in gamma- and alpha-Fe_2O_3 semiconductor nanoparticles, *J. Phys. Chem. B* **102**, 770–776 (1998).
65. L. T. Canham, Silicon quantum wire array fabrication by electrochemical and chemical dissolution of wafers, *Appl. Phys. Lett.* **57**, 1046–1048 (1990).
66. J. R. Heath, A liquid-solution-phase synthesis of crystalline silicon, *Science* **258**, 1131–1133 (1992).
67. J. L. Heinrich, C. L. Curtis, G. M. Credo, K. L. Kavanagh, and M. J. Sailor, Luminescent colloidal silicon suspensions from porous silicon, *Science* **255**, 66–68 (1992).
68. R. A. Bley, S. M. Kauzlarich, J. E. Davis, and H. W. H. Lee, Characterization of silicon nanoparticles prepared from porous silicon, *Chem. Mater.* **8**, 1881–1888 (1996).
69. L. B. Zhang, J. L. Coffer, and T. W. Zerda, Properties of luminescent Si nanoparticles in sol–gel matrices, *J. Sol-Gel Sci. Technol.* **11**, 267–272 (1998).
70. H. Takagi, H. Ogawa, Y. Yamazaki, A. Ishizaki, and T. Nakagiri, Quantum size effects on photoluminescence in ultrafine si particles, *Appl. Phys. Lett.* **56**, 2379–2380 (1990).
71. J. D. Holmes, K. P. Johnston, R. C. Doty, and B. A. Korgel, Control of thickness and orientation of solution-grown silicon nanowires, *Science* **287**, 1471–1473 (2000).
72. J. R. Heath, J. J. Shiang, and A. P. Alivisatos, Germanium quantum dots – optical properties and synthesis, *J. Chem. Phys.* **101**, 1607–1615 (1994).
73. M. W. Peterson and A. J. Nozik, in: *Photoelectrochemistry and Photovoltaics of Layered Semiconductors*, edited by A. Aruchamy (Kluwer Academic Publishers, Dordrecht, 1992), pp. 297–347.
74. C. J. Sandroff, D. M. Hwang, and W. M. Chung, Carrier confinement and spatial crystallite dimensions in layered semiconductor colloids, *Phys. Rev. B: Condens. Matter* **33**, 5953–5955 (1986).
75. O. I. Micic, L. Zongguan, G. Mills, J. C. Sullivan, and D. Meisel, On the formation of small particles of PbI_2, HgI_2, and BiI_3 Layered semiconductors, *J. Phys. Chem.* **91**, 6221–6229 (1987).
76. A. Sengupta, B. Jiang, K. C. Mandal, and J. Z. Zhang, Ultrafast electronic relaxation dynamics in PbI_2 semiconductor colloidal nanoparticles: A femtosecond transient absorption study, *J. Phys. Chem. B* **103**, 3128–3137 (1999).

77. A. Sengupta, K. C. Mandal, and J. Z. Zhang, Ultrafast electronic relaxation dynamics in layered iodide semiconductors: A comparative study of colloidal BiI_3 and PbI_2 nanoparticles, *J. Phys. Chem. B* **104**, 9396–9403 (2000).
78. A. Sengupta and J. Z. Zhang, in *First International Chinese Workshop on Nanoscience and Nanotechnology* (Tsinghua University Press, Beijing), Vol. 1, in press.
79. T. K. Chaudhuri, A. B. Patra, P. K. Basu, R. S. Saraswat, and H. N. Acharya, Preparation of bismuth iodide thin films by a chemical method, *Mater. Lett.* **8**, 361–363 (1989).
80. T. Karasawa, K. Miyata, T. Komatsu, and Y. Kaifu, Resonant Raman scattering and hot luminescence at the indirect exciton in BiI/sub 3, *J. Phys. Soc. Jpn.* **52**, 2592–2602 (1983).
81. C. J. Sandroff, S. P. Kelty, and D. M. Hwang, Clusters in solution: Growth and optical properties of layered semiconductors with hexagonal and honeycombed structures, *J. Chem. Phys.* **85**, 5337–5340 (1986).
82. T. Kamatsu, T. Karasawa, I. Akai, and T. Iida, Optical properties of nanostructured in layered melt tri-iodide crystals, *J. Lumin.* **70**, 448–467 (1996).
83. Z. K. Tang, Y. Nozue, and T. Goto, Quantum size effect on the excited state of HgI_2, PbI_2 and BiI_3 clusters and molecules in zeolite LTA, *J. Phys. Soc. Jpn.* **61**, 2943–2950 (1992).
84. J. H. Fendler and F. C. Meldrum, The colloid chemical approach to nanostructured materials, *Adv. Mater.* **7**, 607–632 (1995).
85. C. P. Collier, T. Vossmeyer, and J. R. Heath, Nanocrystal superlattices, *Ann. Rev. Phys. Chem.* **49**, 371–404 (1998).
86. M. V. Artemyev, U. Woggon, H. Jaschinski, L. I. Gurinovich, and S. V. Gaponenko, Spectroscopic study of electronic states in an ensemble of close-packed CdSe nanocrystals, *J. Phys. Chem. B* **104**, 11617–11621 (2000).
87. H. C. Hamaker, The London–van der Waals attraction between spherical particles, *Physica* **4**, 1058–1072 (1937).
88. W. van Megen and S. M. Underwood, Glass transition in colloidal hard spheres – mode-coupling theory analysis, *Phys. Rev. Lett.* **70**, 2766–2769 (1993).
89. C. B. Murray, D. J. Norris, and M. G. Bawendi, Synthesis and characterization of nearly monodisperse CdE ($E = S$, Se, Te) semiconductor nanocrystallites, *J. Am. Chem. Soc.* **115**, 8706–8715 (1993).
90. C. B. Murray, C. R. Kagan, and M. G. Bawendi, Self-organization of CdSe nanocrystallites into three-dimensional quantum dot superlattices, *Science* **270**, 1335–1338 (1995).
91. C. R. Kagan, C. B. Murray, M. Nirmal, and M. G. Bawendi, Electronic energy transfer in CdSe quantum dot solids, *Phys. Rev. Lett.* **76**, 1517–1520 (1996).
92. C. R. Kagan, C. B. Murray, and M. G. Bawendi, Long-range resonance transfer of electronic excitations in close-packed CdSe quantum-dot solids. *Phys. Rev. B: Condens Matter* **54**, 8633–8643 (1996).
93. R. Ugajin, Mott metal–insulator transition driven by an external electric field in coupled quantum dot arrays and its application to field effect devices, *J. Appl. Phys.* **76**, 2833–2836 (1994).
94. S. Lloyd, A potentially realizable quantum computer, *Science* **261**, 1569–1571 (1993).
95. V. L. Colvin, M. C. Schlamp, and A. P. Alivisatos, Light-emitting diodes made from cadmium selenide nanocrystals and a semiconducting polymer, *Nature* **370**, 354–357 (1994).
96. B. O. Dabbousi, M. G. Bawendi, O. Onitsuka, and M. F. Rubner, Electroluminescence from CdSe quantum-dot polymer composites, *Appl. Phys. Lett.* **66**, 1316–1318 (1995).
97. N. C. Greenham, X. G. Peng, and A. P. Alivisatos, Charge separation and transport in conjugated polymer cadmium selenide nanocrystal composites studied by photoluminescence quenching and photoconductivity, *Synth. Met.* **84**, 545–546 (1997).
98. B. Alen, J. Martinez-Pastor, A. Garcia-Cristobal, L. Gonzalez, and J. M. Garcia, Optical transitions and excitonic recombination in InAs/InP self-assembled quantum wires, *Appl. Phys. Lett.* **78**, 4025–4027 (2001).
99. L. R. Becerra, C. B. Murray, R. G. Griffin, and M. G. Bawendi, Investigation of the surface morphology of capped CdSe nanocrystallites by P-31 nuclear magnetic resonance, *J. Chem. Phys.* **100**, 3297–3300 (1994).

100. J. E. B. Katari, V. L. Colvin, and A. P. Alivisatos, X-ray photoelectron spectroscopy of CdSe nanocrystals with applications to studies of the nanocrystal surface, *J. Phys. Chem.* **98**, 4109–4117 (1994).
101. L. Spanhel, M. Haase, H. Weller, and A. Henglein, Photochemistry of colloidal semiconductors. 20. Surface modification and stability of strong luminescing CdS particles, *J. Am. Chem. Soc.* **109**, 5649–5655 (1987).
102. P. V. Kamat and N. M. Dimitrijevic, Photochemistry in semiconductor particulate systems. 13. Surface modification of CdS semiconductor colloids with diethyldithiocarbamate, *J. Phys. Chem.* **93**, 4259–4263 (1989).
103. P. V. Kamat, M. D. Vanwijngaarden, and S. Hotchandani, Surface modification of CdS colloids with mercaptoethylamine, *Isr. J. Chem.* **33**, 47–51 (1993).
104. C. Luangdilok and D. Meisel, Size control and surface modification of colloidal semiconductor particles, *Isr. J. Chem.* **33**, 53–58 (1993).
105. M. Y. Gao, S. Kirstein, H. Mohwald, A. L. Rogach, A. Kornowski, A. Eychmuller, and H. Weller, Strongly photoluminescent CdTe nanocrystals by proper surface modification, *J. Phys. Chem. B* **102**, 8360–8363 (1998).
106. M. A. Hines and P. Guyot-Sionnest, Synthesis and characterization of strongly luminescing ZnS-capped CdSe nanocrystals, *J. Phys. Chem.* **100**, 468–471 (1996).
107. M. C. Schlamp, X. G. Peng, and A. P. Alivisatos, Improved efficiencies in light emitting diodes made with CdSe(CdS) core shell type nanocrystals and a semiconducting polymer, *J. Appl. Phys.* **82**, 5837–5842 (1997).
108. X. G. Peng, M. C. Schlamp, A. V. Kadavanich, and A. P. Alivisatos, Epitaxial growth of highly luminescent CdSe CdS core shell nanocrystals with photostability and electronic accessibility, *J. Am. Chem. Soc.* **119**, 7019–7029 (1997).
109. P. Palinginis and W. Hailin, High-resolution spectral hole burning in CdSe/ZnS core/shell nanocrystals, *Appl. Phys. Lett.* **78**, 1541–1543 (2001).
110. M. A. Hines and P. Guyot-Sionnest, Bright UV-blue luminescent colloidal ZnSe nanocrystals, *J. Phys. Chem. B* **102**, 3655–3657 (1998).
111. S. A. Blanton, M. A. Hines, M. E. Schmidt, and P. Guyotsionnest, Two-photon spectroscopy and microscopy of II VI semiconductor nanocrystals, *J. Lumin.* **70**, 253–268 (1996).
112. E. Poles, D. C. Selmarten, O. I. Micic, and A. J. Nozik, Anti-Stokes photoluminescence in colloidal semiconductor quantum dots, *Appl. Phys. Lett.* **75**, 971–973 (1999).
113. I. V. Ignatiev, I. E. Kozin, H. W. Ren, S. Sugou, and Y. Matsumoto, Anti-Stokes photoluminescence of InP self-assembled quantum dots in the presence of electric current, *Phys. Rev. B: Condens. Matter* **60**, R14001–R14004 (1999).
114. P. P. Paskov, P. O. Holtz, B. Monemar, J. M. Garcia, W. V. Schoenfeld, and P. M. Petroff, Photoluminescence up-conversion in InAs GaAs self-assembled quantum dots, *Appl. Phys. Lett.* **77**, 812–814 (2000).
115. H. X. Zhang, C. H. Kam, Y. Zhou, X. Q. Han, S. Buddhudu, and Y. L. Lam, Visible up-conversion luminescence in Er^{3+}: $BaTiO_3$ nanocrystals, *Opt. Mater.* **15**, 47–50 (2000).
116. W. Chen, A. G. Joly, and J. Z. Zhang, Anti-Stokes luminescence of Mn^{2+} in ZnS:Mn^{2+} nanoparticles, *Phys. Rev. B* **64**, 141202 (2001).
117. S. Shionoya and W. M. Yen, *Phosphor Handbook* (CRC Press, New York, 1999).
118. T. A. Kennedy, E. R. Glaser, P. B. Klein, and R. N. Bhargava, Symmetry and electronic structure of the Mn impurity in ZnS nanocrystals, *Phys. Rev. B: Condens. Matter* **52**, 14356–14359 (1995).
119. G. Counio, S. Esnouf, T. Gacoin, and J. P. Boilot, CdS : Mn nanocrystals in transparent xerogel matrices: Synthesis and luminescence properties, *J. Phys. Chem.* **100**, 20021–20026 (1996).
120. T. Igarashi, T. Isobe, and M. Senna, EPR study of Mn^{2+} electronic states for the nanosized ZnS:Mn powder modified by acrylic acid, *Phys. Rev. B: Condens. Matter* **56**, 6444–6445 (1997).
121. N. Feltin, L. Levy, D. Ingert, and M. P. Pileni, Magnetic properties of 4-nm Cd1-yMnyS nanoparticles differing by their compositions, *J. Phys. Chem. B* **103**, 4–10 (1999).
122. R. N. Bhargava, D. Gallagher, and T. Welker, Doped nanocrystals of semiconductors – a new class of luminescent materials, *J. Lumin.* **60–61**, 275–280 (1994).

123. U. W. Pohl and H. E. Gumlich, Optical transitions of different Mn–ion pairs in ZnS, *Phys. Rev. B: Condens. Matter* **40**, 1194–1201 (1989).
124. P. Devisschere, K. Neyts, D. Corlatan, J. Vandenbossche, C. Barthou, P. Benalloul, and J. Benoit, Analysis of the luminescent decay of ZnS–Mn electroluminescent thin films, *J. Lumin.* **65**, 211–219 (1995).
125. R. N. Bhargava, Doped nanocrystalline materials—physics and applications, *J. Lumin.* **70**, 85–94 (1996).
126. J. Q. Yu, H. M. Liu, Y. Y. Wang, and W. Y. Jia, Hot luminescence of Mn^{2+} in ZnS nanocrystals, *J. Lumin.* **79**, 191–199 (1998).
127. K. Sooklal, B. S. Cullum, S. M. Angel, and C. J. Murphy, Photophysical properties of ZnS nanoclusters with spatially localized Mn^{2+}, *J. Phys. Chem.* **100**, 4551–4555 (1996).
128. L. D. Sun, C. H. Yan, C. H. Liu, C. S. Liao, D. Li, and J. Q. Yu, Study of the optical properties of Eu^{3+}-doped ZnS nanocrystals, *J. Alloys Compd.* **277**, 234–237 (1998).
129. D. D. Papakonstantinou, J. Huang, and P. Lianos, Photoluminescence of ZnS nanoparticles doped with europium ions in a polymer matrix, *J. Mater. Sci. Lett.* **17**, 1571–1573 (1998).
130. A. A. Khosravi, M. Kundu, L. Jatwa, S. K. Deshpande, U. A. Bhagwat, M. Sastry, and S. K. Kulkarni, Green luminescence from copper doped zinc sulphide quantum particles, *Appl. Phys. Lett.* **67**, 2702–2704 (1995).
131. J. M. Huang, Y. Yang, S. H. Xue, B. Yang, S. Y. Liu, and J. C. Shen, Photoluminescence and electroluminescence of ZnS:Cu nanocrystals in polymeric networks, *Appl. Phys. Lett.* **70**, 2335–2337 (1997).
132. C. M. Jin, J. Q. Yu, L. D. Sun, K. Dou, S. G. Hou, J. L. Zhao, Y. M. Chen, and S. H. Huang, Luminescence of Mn^{2+}-doped ZnS nanocrystallites, *J. Lumin.* **66–7**, 315–318 (1995).
133. R. N. Bhargava, D. Gallagher, X. Hong, and A. Nurmikko, Optical properties of manganese-doped nanocrystals of ZnS, *Phys. Rev. Lett.* **72**, 416–419 (1994).
134. W. G. Becker and A. J. Bard, Photoluminescence and photoinduced oxygen absorption of colloidal zinc sulfide dispersions, *J. Phys. Chem.* **87**, 4888–4893 (1983).
135. G. Counio, T. Gacoin, and J. P. Boilot, Synthesis and photoluminescence of $Cd_{1-x}Mn_xS$ ($x \leq 5\%$) nanocrystals, *J. Phys. Chem. B* **102**, 5257–5260 (1998).
136. J. Q. Yu, H. M. Liu, Y. Y. Wang, F. E. Fernandez, and W. Y. Jia, Optical properties of ZnS:Mn^{2+} nanoparticles in polymer films, *J. Lumin.* **76–77**, 252–255 (1998).
137. A. D. Dinsmore, D. S. Hsu, H. F. Gray, S. B. Qadri, Y. Tian, and B. R. Ratna, Mn-doped ZnS nanoparticles as efficient low-voltage cathodoluminescent phosphors, *Appl. Phys. Lett.* **75**, 802–804 (1999).
138. W. Chen, R. Sammynaiken, and Y. N. Huang, Luminescence enhancement of ZnS : Mn nanoclusters in zeolite, *J. Appl. Phys.* **88**, 5188–5193 (2000).
139. W. Chen, R. Sammynaiken, Y. N. Huang, J. O. Malm, R. Wallenberg, J. O. Bovin, V. Zwiller, and N. A. Kotov, Crystal field, phonon coupling and emission shift of Mn^{2+} in ZnS : Mn nanoparticles, *J. Appl. Phys.* **89**, 1120–1129 (2001).
140. M. Konishi, T. Isobe, and M. Senna, Enhancement of photoluminescence of ZnS : Mn nanocrystals by hybridizing with polymerized acrylic acid, *J. Lumin.* **93**, 1–8 (2001).
141. K. Yan, C. K. Duan, Y. Ma, S. D. Xia, and J. C. Krupa, Photoluminescence lifetime of nanocrystalline ZnS: Mn^{2+}, *Phys. Rev. B: Condens. Matter* **58**, 13585–13589 (1998).
142. A. A. Bol and A. Meijerink, Long-lived Mn^{2+} emission in nanocrystalline ZnS : Mn^{2+}, *Phys. Rev. B: Condens. Matter* **58**, R15997–R16000 (1998).
143. B. A. Smith, J. Z. Zhang, A. Joly, and J. Liu, Luminescence decay kinetics of Mn^{2+}-doped ZnS nanoclusters grown in reverse micelles, *Phys. Rev. B* **62**, 2021–2028 (2000).
144. J. H. Chung, C. S. Ah, and D.-J. Jang, Formation and distinctive times of surface- and lattice-bound Mn^{2+} impurity luminescence in ZnS nanoparticles, *J. Phys. Chem. B* **105**, 4128–4132 (2001).
145. S. R. Cordero, P. J. Carson, R. A. Estabrook, G. F. Strouse, and S. K. Buratto, Photo-activated luminescence of CdSe quantum dot monolayers, *J. Phys. Chem. B* **104**, 12137–12142 (2000).
146. B. C. Hess, I. G. Okhrimenko, R. C. Davis, B. C. Stevens, Q. A. Schulzke, K. C. Wright, C. D. Bass, C. D. Evans, and S. L. Summers, Surface transformation and photoinduced recovery in CdSe nanocrystals, *Phys. Rev. Lett.* **86**, 3132–3135 (2001).

147. D. F. Underwood, T. Kippeny, and S. J. Rosenthal, Ultrafast carrier dynamics in CdSe nanocrystals determined by femtosecond fluorescence upconversion spectroscopy, *J. Phys. Chem. B* **105**, 436–443 (2001).
148. S. Shih, K. H. Jung, J. Yan, D. L. Kwong, M. Kovar, J. M. White, T. George, and S. Kim, Photoinduced luminescence enhancement from anodically oxidized porous Si, *Appl. Phys. Lett.* **63**, 3306–3308 (1993).
149. S. R. Codero, P. J. Carson, R. A. Estabrook, G. F. Strouse, and S. K. Buratto, Photo-activated luminescence of CdSe quantum dot monolayers, *J. Phys. Chem. B* **104**, 12137–12142 (2000).
150. J. Q. Yu, H. M. Liu, Y. Y. Wang, F. E. Fernandez, W. Y. Jia, L. D. Sun, C. M. Jin, D. Li, J. Y. Liu, and S. H. Huang, Irradiation-induced luminescence enhancement effect of ZnS : Mn^{2+} nanoparticles in polymer films, *Opt. Lett.* **22**, 913–915 (1997).
151. A. A. Bol and A. Meijerink, Factors influencing the luminescence quantum efficiency of nanocrystalline ZnS : Mn^{2+}, *Phys. Status Solidi B:Basic Res.* **224**, 291–296 (2001).
152. J. J. Shiang, A. N. Goldstein, and A. P. Alivisatos, Lattice reorganization in electronically excited semiconductor clusters, *J. Chem. Phys.* **92**, 3232–3233 (1990).
153. G. Scamarcio, M. Lugara, and D. Manno, Size-dependent lattice contraction in CdS1-XSe$_x$ nanocrystals embedded in glass observed by Raman scattering, *Phys. Rev. B:Condens. Matter* **45**, 13792–13795 (1992).
154. J. J. Shiang, S. H. Risbud, and A. P. Alivisatos, Resonance Raman studies of the ground and lowest electronic excited state in CdS nanocrystals, *J. Chem. Phys.* **98**, 8432–8442 (1993).
155. M. Abdulkhadar and B. Thomas, Study of Raman spectra of nanoparticles of CdS and ZnS, *Nanostruct. Mater.* **5**, 289–298 (1995).
156. A. A. Sirenko, V. I. Belitsky, T. Ruf, M. Cardona, A. I. Ekimov, and C. TralleroGiner, Spin-flip and acoustic-phonon Raman scattering in CdS nanocrystals, *Phys. Rev. B:Condens. Matter* **58**, 2077–2087 (1998).
157. V. G. Melehin and V. D. Petrikov, Low-frequency Raman scattering from CdS nanocrystals embedded in phosphate glass matrix, *Phys. Low-Dimensional Struct.* **9–10**, 73–83 (1999).
158. A. P. Alivisatos, T. D. Harris, P. J. Carroll, M. L. Steigerwald, and L. E. Brus, Electron-vibration coupling in semiconductor clusters studied by resonance Raman spectroscopy, *J. Chem. Phys.* **90**, 3463–3468 (1989).
159. J. J. Shiang, I. M. Craig, and A. P. Alivisatos, Resonance Raman depolarization in CdSe nanocrystals, *Z. Phys. D:Atoms, Molecules and Clusters* **26**, 358–360 (1993).
160. V. Spagnolo, G. Scamarcio, M. Lugara, and G. C. Righini, Raman scattering in CdTe1-XSe$_x$ and CdS1–XSe$_x$ nanocrystals embedded in glass, *Superlattices Microstruct.* **16**, 51–54 (1994).
161. J. J. Shiang, R. H. Wolters, and J. R. Heath, Theory of size-dependent resonance Raman intensities in InP nanocrystals, *J. Chem. Phys.* **106**, 8981–8994 (1997).
162. V. A. Volodin, M. D. Efremov, V. A. Gritsenko, and S. A. Kochubei, Raman study of silicon nanocrystals formed in SiN$_x$ films by excimer laser or thermal annealing, *Appl. Phys. Lett.* **73**, 1212–1214 (1998).
163. M. D. Efremov, V. V. Bolotov, V. A. Volodin, and S. A. Kochubei, Raman scattering anisotropy in a system of (1 1 0)-oriented silicon nanocrystals formed in a-Si film, *Solid State Commun.* **108**, 645–648 (1998).
164. W. F. A. Besling, A. Goossens, and J. Schoonman, In-situ Raman spectroscopy and laser-induced fluorescence during laser chemical vapor precipitation of silicon nanoparticles, *J. Phys. IV* **9**, 545–550 (1999).
165. G. H. Li, K. Ding, Y. Chen, H. X. Han, and Z. P. Wang, Photoluminescence and Raman scattering of silicon nanocrystals prepared by silicon ion implantion into SiO_2 films, *J. Appl. Phys.* **88**, 1439–1442 (2000).
166. Y. Y. Wang, Y. H. Yang, Y. P. Guo, J. S. Yue, and R. J. Gan, Raman scattering and room-temperature visible photoluminescence from Ge nanocrystals embedded in SiO_2 thin films, *Mater. Lett.* **29**, 159–164 (1996).
167. W. K. Choi, V. Ng, S. P. Ng, H. H. Thio, Z. X. Shen, and W. S. Li, Raman characterization of germanium nanocrystals in amorphous silicon oxide films synthesized by rapid thermal annealing, *J. Appl. Phys.* **86**, 1398–1403 (1999).

168. A. V. Kolobov, Y. Maeda, and K. Tanaka, Raman spectra of Ge nanocrystals embedded into SiO_2, *J. Appl. Phys.* **88**, 3285–3289 (2000).
169. Y. W. Ho, V. Ng, W. K. Choi, S. P. Ng, T. Osipowicz, H. L. Seng, W. W. Tjui, and K. Li, Characterisation of Ge nanocrystals in co-sputtered Ge + SiO_2 system using Raman spectroscopy, RBS and TEM, *Scripta Materialia* **44**, 1291–1295 (2001).
170. R. Rinaldi, R. Cingolani, M. Ferrara, A. C. Maciel, J. Ryan, U. Marti, D. Martin, F. Moriergemoud, and F. K. Reinhart, Modulated reflectance and resonant Raman scattering of GaAs quantum wires grown on nonplanar substrates, *Appl. Phys. Lett.* **64**, 3587–3589 (1994).
171. B. Schreder, A. Materny, W. Kiefer, G. Bacher, A. Forchel, and G. Landwehr, Resonance Raman spectroscopy on strain relaxed CdZnSe/ZnSe quantum wires, *J. Raman Spectrosc.* **31**, 959–963 (2000).
172. H. Dollefeld, H. Weller, and A. Eychmuller, Particle–particle interaction in semiconductor nanocrystale assemblies, *Nano Lett.* **1**, 267–269 (2001).
173. N. M. Dimitrijevic and P. V. Kamat, Transient photobleaching of small CdSe colloids in acetonitrile. Anodic decomposition, *J. Phys. Chem.* **91**, 2096–2099 (1987).
174. M. Haase, H. Weller, and A. Henglein, Photochemistry of colloidal semiconductors. 26. Photoelectron emission from CdS particles and related chemical effects, *J. Phys. Chem.* **92**, 4706–4712 (1988).
175. E. F. Hilinski, P. A. Lucas, and W. Ying, A picosecond bleaching study of quantum-confined cadmium sulfide microcrystallites in a polymer film, *J. Chem. Phys.* **89**, 3534–3541 (1988).
176. P. V. Kamat, N. M. Dimitrijevic, and A. J. Nozik, Dynamic Burstein–Moss shift in semiconductor colloids, *J. Phys. Chem.* **93**, 2873–2875 (1989).
177. Y. Wang, A. Suna, J. McHugh, E. F. Hilinski, P. A. Lucas, and R. D. Johnson, Optical transient bleaching of quantum-confined CdS clusters — the effects of surface-trapped electron hole pairs, *J. Chem. Phys.* **92**, 6927–6939 (1990).
178. T. Vossmeyer, L. Katsikas, M. Giersig, I. G. Popovic, K. Diesner, A. Chemseddine, A. Eychmuller, and H. Weller, CdS Nanoclusters—Synthesis, characterization, size dependent oscillator strength, temperature shift of the excitonic transition energy, and reversible absorbance shift, *J. Phys. Chem.* **98**, 7665–7673 (1994).
179. A. Henglein, A. Kumar, E. Janata, and H. Weller, Photochemistry and radiation chemistry of semiconductor colloids: Reaction of the hydrated electron with CdS and Non-linear optical effects, *Chem. Phys. Lett.* **132**, 133–136 (1986).
180. K. I. Kang, A. D. Kepner, S. V. Gaponenko, S. W. Koch, Y. Z. Hu, and N. Peyghambarian, Confinement-enhanced biexciton binding energy in semiconductor quantum dots, *Phys. Rev. B: Condens. Matter* **48**, 15449–15452 (1993).
181. V. Klimov, S. Hunsche, and H. Kurz, Biexciton effects in femtosecond nonlinear transmission of semiconductor quantum dots, *Phys. Rev. B: Condens. Matter* **50**, 8110–8113 (1994).
182. V. Dneprovskii, N. Gushina, O. Pavlov, V. Poborchii, I. Salamatina, and E. Zhukov, Nonlinear optical absorption of GaAs quantum wires, *Phys. Lett. A* **204**, 59–62 (1995).
183. N. V. Gushchina, V. S. Dneprovskii, E. A. Zhukov, O. V. Pavlov, V. V. Poborchii, and I. A. Salamatina, Nonlinear optical properties of Gaas quantum wires, *JETP Lett.* **61**, 507–510 (1995).
184. V. Dneprovskii, A. Eev, N. Gushina, D. Okorokov, V. Panov, V. Karavanskii, A. Maslov, V. Sokolov, and E. Dovidenko, Strong optical nonlinearities in quantum wires and dots of porous silicon, *Phys. Status Solidi B: Basic Res.* **188**, 297–306 (1995).
185. V. Dneprovskii, N. Gushina, D. Okorokov, V. Karavanskii, and E. Dovidenko, Saturation of optical transitions in quantum dots and wires of porous silicon, *Superlattices Microstruct.* **17**, 41–45 (1995).
186. F. Z. Henari, K. Morgenstern, W. J. Blau, V. A. Karavanskii, and V. S. Dneprovskii, Third-order optical nonlinearity and all-optical switching in porous silicon, *Appl. Phys. Lett.* **67**, 323–325 (1995).
187. B. L. Yu, C. S. Zhu, and F. X. Gan, Optical nonlinearity of Bi_2O_3 nanoparticles studied by Z-scan technique, *J. Appl. Phys.* **82**, 4532–4537 (1997).
188. Y. Wang, Nonlinear optical properties of nanometer-sized semiconductor clusters, *Acc. chem. Res.* **24**, 133–139 (1991).

189. T. Dannhauser, M. O'Neil, K. Johanseon, D. Whitter, and G. McLendon, Photophysics of quantized colloidal semiconductor: Dramatic luminescence enhancement by binding of simple amines, *J. Phys. Chem.* **90**, 6074–6076 (1986).
190. Y. R. Shen, *The Principles of Nonlinear Optics* (John Wiley, New York, 1984).
191. T. Yamaki, K. Asai, K. Ishigure, K. Sano, and K. Ema, DFWM Study of thin films containing surface-modified CdS nanoparticles, *Synth. Met.* **103**, 2690–2691 (1999).
192. M. Jacobsohn and U. Banin, Size dependence of second harmonic generation in CdSe nanocrystal quantum dots, *J. Phys. Chem. B* **104**, 1–5 (2000).
193. Z. Yu, F. Dgang, W. Xin, L. Juzheng, and L. Zuhong, Optical and nonlinear optical properties of surface-modified CdS nanoparticles, *Colloids Surf. A: Phyiscochem. Eng. Aspects* **181**, 145–149 (2001).
194. J. Lenglet, A. Bourdon, J. C. Bacri, R. Perzynski, and G. Demouchy, Second-harmonic generation in magnetic colloids by orientation of the nanoparticles, *Phys. Rev. B: Condens. Matter* **53**, 14941–14956 (1996).
195. T. Takagahara, Dependence on dimensionality of excitonic optical nonlinearity in quantum confined structures, *Solid State Commun.* **78**, 279–282 (1991).
196. T. Takagahara, Quantum dot lattice and enhanced excitonic optical nonlinearity, *Surf. Sci.* **267**, 310–314 (1992).
197. Y. Kayanuma, Resonant interaction of photons with a random array of quantum dots, *J. Phys. Soc. Jpn.* **62**, 346–356 (1993).
198. B. Clerjaud, F. Gendron, and C. Porte, Chromium-induced up conversion in GaP, *Appl. Phys. Lett.* **38**, 212–214 (1981).
199. Y. Mita, Other phosphors. Infrared up-conversion phosphors, in: *Phosphor Handbook*, edited by S. A. Y. Shionoya and W. M. Yen (CRC Press, New York, 1999), p. 643.
200. L. G. Quagliano and H. Nather, Up conversion of luminescence via deep centers in high purity GaAs and GaAlAs epitaxial layers, *Appl. Phys. Lett.* **45**, 555–557 (1984).
201. E. J. Johnson, J. Kafalas, R. W. Davies, and W. A. Dyes, Deep center EL2 and anti-Stokes luminescence in semi-insulating GaAs, *Appl. Phys. Lett.* **40**, 993–995 (1982).
202. Y. H. Cho, D. S. Kim, B. D. Choe, H. Lim, J. I. Lee, and D. Kim, Dynamics of anti-Stokes photoluminescence in type-II $Al_xGa_{1-x}As$–$GaInP_2$ heterostructures: The important role of long-lived carriers near the interface, *Phys. Rev. B:Condens. Matter* **56**, R4375–R4378 (1997).
203. M. Potemski, R. Stepniewski, J. C. Maan, G. Martinez, P. Wyder, and B. Etienne, Auger recombination within Landau levels in a 2-dimensional electron gas, *Phys. Rev. Lett.* **66**, 2239–2242 (1991).
204. P. Vagos, P. Boucaud, F. H. Julien, J. M. Lourtioz, and R. Planel, Photoluminescence up-conversion induced by intersubband absorption in asymmetric coupled quantum wells, *Phys. Rev. Lett.* **70**, 1018–1021 (1993).
205. W. Seidel, A. Titkov, J. P. Andre, P. Voisin, and M. Voos, High-efficiency energy up-conversion by an Auger fountain At an Inp-Alinas type-Ii heterojunction, *Phys. Rev. Lett.* **73**, 2356–2359 (1994).
206. F. Driessen, H. M. Cheong, A. Mascarenhas, S. K. Deb, P. R. Hageman, G. J. Bauhuis, and L. J. Giling, Interface-induced conversion of infrared to visible light at semiconductor interfaces, *Phys. Rev. B: Condens. Matter* **54**, R5263–R5266 (1996).
207. R. Hellmann, A. Euteneuer, S. G. Hense, J. Feldmann, P. Thomas, E. O. Gobel, D. R. Yakovlev, A. Waag, and G. Landwehr, Low-temperature anti-stokes luminescence mediated by disorder in semiconductor quantum-well structures, *Phys. Rev. B: Condens. Matter* **51**, 18053–18056 (1995).
208. Z. P. Su, K. L. Teo, P. Y. Yu, and K. Uchida, Mechanisms of photoluminescence upconversion at the Gaas/(ordered) Gainp2 interface, *Solid State Commun.* **99**, 933–936 (1996).
209. J. Zeman, G. Martinez, P. Y. Yu, and K. Uchida, Band alignment and photoluminescence up-conversion at the GaAs/(ordered)$GaInP_2$ heterojunction, *Phys. Revi. B: Condens. Matter* **55**, 13428–13431 (1997).
210. L. Schrottke, H. T. Grahn, and K. Fujiwara, Enhanced anti-Stokes photoluminescence in a GaAs/Al0.17Ga0.83As single quantum well with growth islands, *Phys. Rev. B: Condens. Matter* **56**, 15553–15556 (1997).
211. H. M. Cheong, B. Fluegel, M. C. Hanna, and A. Mascarenhas, Photoluminescence up-conversion in GaAs/$Al_xGa_{1-x}As$ heterostructures, *Phys. Rev. B: Condens. Matter* **58**, R4254–R4257 (1998).

212. Z. Chine, B. Piriou, M. Oueslati, T. Boufaden, and B. El Jani, Anti-Stokes photoluminescence of yellow band in GaN: Evidence of two-photon excitation process, *J. Lumin.* **82**, 81–84 (1999).
213. T. Kita, T. Nishino, C. Geng, F. Scholz, and H. Schweizer, Dynamic process of anti-Stokes photoluminescence at a long-range-ordered $Ga_{0.5}In_{0.5}P$/GaAs heterointerface, *Phys. Rev. B: Condens. Matter* **59**, 15358–15362 (1999).
214. S. C. Hohng and D. S. Kim, Two-color picosecond experiments on anti-Stokes photoluminescence in GaAs/AlGaAs asymmetric double quantum wells, *Appl. Phys. Lett.* **75**, 3620–3622 (1999).
215. L. Schrottke, R. Hey, and H. T. Grahn, Electric-field-induced anti-Stokes photoluminescence in an asymmetric GaAs/(Al,Ga)As double quantum well superlattice, *Phys. Rev. B: Condens. Matter* **60**, 16635–16639 (1999).
216. W. Heimbrodt, M. Happ, and F. Henneberger, Giant anti-Stokes photoluminescence from semimagnetic heterostructures, *Phys. Rev. B: Condens. Matter* **60**, R16326–R16329 (1999).
217. T. Kita, T. Nishino, C. Geng, F. Scholz, and H. Schweizer, Time-resolved observation of anti-Stokes photoluminescence at ordered $Ga_{0.5}In_{0.5}P$ and GaAs interfaces, *J. Lumin.* **87–89**, 269–271 (2000).
218. A. Satake, Y. Masumoto, T. Miyajima, T. Asatsuma, and T. Hino, Ultraviolet anti-Stokes photoluminescence in $In_xGa_{1-x}N$/GaN quantum-well structures, *Phys. Rev. B* **61**, 12654–12657 (2000).
219. G. G. Zegrya and V. A. Kharchenko, A new mechanism for Auger recombination of non-equilibrium current carriers in semiconducting heterostructures, *Zhurnal Eksperimentalnoi I Teoreticheskoi Fiziki* **101**, 327–343 (1992).
220. A. G. Joly, W. Chen, J. Roark, and J. Z. Zhang, Temperature dependence of u-conversion luminescence and photoluminescence of Mn^{2+} in ZnS : Mn^{2+} Nanoparticles, *J. Nanosci. Nanotechnol.* **1**, 295–301 (2001).
221. A. P. Alivisatos, A. L. Harris, N. J. Levinos, M. L. Steigerwald, and L. E. Brus, Electronic states of semiconductor clusters: Homogeneous and inhomogeneous broadening of the optical spectrum, *J. Chem. Phys.* **89**, 4001–4011 (1988).
222. D. J. Norris and M. G. Bawendi, Structure in the lowest absorption feature of CdSe quantum dots, *J. Chem. Phys.* **103**, 5260–5268 (1995).
223. D. M. Mittleman, R. W. Schoenlein, J. J. Shiang, V. L. Colvin, A. P. Alivisatos, and C. V. Shank, Quantum size dependence of femtosecond electronic dephasing and vibrational dynamics in CdSe nanocrystals, *Phys. Rev. B: Condens. Matter* **49**, 14435–14447 (1994).
224. U. Woggon, S. Gaponenko, W. Langbein, A. Uhrig, and C. Klingshirn, Homogeneous linewidth of confined electron–hole-pair States in II–VI-quantum dots, *Phys. Rev. B:Condens. Matter* **47**, 3684–3689 (1993).
225. H. Giessen, B. Fluegel, G. Mohs, N. Peyghambarian, J. R. Sprague, O. I. Micic, and A. J. Nozik, Observation of the quantum confined ground state in InP quantum dots at 300 K, *Appl. Phys. Lett.* **68**, 304–306 (1996).
226. V. Jungnickel and F. Henneberger, Luminescence related processes in semiconductor nanocrystals – the strong confinement regime, *J. Lumin.* **70**, 238–252 (1996).
227. W. E. Moerner, Examining nanoenvironments in solids on the scale of a single, isolated impurity molecule, *Science* **265**, 46–53 (1994).
228. T. Basche, W. E. Moerner, M. Orrit, and U. P. Wild, *Single-Molecule Optical Detection, Imaging and Spectroscopy* (VCH, Weinheim; Cambridge, 1997), p. 250.
229. J. Tittel, W. Gohde, F. Koberling, T. Basche, A. Kornowski, H. Weller, and A. Eychmuller, Fluorescence spectroscopy on single CdS nanocrystals, *J. Phys. Chem. B* **101**, 3013–3016 (1997).
230. S. A. Blanton, A. Dehestani, P. C. Lin, and P. Guyot-Sionnest, Photoluminescence of single semiconductor nanocrystallites by two-photon excitation microscopy, *Chem. Phys. Lett.* **229**, 317–322 (1994).
231. S. A. Empedocles, D. J. Norris, and M. G. Bawendi, Photoluminescence spectroscopy of single CdSe nanocrystallite quantum dots, *Phys. Rev. Lett.* **77**, 3873–3876 (1996).
232. M. Nirmal, B. O. Dabbousi, M. G. Bawendi, J. J. Macklin, J. K. Trautman, T. D. Harris, and L. E. Brus, Fluorescence intermittency in single cadmium selenide nanocrystals, *Nature* **383**, 802–804 (1996).

233. S. A. Blanton, M. A. Hines, and P. Guyot-Sionnest, Photoluminescence wandering in single CdSe nanocrystals, *Appl. Phys. Lett.* **69**, 3905–3907 (1996).
234. S. A. Empedocles and M. G. Bawendi, Quantum-confined stark effect in single CdSe nanocrystallite quantum dots, *Science* **278**, 2114–2117 (1997).
235. S. A. Empedocles, R. Neuhauser, K. Shimizu, and M. G. Bawendi, Photoluminescence from single semiconductor nanostructures, *Adv. Mater.* **11**, 1243–1256 (1999).
236. S. A. Empedocles and M. G. Bawendi, Influence of spectral diffusion on the line shapes of single CdSe nanocrystallite quantum dots, *J. Phys. Chem. B* **103**, 1826–1830 (1999).
237. S. Empedocles and M. Bawendi, Spectroscopy of single CdSe nanocrystallites, *Acc. Chem. Res.* **32**, 389–396 (1999).
238. F. Koberling, A. Mews, and T. Basche, Single-dot spectroscopy of CdS nanocrystals and CdS/HgS heterostructures, *Phys. Rev. B: Condens. Matter* **60**, 1921–1927 (1999).
239. A. L. Efros, M. Rosen, M. Kuno, M. Nirmal, D. J. Norris, and M. Bawendi, Band-edge exciton in quantum dots of semiconductors with a degenerate valence band – dark and bright exciton states, *Phys. Rev. B:Condens. Matter* **54**, 4843–4856 (1996).
240. U. Banin, M. Bruchez, A. P. Alivisatos, T. Ha, S. Weiss, and D. S. Chemla, Evidence for a thermal contribution to emission intermittency in single CdSe/CdS core/shell nanocrystals, *J. Chem. Phys.* **110**, 1195–1201 (1999).
241. M. Ono, K. Matsuda, T. Saiki, K. Nishi, T. Mukaiyama, and M. Kuwata-Gonokami, Time-resolved emission from self-assembled single quantum dots using scanning near-field optical microscope, *Jpn. J. Appl. Phys., Part 2: Lett.* **38**, L1460-L1462 (1999).
242. K. Matsuda, T. Saiki, H. Saito, and K. Nishi, Room-temperature photoluminescence spectroscopy of self-assembled $In_{0.5}Ga_{0.5}As$ single quantum dots by using highly sensitive near-field scanning optical microscope, *Appl. Phys. Lett.* **76**, 73–75 (2000).
243. T. Matsumoto, M. Ohtsu, K. Matsuda, T. Saiki, H. Saito, and K. Nishi, Low-temperature near-field nonlinear absorption spectroscopy of InGaAs single quantum dots, *Appl. Phys. Lett.* **75**, 3246–3248 (1999).
244. R. Mirin, A. Gossard, and J. Bowers, Room temperature lasing from in Gaas quantum dots, *Electron. Lett.* **32**, 1732–1734 (1996).
245. S. Fafard, K. Hinzer, A. J. Springthorpe, Y. Feng, J. McCaffrey, S. Charbonneau, and E. M. Griswold, Temperature effects in semiconductor quantum dot lasers, *Mater. Sci. Eng. B: Solid State Mater. Adv. Technol.* **51**, 114–117 (1998).
246. K. Hinzer, C. N. Allen, J. Lapointe, D. Picard, Z. R. Wasilewski, S. Fafard, and A. J. S. Thorpe, Widely tunable self-assembled quantum dot lasers, *J. Vac. Sci. Technol. A: Vacuum Surfaces and Films* **18**, 578–581 (2000).
247. K. Hinzer, J. Lapointe, Y. Feng, A. Delage, S. Fafard, A. J. SpringThorpe, and E. M. Griswold, Short-wavelength laser diodes based on AlInAs/AlGaAs self-assembled quantum dots, *J. Appl. Phys.* **87**, 1496–1502 (2000).
248. S. Tanaka, H. Hirayama, Y. Aoyagi, Y. Narukawa, Y. Kawakami, and S. Fujita, Stimulated emission from optically pumped GaN quantum dots, *Appl. Phys. Lett.* **71**, 1299–1301 (1997).
249. M. K. Zundel, K. Eberl, N. Y. Jin-Phillipp, F. Phillipp, T. Riedl, E. Fehrenbacher, and A. Hangleiter, Self-assembled InP quantum dots for red LEDs on Si and injection lasers on GaAs, *J. Cryst. Growth* **202**, 1121–1125 (1999).
250. V. I. Klimov, A. A. Mikhailovsky, S. Xu, A. Malko, J. A. Hollingsworth, C. A. Leatherdale, H. J. Eisler, and M. G. Bawendi, Optical gain and stimulated emission in nanocrystal quantum dots, *Science* **290**, 314–317 (2000).
251. F. Hide, B. J. Schwartz, M. A. Diazgarcia, and A. J. Heeger, Laser emission from solutions and films containing semiconducting polymer and titanium dioxide nanocrystals, *Chem. Phys. Lett.* **256**, 424–430 (1996).
252. M. H. Huang, S. Mao, H. Feick, H. Q. Yan, Y. Y. Wu, H. Kind, E. Weber, R. Russo, and P. D. Yang, Room-temperature ultraviolet nanowire nanolasers, *Science* **292**, 1897–1899 (2001).
253. S. A. Carter, J. C. Scott, and P. J. Brock, Enhanced luminance in polymer composite light emitting devices, *Appl. Phys. Lett.* **71**, 1145–1147 (1997).
254. Y. Wang and N. Herron, Semiconductor nanocrystals in carrier-transporting polymers – charge generation and charge transport, *J. Lumin.* **70**, 48–59 (1996).

255. N. C. Greenham, X. G. Peng, and A. P. Alivisatos, Charge separation and transport in conjugated-polymer/semiconductor-nanocrystal composites studied by photoluminescence quenching and photoconductivity, *Phys. Rev. B: Condens. Matter* **54**, 17628–17637 (1996).
256. S. Nakamura, K. Kitamura, H. Umeya, A. Jia, M. Kobayashi, A. Yoshikawa, M. Shimotomai, and K. Takahashi, Bright electroluminescence from CdS quantum dot LED structures, *Electron. Lett.* **34**, 2435–2436 (1998).
257. C. A. Leatherdale, C. R. Kagan, N. Y. Morgan, S. A. Empedocles, M. A. Kastner, and M. G. Bawendi, Photoconductivity in CdSe quantum dot solids, *Phys. Rev. B* **62**, 2669–2680 (2000).
258. H. Mattoussi, A. W. Cumming, C. B. Murray, M. G. Bawendi, and R. Ober, Properties of CdSe nanocrystal dispersions in the dilute regime: Structure and interparticle interactions, *Phys. Rev. B:Condens. Matter* **58**, 7850–7863 (1998).
259. Z. L. Wang, M. B. Mohamed, S. Link, and M. A. El-Sayed, Crystallographic facets and shapes of gold nanorods of different aspect ratios, *Surf. Sci.* **440**, L809–L814 (1999).
260. W. U. Huynh, X. G. Peng, and A. P. Alivisatos, CdSe nanocrystal rods/poly(3-hexylthiophene) composite photovoltaic devices, *Adv. Mater.* **11**, 923–927, 886 (1999).
261. L. Manna, E. C. Scher, and A. P. Alivisatos, Synthesis of soluble and processable rod-, arrow-, teardrop-, and tetrapod-shaped CdSe nanocrystals, *J. Am. Chem. Soc.* **122**, 12700–12706 (2000).
262. Z. Adam and X. Peng, Formation of high-quality CdTe, CdSe, and CdS nanocrystals using CdO as precursor, *J. Am. Chem. Soc.* **123**, 183–184 (2001).
263. Z. W. Pan, Z. R. Dai, and Z. L. Wang, Nanobelts of semiconducting oxides, *Science* **291**, 1947–1949 (2001).
264. D. M. Chapin, C. S. Fuller, and G. L. Pearson, A new silicon p–n junction photocell for converting solar radiation into electrical power, *J. Appl. Phys.* **74**, 230 (1954).
265. A. Goetzberger, J. Knobloch, and B. Voss, *Crystalline Silicon Solar Cells* (Wiley, Chichester, 1998).
266. M. K. Nazeeruddin, A. Kay, I. Rodicio, R. Humphrybaker, E. Muller, P. Liska, N. Vlachopoulos, and M. Gratzel, Conversion of light to electricity by Cis-X2bis(2,2′-bipyridyl-4,4′-dicarboxylate) ruthenium(II) charge-transfer sensitizers ($X = Cl^-$, Br^-, I^-, Cn^-, and Scn^-) On Nanocrystalline TiO_2 Electrodes, *J. Am. Chem. Soc.* **115**, 6382–6390 (1993).
267. U. Bach, D. Lupo, P. Comte, J. E. Moser, F. Weissortel, J. Salbeck, H. Spreitzer, and M. Gratzel, Solid-state dye-sensitized mesoporous TiO_2 solar cells with high photon-to-electron conversion efficiencies, *Nature* **395**, 583–585 (1998).
268. B. Oregan, D. T. Schwartz, S. M. Zakeeruddin, and M. Gratzel, Electrodeposited nanocomposite n–p heterojunctions for solid-state dye-sensitized photovoltaics, *Adv. Mater.* **12**, 1263–1267,1232 (2000).
269. S. Spiekermann, G. Smestad, J. Kowalik, L. M. Tolbert, and M. Gratzel, Poly(4-undecyl-2,2′-bithiophene) as a hole conductor in solid state dye sensitized titanium dioxide solar cells, *Synth. Met.* **121**, 1603–1604 (2001).
270. C. Grant, A. Schwartzberg, G. P. Smestad, J. Kowalik, L. M. Tolbert, and J. Z. Zhang, Characterization of nanocrystalline and thin film TiO_2 solar cells with poly(3-undecyl-2,2′-bithiophene) as a sensitizer and hole conductor, *J. Electroanalytical. Chem.* **522**, 40–48 (2002).
271. G. Yu, G. Pakbaz, and A. J. Heeger, Semiconducting polymer diodes: Large size, low cost photodetectors with excellent visible–ultraviolet sensitivity, *Appl. Phys. Lett.* **64**, 3422–3424 (1994).
272. G. Yu and A. J. Heeger, Charge separation and photovoltaic conversion in polymer composites with internal donor acceptor heterojunctions, *J. Appl. Phys.* **78**, 4510–4515 (1995).
273. T. J. Savenije, J. M. Warman, and A. Goossens, Visible light sensitisation of titanium dioxide using a phenylene vinylene polymer, *Chem. Phys. Lett.* **287**, 148–153 (1998).
274. A. C. Arango, S. A. Carter, and P. J. Brock, Charge transfer in photovoltaics consisting of interpenetrating networks of conjugated polymer and TiO_2 nanoparticles, *Appl. Phys. Lett.* **74**, 1698–1700 (1999).
275. A. C. Arango, L. R. Johnson, V. N. Bliznyuk, Z. Schlesinger, S. A. Carter, and H. H. Horhold, Efficient titanium oxide/conjugated polymer photovoltaics for solar energy conversion, *Adv. Mater.* **12**, 1689–1692, 1642 (2000).
276. D. Gebeyehu, C. J. Brabec, F. Padinger, T. Fromherz, S. Spiekermann, N. Vlachopoulos, F. Kienberger, H. Schindler, and N. S. Sariciftci, Solid state dye-sensitized TiO_2 solar cells with poly(3-octylthiophene) as hole transport layer, *Synth. Met.* **121**, 1549–1550 (2001).

277. T. Fromherz, F. Padinger, D. Gebeyehu, C. Brabec, J. C. Hummelen, and N. S. Sariciftci, Comparison of photovoltaic devices containing various blends of polymer and fullerene derivatives, *Sol. Energy Mater. Sol. Cells* **63**, 61–68 (2000).
278. D. Godovsky, L. C. Chen, L. Pettersson, O. Inganas, M. R. Andersson, and J. C. Hummelen, The use of combinatorial materials development for polymer solar cells, *Adv. Mater. Opt. Electron.* **10**, 47–54 (2000).
279. B. Zhenan, A. Dodabalapur, and A. J. Lovinger, Soluble and processable regioregular poly(3-hexylthiophene) for thin film field-effect transistor applications with high mobility, *Appl. Phys. Lett.* **69**, 4108–4110 (1996).
280. H. Sirringhaus, N. Tessler, and R. H. Friend, Integrated optoelectronic devices based on conjugated polymers, *Science* **280**, 1741–1744 (1998).
281. N. F. Borrelli, D. W. Hall, H. J. Holland, and D. W. Smith, Quantum confinement effects of semiconducting microcrystallites in glass, *J. Appl. Phys.* **61**, 5399–5409 (1987).
282. I. Okur and P. D. Townsend, Waveguide formation by He^+ and H^+ ion implantation in filter glass containing nanoparticles, *Nucl. Instrum. Methods Phys. Res. Sect. B – Beam Interactions with Materials and Atoms* **124**, 76–80 (1997).
283. S. H. Park and Y. N. Xia, Assembly of mesoscale particles over large areas and its application in fabricating tunable optical filters, *Langmuir* **15**, 266–273 (1999).
284. Y. N. Xia, J. A. Rogers, K. E. Paul, and G. M. Whitesides, Unconventional methods for fabricating and patterning nanostructures, *Chem. Rev.* **99**, 1823–1848 (1999).
285. P. L. Flaugh, S. E. O'Donnel, and S. A. Asher, Development of a new optical wavelength rejection filter; demonstration of its utility in Raman spectroscopy, *Appl. Spectrosc.* **38**, 847–850 (1984).
286. R. J. Spry and D. J. Kosan, Theoretical analysis of the crystalline colloidal array filter, *Appl. Spectrosc.* **40**, 782–784 (1986).
287. J. H. Holtz and S. A. Asher, Polymerized colloidal crystal hydrogel films as intelligent chemical sensing materials, *Nature* **389**, 829–832 (1997).
288. J. H. Holtz, J. S. W. Holtz, C. H. Munro, and S. A. Asher, Intelligent polymerized crystalline colloidal arrays: Novel chemical sensor materials, *Anal. Chem.* **70**, 780–791 (1998).
289. J. M. Weissman, H. B. Sunkara, A. S. Tse, and S. A. Asher, Thermally switchable periodicities and diffraction from mesoscopically ordered materials, *Science* **274**, 959–960 (1996).
290. Tarhan, II and G. H. Watson, Photonic band structure of fcc colloidal crystals, *Phys. Rev. Lett.* **76**, 315–318 (1996).
291. W. L. Vos, M. Megens, C. M. Vankats, and P. Bosecke, Transmission and diffraction by photonic colloidal crystals, *J. Phys.:Condens. Matter* **8**, 9503–9507 (1996).
292. H. Miguez, F. Meseguer, C. Lopez, A. Blanco, J. S. Moya, J. Requena, A. Mifsud, and V. Fornes, Control of the photonic crystal properties of fcc-packed submicrometer SiO_2 spheres by sintering, *Adv. Mater.* **10**, 480 (1998).
293. J. Zhou, Y. Zhou, S. Buddhudu, S. L. Ng, Y. L. Lam, and C. H. Kam, Photoluminescence of ZnS: Mn embedded in three-dimensional photonic crystals of submicron polymer spheres, *Appl. Phys. Lett.* **76**, 3513–3515 (2000).
294. Y. Nakatani and M. Matsuoka, Effects of sulfate ion on gas sensitive properties of alpha-Fe/sub 2/O/sub 3/ceramics, *Jpn. J. Appl. Phys., Part 1 (Regular Papers & Short Notes)* **21**, L758–760 (1982).
295. H. Kanal, H. Mizulani, T. Tanaka, T. Funabiki, S. Yoshida, and M. Takano, X-ray absorption study on the local structures of fine particles of alfa-Fe_2O_3–SnO_2 gas sensors, *J. Mater. Chem.* **2**, 703–707 (1992).
296. R. DeMeis, Detector materials — economical nanocrystal fabrication demonstrated, *Laser Focus World* **33**, 46–47 (1997).
297. Y. Chen, H. L. W. Chan, N. M. Hui, Y. W. Wong, and C. I. Choy, Response of nanocomposite pyroelectric detectors, *Sens. Actuators A: Phys.* **69**, 156–165 (1998).
298. D. L. Feldheim and C. D. Keating, Self-assembly of single electron transistors and related devices, *Chem. Soc. Rev.* **27**, 1 (1998).
299. E. H. Lee and K. Park, Potential applications of nanoscale semiconductor quantum devices for information and telecommunications technologies, *Mater. Sc. Eng. B: Solid State Mater. Adv. Technol.* **74**, 1–6 (2000).

300. Y. Masumoto, Persistent hole burning in semiconductor nanocrystals, *J. Lumin.* **70**, 386–399 (1996).
301. J. Valenta, J. Dian, J. Hala, P. Gilliot, and R. Levy, Persistent spectral hole-burning and hole-filling in CuBr semiconductor nanocrystals, *J. Chem. Phys.* **111**, 9398–9405 (1999).
302. R. W. Schoenlein, D. M. Mittleman, J. J. Shiang, A. P. Alivisatos, and C. V. Shank, Investigation of femtosecond electronic dephasing in CdSe nanocrystals using quantum-beat-suppressed photon echoes, *Phys. Rev. Lett.* **70**, 1014–1017 (1993).
303. U. Banin, G. Cerullo, A. A. Guzelian, A. P. Alivisatos, and C. V. Shank, Quantum confinement and ultrafast dephasing dynamics in InP nanocrystals, *Phys. Rev. B: Condens. Matter* **55**, 7059–7067 (1997).
304. H. Giessen, B. Fluegel, G. Mohs, Y. Z. Hu, N. Peyghambarian, U. Woggon, C. Klingshirn, P. Thomas, and S. W. Koch, Dephasing in the gain region of II–VI semiconductor nanocrystals, *J. Opt. Soc. Am. B: Opt. Phys.* **13**, 1039–1044 (1996).
305. A. Nakamura and H. Ohmura, Coherent transient phenomena of confined excitons in CuCl nanocrystals: Non-Markovian dephasing and quantum beats, *J. Lumin.* **83–84**, 97–103 (1999).
306. U. Bockelmann, Electronic relaxation in quasi-one-dimensional and zero-dimensional structures, *Semicond. Sci. Technol.* **9**, 865–870 (1994).
307. D. E. Skinner, D. P. Colombo, J. J. Cavaleri, and R. M. Bowman, Femtosecond investigation of electron trapping in semiconductor nanoclusters, *J. Phys. Chem.* **99**, 7853–7856 (1995).
308. M. Oneil, J. Marohn, and G. McLendon, Picosecond measurements of exciton trapping in semiconductor clusters, *Chem. Phys. Lett.* **168**, 208–210 (1990).
309. V. Klimov, P. H. Bolivar, and H. Kurz, Ultrafast carrier dynamics in semiconductor quantum dots, *Phys. Rev. B: Condens. Matter* **53**, 1463–1467 (1996).
310. S. Logunov, T. Green, S. Marguet, and M. A. El-Sayed, Interfacial carriers dynamics of CdS nanoparticles, *J. Phys. Chem. A* **102**, 5652–5658 (1998).
311. N. P. Ernsting, M. Kaschke, H. Weller, and L. Katsikas, Colloidal $Zn_{1-x}Cd_xS$—optical saturation of the exciton band and primary photochemistry studied by subpicosecond laser flash photolysis, *J. Opt. Soc. Am. B:Opt. Phys.* **7**, 1630–1637 (1990).
312. J. Z. Zhang, R. H. O'Neil, T. W. Roberti, J. L. McGowen, and J. E. Evans, Femtosecond studies of trapped electrons at the liquid–solid interface of aqueous CdS colloids, *Chem. Phys. Lett.* **218**, 479–484 (1994).
313. J. P. Zheng and H. S. Kwok, Exciton and biexciton recombination in semiconductor nanocrystals, *Appl. Phys. Lett.* **65**, 1151–1153 (1994).
314. J. Z. Zhang, Ultrafast studies of electron dynamics in semiconductor and metal colloidal nanoparticles: Effects of size and surface, *Acc. chem. Res.* **30**, 423–429 (1997).
315. F. Wu, J. Z. Zhang, R. Kho, and R. K. Mehra, Radiative and nonradiative lifetimes of band edge states and deep trap states of CdS nanoparticles determined by time-correlated single photon counting, *Chem. Phys. Lett.* **330**, 237–242 (2000).
316. M. Nirmal, C. B. Murray, and M. G. Bawendi, Fluorescence-line narrowing in CdSe quantum dots — surface localization of the photogenerated exciton, *Phys. Rev. B:Condens. Matter* **50**, 2293–2300 (1994).
317. O. I. Micic, M. Meglic, D. Lawless, D. K. Sharma, and N. Serpone, Semiconductor photophysics. 5. Charge carrier trapping in ultrasmall silver iodide particles and kinetics of formation of silver atom clusters, *Langmuir* **6**, 487–492 (1990).
318. D. P. Colombo, K. A. Roussel, J. Saeh, D. E. Skinner, J. J. Cavaleri, and R. M. Bowman, Femtosecond study of the intensity dependence of electron hole dynamics in TiO_2 nanoclusters, *Chem. Phys. Lett.* **232**, 207–214 (1995).
319. J. J. Cavaleri, D. E. Skinner, D. P. Colombo, and R. M. Bowman, Femtosecond study of the size-dependent charge carrier dynamics in ZnO nanocluster solutions, *J. Chem. Phys.* **103**, 5378–5386 (1995).
320. V. I. Klimov, C. J. Schwarz, D. W. McBranch, and C. W. White, Initial carrier relaxation dynamics in ion-implanted Si nanocrystals: Femtosecond transient absorption study, *Appl. Phys. Lett.* **73**, 2603–2605 (1998).
321. R. Doolen, R. Laitinen, F. Parsapour, and D. F. Kelley, Trap state dynamics in MoS_2 Nanoclusters, *J. Phys. Chem. B* **102**, 3906–3911 (1998).

322. H. E. Gumlich, Electro- and photoluminescence properties of Mn(II) in ZnS and ZnCdS, *J. Lumin.* **23**, 73–99 (1981).
323. H. Ito, T. Takano, T. Kuroda, F. Minami, and H. Akinaga, Two-dimensional confinement effect on Mn^{2+} intraionic transition, *J. Lumin.* **72–14**, 342–343 (1997).
324. N. Murase, R. Jagannathan, Y. Kanematsu, M. Watanabe, A. Kurita, K. Hirata, T. Yazawa, and T. Kushida, Fluorescence and EPR characteristics of Mn^{2+}-doped ZnS nanocrystals prepared by aqueous colloidal method, *J. Phys. Chem. B* **103**, 754–760 (1999).
325. J. Christen, E. Kapon, M. Grundmann, D. M. Hwang, M. Joschko, and D. Bimberg, 1D charge carrier dynamics in Gaas quantum wires — carrier capture, relaxation, and recombination, *Phys. Status Solidi B:Basic Res.* **173**, 307–321 (1992).
326. L. Rota, F. Rossi, P. Lugli, and E. Molinari, An investigation of carrier dynamics in semiconductor quantum wires following femtosecond laser excitation, *Semicond. Sci. Technol.* **9**, 871–874 (1994).
327. R. Cingolani, R. Rinaldi, M. Ferrara, G. C. Larocca, H. Lage, D. Heitmann, and H. Kalt, Electron–hole plasma spectroscopy of GaAs quantum wires, *Semicond. Sci. Technol.* **9**, 875–877 (1994).
328. R. J. D. Miller, G. L. McLendon, A. J. Nozik, W. Schmickler, and F. Willig, *Surface Electron Transfer Process* (VCH, New York, 1995), p. 167.
329. A. J. Nozik and R. Memming, Physical chemistry of semiconductor–liquid interfaces, *J. Phys. Chem.* **100**, 13061–13078 (1996).
330. P. V. Kamat and D. Meisel, *Semiconductor Nanoclusters—Physical, Chemical, and Catalytic Aspects* (Elsevier, New York, 1997).
331. J. E. Moser, P. Bonnote, and M. Gratzel, Molecular photovoltaics, *Coord. Chem. Rev.* **171**, 245–250 (1998).
332. N. Serpone, A decade of heterogeneous photocatalysis in our laboratory — pure and applied studies in energy production and environmental detoxification, *Res. Chem. Intermed.* **20**, 953–992 (1994).
333. P. V. Kamat, Interfacial charge transfer processes in colloidal semiconductor systems, *Prog. React. Kinet.* **19**, 277–316 (1994).
334. N. Serpone and E. Pelizzetti, *Photocatalysis: Fundamentals and Applications* (Wiley, New York, 1989).
335. K. Kalyanasundaram and M. Gratzel, Applications of functionalized transition metal complexes in photonic and optoelectronic devices, *Coord. Chem. Rev.* **177**, 347–414 (1998).
336. Y. Q. Gao and R. A. Marcus, On the theory of electron transfer reactions at semiconductor/liquid interfaces. II. A free electron model, *J. Chem. Phys.* **113**, 6351–6360 (2000).
337. S. Gosavi and R. A. Marcus, Nonadiabatic electron-transfer at metal surfaces, *J. Phys. Chem. B* **104**, 2067–2072 (2000).
338. Y. Q. Gao, Y. Georgievskii, and R. A. Marcus, On the theory of electron transfer reactions at semiconductor electrode/liquid interfaces, *J. Chem. Phys.* **112**, 3358–3369 (2000).
339. J. B. Asbury, E. Hao, Y. Q. Wang, H. N. Ghosh, and T. Q. Lian, Ultrafast electron transfer dynamics from molecular adsorbates to semiconductor nanocrystalline thin films, *J. Phys. Chem. B* **105**, 4545–4557 (2001).
340. J. M. Rehm, G. L. McLendon, Y. Nagasawa, K. Yoshihara, J. Moser, and M. Grätzel, Femtosecond electron-transfer dynamics at a sensitizing dye-semiconductor (TiO_2) interface, *J. Phys. Chem.* **100**, 9577 (1996).
341. T. Hannappel, B. Burfeindt, W. Storck, and F. Willig, Measurement of ultrafast photoinduced electron transfer from chemically anchored Ru-dye molecules into empty electronic states in a colloidal anatase TiO_2 film, *J. Phys. Chem. B* **101**, 6799–6802 (1997).
342. J. E. Moser, D. Noukakis, U. Bach, Y. Tachibana, D. R. Klug, J. R. Durrant, R. HumphryBaker, and M. Gratzel, Comment on "Measurement of ultrafast photoinduced electron transfer from chemically anchored Ru-dye molecules into empty electronic states in a colloidal anatase TiO_2 film", *J. Phys. Chem. B* **102**, 3649–3650 (1998).
343. T. Hannappel, C. Zimmermann, B. Meissner, B. Burfeindt, W. Storck, and F. Willig, Reply to comment on "measurement of ultrafast photoinduced electron transfer from chemically anchored Ru-dye molecules into empty electronic states in a colloidal anatase TiO_2 film", *J. Phys. Chem. B* **102**, 3651–3652 (1998).

344. T. A. Heimer and E. J. Heilweil, Direct time-resolved infrared measurement of electron injection in dye-sensitized titanium dioxide films, *J. Phys. Chem. B* **101**, 10990–10993 (1997).
345. R. J. Ellingson, J. B. Asbury, S. Ferrere, H. N. Ghosh, J. R. Sprague, T. Q. Lian, and A. J. Nozik, Dynamics of electron injection in nanocrystalline titanium dioxide films sensitized with [Ru(4,4'-dicarboxy-2,2'-bipyridine)(2)(NCS)(2)] by infrared transient absorption, *J. Phys. Chem. B* **102**, 6455–6458 (1998).
346. N. J. Cherepy, G. P. Smestad, M. Gratzel, and J. Z. Zhang, Ultrafast electron injection: Implications for a photoelectrochemical cell utilizing an anthocyanin dye-sensitized TiO_2 nanocrystalline electrode, *J. Phys. Chem. B* **101**, 9342–9351 (1997).
347. P. V. Kamat, Native and surface modified semiconductor nanoclusters, *Prog. Inorg. Chem.* **44**, 273–343 (1997).
348. I. Martini, J. H. Hodak, and G. V. Hartland, Effect of structure on electron transfer reactions between anthracene dyes and TiO_2 nanoparticles, *J. Phys. Chem. B* **102**, 9508–9517 (1998).
349. H. N. Ghosh, J. B. Asbury, Y. X. Weng, and T. Q. Lian, Interfacial electron transfer between Fe(II)(CN)(6)(4 –) and TiO_2 nanoparticles: Direct electron injection and nonexponential recombination, *J. Phys. Chem. B* **102**, 10208–10215 (1998).
350. H. N. Ghosh, J. B. Asbury, and T. Q. Lian, Direct observation of ultrafast electron injection from coumarin 343 to TiO_2 nanoparticles by femtosecond infrared spectroscopy, *J. Phys. Chem. B* **102**, 6482–6486 (1998).
351. J. B. Asbury, Y. Q. Wang, and T. Q. Lian, Multiple-exponential electron injection in Ru(dcbpy)(2)(SCN)(2) sensitized ZnO nanocrystalline thin films, *J. Phys. Chem. B* **103**, 6643–6647 (1999).
352. J. B. Asbury, R. J. Ellingson, H. N. Ghosh, S. Ferrere, A. J. Nozik, and T. Q. Lian, Femtosecond IR study of excited-state relaxation and electron-injection dynamics of Ru(dcbpy)(2)(NCS)(2) in solution and on nanocrystalline TiO_2 and Al_2O_3 thin films, *J. Phys. Chem. B* **103**, 3110–3119 (1999).
353. J. B. Asbury, E. C. Hao, Y. Q. Wang, and T. Q. Lian, Bridge length-dependent ultrafast electron transfer from Re polypyridyl complexes to nanocrystalline TiO_2 thin films studied by femtosecond infrared spectroscopy, *J. Phys. Chem. B* **104**, 11957–11964 (2000).
354. Y. X. Weng, Y. Q. Wang, J. B. Asbury, H. N. Ghosh, and T. Q. Lian, Back electron transfer from TiO_2 nanoparticles to Fe-III(CN)(6)(3 –): Origin of non-single-exponential and particle size independent dynamics, *J. Phys. Chem. B* **104**, 93–104 (2000).
355. C. Nasr, D. Liu, S. Hotchandani, and P. V. Kamat, Dye-capped semiconductor nanoclusters—excited state and photosensitization aspects of rhodamine 6g H-aggregates bound to SiO_2 and SnO_2 colloids, *J. Phys. Chem.* **100**, 11054–11061 (1996).
356. D. Liu, R. W. Fessenden, G. L. Hug, and P. V. Kamat, Dye capped semiconductor nanoclusters. Role of back electron transfer in the photosensitization of SnO_2 nanocrystallites with cresyl violet aggregates, *J. Phys. Chem. B* **101**, 2583–2590 (1997).
357. I. Martini, G. V. Hartland, and P. V. Kamat, Ultrafast photophysical investigation of cresyl violet aggregates adsorbed onto nanometer-sized particles of SnO_2 and SiO_2, *J. Phys. Chem. B* **101**, 4826–4830 (1997).
358. C. Burda, T. C. Green, S. Link, and M. A. El-Sayed, Electron shuttling across the interface of CdSe nanoparticles monitored by femtosecond laser spectroscopy, *J. Phys. Chem. B* **103**, 1783–1788 (1999).

9

Optical, Electronic, and Dynamic Properties of Metal Nanomaterials

9.1. STATIC ABSORPTION PROPERTIES OF METAL NANOPARTICLES AND ASSEMBLIES

Metal nanoparticles resemble and differ from semiconductor nanoparticles. As a consequence of their difference in electronic band structures, the quantum size confinement effect is expected to occur at much smaller sizes for metal than for semiconductor nanoparticles. This difference has fundamental consequences on their optical, electronic, and dynamic properties. Reducing size is expected to eventually lead to metal to nonmetal transition.[1–4]

One notable feature in the optical absorption spectra of many metal nanoparticles is the surface plasmon band. The surface plasmon band is due to collective electron oscillation around the surface mode of the particle. For example, Ag nanoparticles usually have a strong absorption around 390 nm while Au nanoparticles have the plasmon band around 520 nm (Fig. 9.1). Pt does not show a noticeable band in the visible–UV region (Fig. 9.1). The peak position and bandwidth of the surface plasmon absorption are related to the particle size and shape. For spherical particles with radius R smaller than $\lambda/2\pi$ (where λ is the wavelength of light in the media) and the electric dipole term dominating, the dependence of plasmon band position and width on particle size can be accounted for by the classic electrodynamics Mie theory.[5] According to the Mie theory, the cross section of optical absorption K can be written as[6]

$$K = \frac{18\pi f \varepsilon_m^{3/2}}{\lambda} \frac{\varepsilon_2}{(2\varepsilon_m + \varepsilon_1)^2 + \varepsilon_2^2} \tag{9.1}$$

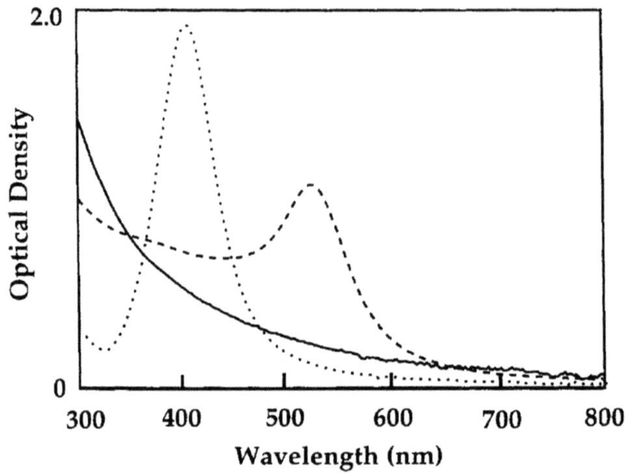

FIGURE 9.1. Electronic absorption spectra of metal NPs: Au (- - -); Ag (· · ·); and Pt (—). Pt NPs lacks a surface plasmon band in this spectral region.

where f is the volume-filling fraction of the metal, ε_m is the dielectric constant of the medium in which the particles are embedded, ε_1 and ε_2 are the real and imaginary part of the complex dielectric constant of the metal particle with a diameter d or radius R. The Mie theory had been modified by others to incorporate other effects including quantum effect later.[6,7]

According to Eq. (9.1), when ε_2 is small or does not change so much around the band, the absorption coefficient has a maximum value at a resonance frequency, where

$$\varepsilon_1 = -2\varepsilon_m \tag{9.2}$$

The wavelength of this plasmon resonance is, therefore, determined by the wavelength dependence of $\varepsilon_1(\omega)$, while the width of the plasmon band is determined by $\varepsilon_2(\omega)$ and ε_m.

If the particle size becomes comparable or smaller than the mean free path of the conduction band electrons, the collisions of the electrons with the particle surface becomes important and the effective mean free path is less than that in bulk materials. This usually results in broadening and blueshift of the plasmon band for particles smaller than about 10 nm.[8–10] From a quantum viewpoint, for particles containing about <100 atoms the electronic energy bands are quantized and energy level spacing may become comparable to thermal energy kT. This affects intraband transitions of the conduction electrons and leads to a damping of electron motion, which corresponds to the free path effect in the classical approach.[11] This damping influences the dielectric constant.

One way to account for the size or surface effect is to divide the contribution to the dielectric constant into two parts: one from interband transitions and the second from intraband transitions including the surface effect:[9,12,13]

$$\begin{aligned} \varepsilon_1(\omega) &= \varepsilon_1^{\text{intra}}(\omega) + \varepsilon_1^{\text{inter}}(\omega) \\ \varepsilon_2(\omega) &= \varepsilon_2^{\text{intra}}(\omega) + \varepsilon_2^{\text{inter}}(\omega) \end{aligned} \tag{9.3}$$

The intraband contribution can be calculated using the Drude model for the nearly free electrons:[13]

$$\begin{aligned} \varepsilon_1^{\text{intra}}(\omega) &= 1 - \frac{\omega_p^2}{\omega^2 + \Gamma^2} \\ \varepsilon_2^{\text{intra}}(\omega) &= \frac{\omega_p^2 \Gamma}{\omega(\omega^2 + \Gamma^2)} \end{aligned} \tag{9.4}$$

where ω_p is the plasmon frequency, which is related to the free electron density N, and Γ is the damping constant. For metals such as Al and Ag, where the onset of interband transitions is well separated from the plasmon band, the plasmon bandwidth is controlled by Γ.[8] For metals such as Au and Cu, where the interband transitions occur in the same spectral region as the plasmon band, the bandwidth is determined by both Γ and $\varepsilon^{\text{inter}}(\omega)$, which makes analysis more difficult. In either case, however, the dependence of the bandwidth on particle size can be accounted for by expressing:[8,10,12,14]

$$\Gamma = \Gamma_0 + A\frac{v_F}{R} \tag{9.5}$$

where R is the particle radius, v_F is the Fermi velocity of the electrons and is 1.4×10^8 cm/s for Au and Ag,[9] A is a constant dependent on the electron–surface interaction and is usually on the order of unity,[9,10] and Γ_0 stands for the frequency of inelastic collisions of free electrons within the bulk metal, for example, electron–phonon coupling or defects, and is on the order of hundredths of an electron volt.[9] For small size particles, the second term in Eq. (9.5), which accounts for dephasing due to electron–surface scattering, can greatly exceed the bulk scattering frequency Γ_0, which accounts for bulk contribution to the electronic dephasing. For example, it has been determined experimentally for Au nanoparticles that $A = 0.43$ and $\Gamma_0 = 4.4 \times 10^{14}$ Hz, which implies an intrinsic, bulk electronic dephasing time of 2.3 fs.[10] It is clear from Eq. (9.5) that the smaller the particle size, the higher is the surface collision frequency, which usually leads to a blueshift of the plasmon peak.

Similar to semiconductor nanoparticles, the shape of metal nanoparticles significantly affects their optical absorption and emission properties. For example, for gold nanoparticles with approximately spherical shapes, the surface plasmon band centers around 520 nm. For gold nanorods, however, the absorption spectrum changes dramatically. As shown in Fig. 9.2, in addition to the band centered around 520 nm, there is a strong absorption band in the near IR region with peak position dependent on the aspect ratio of the nanorod.[15–18] The gold nanorods can be aligned by electric fields.[15] The absorption spectra strongly dependent on the degree of orientational order of the rods and the direction of the polarization of the incident light with respect to the applied electric field.

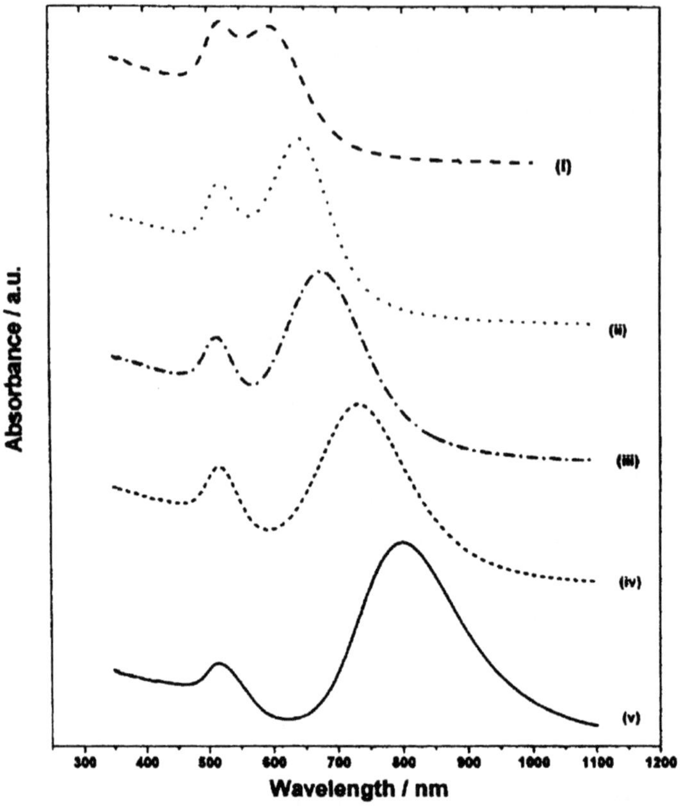

FIGURE 9.2. Electronic absorption spectra of Au nanorods of average aspect ratio of (i) 2.0, (ii) 2.6, (iii) 3.3, (iv) 4.3, and (v) 5.4, respectively. The longitudinal surface plasmon absorption band of gold nanorods mainly depends on the average aspect ratio. (Reproduced from Fig. 2 of Ref. 17 with permission.)

Compared to isolated metal particles, assemblies or aggregates of metal particles exhibit new optical absorption properties. For instance, when silver nanoparticles form monolayers organized in a hexagonal network, their plasmon peak shifts toward lower energy or longer wavelength with an increase in bandwidth compared to that observed for isolated, coated particles dispersed in hexane solution.[11,19] The spectral shift was attributed to an increase in the dielectric constant of the matrix environment of the nanoparticles. When the particles form a three-dimensional (3D) superlattice with a face-centered cubic (fcc) structure, the plasmon peak shifts further toward lower energy but the bandwidth remains about the same as that observed for free nanoparticles in hexane. The optical absorption properties of the 3D superlattices are interpreted as an increase in the mean free path of the conduction electrons, which could indicate the presence of tunneling electrons across the double layers due to the coating of the particles with dodecanethiol.

For spherical gold particles, the SPR occurs as a band centered at ~520 nm. When the particles deviate from spherical geometry, the transverse and longitudinal dipole polarizability no longer produce equivalent resonances. As a consequence two plasma resonances appear, a broadened and redshifted longitudinal plasma resonance in the range of 600–900 nm, and a transverse plasma resonance, whose absorbance peak remains essentially unchanged from that of the spherical particles. Several asymmetric colloidal gold systems, which typify this phenomenon, have been reported, including gold nanorods,[20] gold aggregates,[21,22] and gold nanoshells.[23] The gold nanorods clearly display optical properties predicted by theory. As the ratio of the length to the diameter increases, there is a redshift in the longitudinal plasma resonance while the position of the transverse resonance remains unchanged. Gold shells have been grown on SiO_2 nanoparticles and other colloidal materials. When grown on SiO_2, the gold optical absorption shows distinct resonances for the transverse and longitudinal plasmon. The position and width of these resonances have been shown to be dependent upon the diameter of the SiO_2 nanoparticle and the shell thickness.[23] Gold nanoparticle aggregates show similar behavior. Aggregation causes a coupling of the gold nanoparticle's plasma modes. In the optical spectrum, this appears as a redshift and broadening of the longitudinal plasma resonance,[22,23] as shown in Fig. 9.3.

Similar to semiconductor nanoparticles, metal nanoparticles exhibit non-linear optical properties.[24] The third-order non-linear susceptibility, $\chi^{(3)}$, of core/shell Ag/SiO_2 nanocomposite materials have been studied using DFWM techniques and found to be enhanced by two orders of magnitude with the SiO_2 shell compared to Ag particles without the shell.[25] The enhancement was attributed to concentration of the electric field in the interior of the particle and further magnification at the surface plasmon resonance. Similar technique has

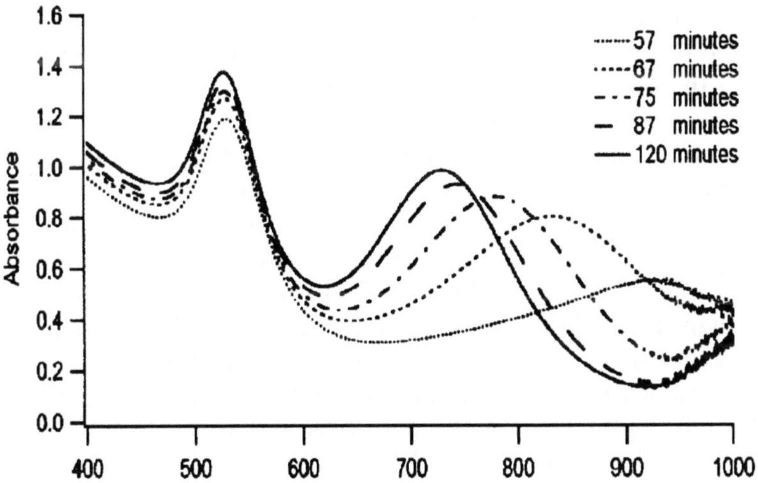

FIGURE 9.3. Time evolution of the optical absorption spectra of Au nanoparticles produced following reduction of $AuCl_4^-$ by Na_2S. The near IR absorption in the 700–900 nm region is believed to be due to aggregate of Au NPs.[22]

been applied to study the third-order non-linearities of glasses doped with copper or silver particles and $\chi^{(3)}$, with a maximum value of the order of 10^{-7} esu, was found to increase with an increase in particle radius.[26] Wavelength dispersive femtosecond DFWM has been applied to study the third-order non-linear optical properties of polyelectrolyte films containing gold nanoparticles and a wavelength dependence around the surface plasmon region was found.[27] Resonantly enhanced $\chi^{(3)}$ for colloidal gold nanoparticles has been found to be around 10^{-16} m^2 V^{-2} based on DFWM using incoherent light.[28] SHG has been realized in composite films of spheroidal metal particles such as silver.[29] SHG based on femtosecond lasers has also been applied to study the plasmon dephasing of Au and Ag nanoparticles.[30] In this case, the particles produced by electron beam lithography were designed to have specific shapes without centrosymmetry to enhance the SHG efficiency.

9.2. EMISSION OF METAL PARTICLES

Metal is usually non-luminescent or weakly luminescent. Recently, there have been a number of reports on light emission from metal nanoparticles. Some claimed significantly enhanced fluorescence.[17,31,32] The origin of the emission is intriguing and is still the subject of discussion. In the case of Au nanorods, fluorescence enhancement over a million compared to the gold metal was observed.[17] The emission with a quantum yield of 10^{-4}–10^{-3} was assigned to electron–hole interband d–sp recombination and the huge enhancement was proposed to result from the enhancement effect of the incoming and outgoing electric fields via coupling to the surface plasmon resonance in the nanorods, similar to the previously proposed fluorescence and Raman enhancement on rough surfaces of noble metals except the much lower yield ($\sim 10^{-11}$–10^{-10}).[33] It was observed that for rods with fixed width, the emission wavelength maximum increases with the rod length and the emission quantum yield increases with the square of the rod length. Using the model developed to explain the enhancement of fluorescence from rough noble metal surfaces,[34] the observed fluorescence from nanorods can be simulated as a function of the aspect ratio.[17,18]

Near IR (1.1–1.6 μm) photoluminescence spectra with 1.06 μm excitattion have also been observed at room temperature for 1.1 and 1.7 nm Au nanocrystals.[35] The quantum yield for 1.7 nm nanocrystals is of the order of 10^{-5}, more than five orders of magnitude greater than that of bulk gold. The photoluminescence was attributed to sp to sp-like transitions, however, the exact mechanism is unknown. Similarly, efficient luminescence (quantum yield about 0.3%) in the 700–800 nm region has been observed with 451 nm excitation for monolayer-protected Au clusters with 1.8 nm diameter cores.[36] The luminescence was hypothesized to be associated with interband transitions between the filled $5d^{10}$ band and $6(sp)^1$ conduction band.

In other cases, the emission is most likely either from very small metal clusters with significant surface contribution or from metal compounds or complexes such as Ag_2O.[31] For example, surface species or very small metal ions on the surface of the colloidal particles could be highly luminescent. Enhanced light emission from gold- and silver-dendrimer composites have also been reported recently.[32] The enhancement was suggested to be due to local field enhancement in the elongated metal-dendrimer nanoparticles, although it could also be due to nanoparticle aggregates or structures resembling percolation clusters. Further investigation is needed to clarify the mechanism of the acclaimed strong luminescence of metal particles. The effect of surface is a difficult one to determine but could be a key to answering the question.

9.3. SURFACE-ENHANCED RAMAN SCATTERING (SERS)

One interesting optical property associated with metal nanoparticles is surface-enhanced Raman scattering (SERS) of molecules on the particle surface.[37–44] The effect of SERS is similar to that first observed on rough metal surfaces.[45] Although not all the fine points of the enhancement mechanism have been clarified, the majority view is that the largest contributor to the intensity amplification results from the electric field enhancement that occurs in the vicinity of small, interacting metal particles that are illuminated with light resonant or near resonant with the localized surface plasmon frequency of the metal structure.[37] Metal nanoparticles have extremely large surface area relative to volume compared to bulk and thereby provide a convenient system for studying the SERS effect. The SERS technique based on metal nanoparticles has various potential applications in analytical chemistry[40,46] and biochemistry.[41,47] For example, it has been considered for use in immunoassays[41] and for studying protein–metal interactions.[39]

9.4. SPECTRAL LINE WIDTHS AND ELECTRONIC DEPHASING

Similar to atoms and molecules or semiconductor nanoparticles, the spectral line width of metal nanoparticles is determined by homogeneous and inhomogeneous broadening. The spectral line width experimentally observed for an ensemble average of particles is usually much broader than the homogeneous line width, due to inhomogeneous broadening caused by a distribution of particle size, shape and local environment. If inhomogeneous broadening is removed, the homogeneous line width is related to the dephasing time. The homogeneous line width of electronic absorption spectrum is usually broad and the dephasing time is very short.

For example, for a single gold nanoparticle, the dephasing time has been determined to be around 8 fs using a near-field optical antenna effect.[48] The measured dephasing time agrees with calculations based on the Mie theory with surface effects neglected. The dephasing time was also found to change for individual particles due to variations in the local environment. A dephasing time of <20 fs has been reported for gold nanoparticles based on measurement using incoherent light.[28] The dephasing time for silver island films has been reported to be around 40 fs.[49] Persistent spectral hole burning studies of supported oblate Ag particles with radii of 7.5 nm found a homogeneous line width of 260 meV corresponding to a dephasing time of 4.8 fs.[50] This time is shorter than expected for damping by bulk electron scattering, and, therefore, it was concluded that additional damping mechanisms have been observed, reflecting confinement of the electrons in nanoparticles with size <10 nm. A dephasing time of 10 fs has recently been reported for Ag nanoparticles based on two-pulse second-order interferometry measurements.[51] SHG studies of plasmon dephasing in Au and Ag nanoparticles found a dephasing time of 12 and 14 fs for Au and Ag, respectively.[30] All the studies show that the plasmon dephasing time in metal particles such as Au and Ag is extremely fast, on the order of 10 fs.

When particles are brought together such as in superlattice films, interparticle interaction becomes important. This interaction typically results in broadening of the spectral line width and faster dephasing time compared to isolated individual particles. Most studies of dephasing times have been conducted on isolated particles so far. Direct studies of the dephasing time for metal particles with strong interactions should be interesting and are yet to be carried out.

9.5. ELECTRONIC RELAXATION DYNAMICS

The electronic relaxation dynamics in metal nanoparticles occur on ultrafast timescales. Direct probe of the relaxation is made possible by using femtosecond laser spectroscopy. Following the initial work by Zhang *et al.* on Ag[52] and Au[53] and Bigot *et al.* on Cu,[54] a number of studies have been reported on the relaxation dynamics of metal nanoparticles including Au,[55–61] Ag,[61] Sn,[62] Pt,[56,63] and Ga.[64] One issue of fundamental interest is the size dependence of the relaxation time. Earlier work by Zhang *et al.* based on transient absorption measurements indicated a possible size dependence of the relaxation time for particles studied in the size range of 1–40 nm.[56] Subsequent work by El-Sayed and Hartland found no size dependence down to 2.5 nm.[57,59–61] The relaxation time was claimed to be the same in Au nanoparticles as in bulk Au. There are some complications in the study of the size effect. First, the dynamics have been found to be excitation intensities dependent[59,60] and solvent dependent.[63] Due to the small transient absorption signal, the excitation intensity used in the transient absorption measurements by Zhang *et al.* was likely to be much higher than that used in the transient bleach measurements by El-Sayed and Hartland. This could be responsible for the difference in measured lifetimes. Another possible cause for the difference in the surface or solvent environment. As found by Zhang *et al.*, the relaxation time is dependent on solvent, 7 ps in water and 5 ps in cyclohexane.[53,63] The effect of surface or solvent on the relaxation time is an issue worthy of further investigation.

Electronic relaxation dynamics in gold nanoshells have also been reported.[65] The relaxation time was found to be 1.65 ps, corresponding to an electron–phonon coupling constant of 2.2×10^{16} W m^{-3} K^{-1}, smaller than that observed for bulk gold films.[66]

The relaxation of the hot electrons in metal NPs can be understood by the two-temperature model that describes the energy exchange between the electrons and phonons by the two coupled equations:[67–69]

$$
\begin{aligned}
C_e(T_e)C_e(T_e)\frac{\partial T_e}{\partial t} &= -g(T_e - T_l) \\
C_l\frac{\partial T_l}{\partial t} &= g(T_e - T_l)
\end{aligned}
\qquad (9.6)
$$

where T_e and T_l are the electronic and lattice temperatures, $C_e(T_e) = gT_e$ is the temperature-dependent electronic heat capacity, $\gamma = 66$ J m^{-3} K^{-2} for Au,[13] C_l is the lattice heat capacity, and g is the electron–phonon coupling constant. The equations indicate that the relaxation time should increase with higher excitation laser power due to higher initial electronic temperature.[10]

One interesting recent observation is oscillations in the electronic relaxation dynamics of Au and bimetallic core/shell particles composed of Au, Ag, Pt, and Pd, which are attributed to coherent excitation of breathing vibrational modes of the metal particles.[10,70,71] As shown in Fig. 9.4(a), femtosecond transient absorption data for 60 nm diameter Au particles probed at 550 nm following 400 nm pulse excitation show clearly modulations with a period of about 16 ps. The period or frequency of modulation was found to be strongly dependent on particle size [Fig. 9.4(b)] with the frequency increasing linearly with decreasing particle size in the range of 120 to 8 nm. It was also interesting to note that

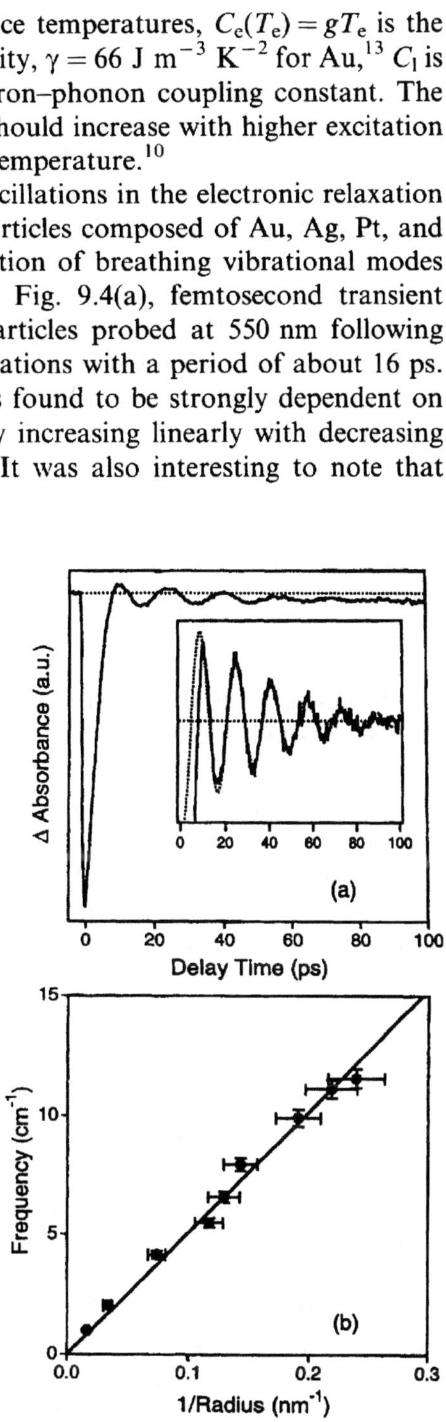

FIGURE 9.4. (a) Transient absorption data for 60 nm diameter Au particles recorded with 400 pump and 550 nm probe pulses. Modulations due to the coherently excited breathing motion can be seen clearly. (b) Plot of the measured modulation frequencies versus $1/R$, R being the particle radius, for different sized Au samples. A straight line fit to the data (b) is also shown. (Reproduced from Fig. 6 of Ref. 10 with permission.)

the observed "beat" signal can be well understood by classical mechanics. Similar observation has been made on Sn and Ga nanoparticles.[70]

9.6. ELECTRON–PHONON INTERACTION

The electronic relaxation in metal particles is dominated by electron–phonon interaction. The relaxation time measured thus reflects how strongly the electron interacts with the phonons. With decreasing size, the electron–phonon interaction is expected to decrease since the electron and phonon spectral densities as well as spectral overlap are expected to decrease. Tomchuk *et al.* have suggested a size dependence of the electron–phonon interaction based on theoretical studies[72,73] and experimental studies of Ag islands.[74] However, the relaxation time determined for a broad size range of Au particles seems to indicate that the electron–phonon interaction is independent of size. This had led to the suggestion that the electron–phonon coupling constant is the same for particle as for bulk, at least for particles down to 2 nm in diameter.[10] However, the issue of surface effect has not been well addressed, especially given that there is an indication of solvent or surface dependence of the electronic relaxation time for Au nanoparticles.[63] The "surface" effect may become even more significant when the particles are smaller than 1–2 nm and contains only a few tens of atoms. Such very small clusters are usually not very stable, making optical and dynamic studies difficult.[56]

9.7. SINGLE PARTICLE SPECTROSCOPY OF METAL NANOPARTICLES

Similar to semiconductor nanoparticles, single particle spectroscopy has been applied to study optical properties of single metal nanoparticles. For example, using near-field optics based on tapered single-mode fiber probes, surface-enhanced Raman spectra of dye molecules on single silver colloidal particles have been measured.[75–77] The homogeneous line width of the surface plasmon band of a single gold nanoparticle has been determined to be 170 meV based on CW transmission experiments performed with a scanning near-field optical microscope (NSOM).[78]

9.8. APPLICATIONS OF METAL NANOPARTICLES

Similar to semiconductor nanoparticles, metal nanoparticles find applications in numerous areas of science and technology, ranging from medicine to optics and biological labeling and imaging.[79] For example, metal nanoparticles such as silver and gold have been used to enhance the non-linearities of molecular probes that are potentially useful for selectively imaging the structure and physiology of nanometric regions in cellular systems.[80] As mentioned earlier, the SERS technique based on metal nanoparticles has potential applications in analytical chemistry[40,46] and biochemistry.[41,47] It has also been considered for application in immunoassays[41] and for studying protein–metal interactions.[39] SERS can also be used to determine vibrational spectrum of molecules that are otherwise difficult to do.[40,43]

One important class of metal nanoparticles is magnetic metal nanoparticle that are critical to magnetic recording industry. Examples include Co, Fe, and Ni.[81–86] These nanoparticles can be made with disk or rod shapes that can be used for magnetic recording applications. A detailed discussion of magnetic nanoparticles is beyond the scope of this book.

REFERENCES

1. V. Vijayakrishnan, A. Chainani, D. D. Sarma, and C. N. R. Rao, Metal insulator transitions in metal clusters—a high-energy spectroscopy study of Pd and Ag clusters, *J. Phys. Chem.* **96**, 8679–8682 (1992).
2. D. R. Huffman, in: *Optical Effects Associated with Small Particles*, edited by P. W. Barber and R. K. Chang (World Scientific, Singapore, 1988), pp. 279–324.
3. U. Kreibig, Electronic properties of small silver particles: the optical constants and their temperature dependence, *J. Phys. F (Metal Phys.)* **4**, 999–1014 (1974).
4. C. P. Collier, R. J. Saykally, J. J. Shiang, S. E. Henrichs, and J. R. Heath, Reversible tuning of silver quantum dot monolayers through the metal–insulator transition, *Science* **277**, 1978–1981 (1997).
5. G. Mie, Contribution to optical properties of turbulent media, specifically colloidal metal dispersions, *Annalen der Physik* **25**, 377–445 (1908).
6. W. P. Halperin, Quantum size effects in metal particles, *Rev. Mod. Phys.* **58**, 533–606 (1986).
7. L. Genzel, T. P. Martin, and U. Kreibig, Dielectric function and plasma resonances of small metal particles, *Z. Physik B* **21**, 339–346 (1975).
8. U. Kreibig and M. Vollmer, *Optical Properties of Metal Clusters* (Springer, Berlin, 1995).
9. M. M. Alvarez, J. T. Khoury, T. G. Schaaff, M. N. Shafigullin, I. Vezmar, and R. L. Whetten, Optical absorption spectra of nanocrystal gold molecules, *J. Phys. Chem. B* **101**, 3706–3712 (1997).
10. J. H. Hodak, A. Henglein, and G. V. Hartland, Photophysics of nanometer sized metal particles: electron–phonon coupling and coherent excitation of breathing vibrational modes, *J. Phys. Chem. B* **104**, 9954–9965 (2000).
11. A. Taleb, C. Petit, and M. P. Pileni, Optical properties of self-assembled 2D and 3D superlattices of silver nanoparticles, *J. Phys. Chem. B* **102**, 2214–2220 (1998).
12. P. B. Johnson and R. W. Christy, Optical constants of the noble metals, *Phys. Rev. B: Solid State* **6**, 4370–4379 (1972).
13. N. W. Ashcroft and N. D. Mermin, *Solid State Physics* (Saunders College, Philadelphia, 1976).
14. D. M. Wood and N. W. Ashcroft, Quantum size effects in the optical properties of small metallic particles, *Phys. Rev. B: Condens. Matter* **25**, 6255–6274 (1982).
15. B. M. I. van der Zande, G. J. M. Koper, and H. N. W. Lekkerkerker, Alignment of rod-shaped gold particles by electric fields, *J. Phys. Chem. B* **103**, 5754–5760 (1999).
16. B. M. I. van der Zande, L. Pages, R. A. M. Hikmet, and A. van Blaaderen, Optical properties of aligned rod-shaped gold particles dispersed in poly(vinyl alcohol) films, *J. Phys. Chem. B* **103**, 5761–5767 (1999).
17. M. B. Mohamed, V. Volkov, S. Link, and M. A. El-Sayed, The 'lightning' gold nanorods: fluorescence enhancement of over a million compared to the gold metal, *Chem. Phys. Lett.* **317**, 517–523 (2000).
18. M. A. El-Sayed, Some interesting properties of metals confined in time and nanometer space of different shapes, *Acc. Chem. Res.* **34**, 257–264 (2001).
19. A. Taleb, V. Russier, A. Courty, and M. P. Pileni, Collective optical properties of silver nanoparticles organized in two-dimensional superlattices, *Phys. Rev. B: Condens. Matter* **59**, 13350–13358 (1999).

20. Y. Y. Yu, S. S. Chang, C. L. Lee, and C. R. C. Wang, Gold nanorods: electrochemical synthesis and optical properties, *J. Phys. Chem. B* **101**, 6661–6664 (1997).
21. A. N. Shipway, M. Lahav, R. Gabai, and I. Willner, Investigations into the electrostatically induced aggregation of Au nanoparticles, *Langmuir* **16**, 8789–8795 (2000).
22. J. Norman, T., C. D. Grant, D. Magana, D. Cao, F. Bridges, J. Liu, A. van Buuren, and J. Z. Zhang, Near infrared optical absorption of gold nanoparticle aggregates, *J. Phys. Chem. B*, in press (2002).
23. S. J. Oldenburg, R. D. Averitt, S. L. Westcott, and N. J. Halas, Nanoengineering of optical resonances, *Chem. Phys. Lett.* **288**, 243–247 (1998).
24. D. Ricard, P. Roussignol, and C. Flytzanis, Surface-mediated enhancement of optical phase conjugation in metal colloids, *Opt. Lett.* **10**, 511–513 (1985).
25. J. H. Adair, T. Li, T. Kido, K. Havey, J. Moon, J. Mecholsky, A. Morrone, D. R. Talham, M. H. Ludwig, and L. Wang, Recent developments in the preparation and properties of nanometer-size spherical and platelet-shaped particles and composite particles, *Mater. Sci. Eng. R.* **23**, 139–242 (1998).
26. K. Uchida, S. Kaneko, S. Omi, C. Hata, H. Tanji, Y. Asahara, A. J. Ikushima, T. Tokizaki, and A. Nakamura, Optical nonlinearities of a high concentration of small metal particles dispersed in glass: copper and silver particles, *J. Opt. Soc. Am. B (Opt. Phys.)* **11**, 1236–1243 (1994).
27. W. Schrof, S. Rozouvan, E. van Keuren, D. Horn, J. Schmitt, and G. Decher, Nonlinear optical properties of polyelectrolyte thin films containing gold nanoparticles investigated by wavelength dispersive femtosecond degenerate four wave mixing (DFWM), *Adv. Mater.* **10**, 338–341 (1998).
28. K. Puech, F. Z. Henari, W. J. Blau, D. Duff, and G. Schmid, Investigation of the ultrafast dephasing time of gold nanoparticles using incoherent light, *Chem. Phys. Lett.* **247**, 13–17 (1995).
29. G. Berkvic and S. Efrima, Second harmonic generation from composite films of spheroidal metal particles, *Langmuir* **9**, 35–357 (1993).
30. B. Lamprecht, A. Leitner, and F. R. Aussenegg, SHG studies of plasmon dephasing in nanoparticles, *Appl. Phys. B—Lasers and Optics* **68**, 419–423 (1999).
31. L. A. Peyser, A. E. Vinson, A. P. Bartko, and R. M. Dickson, Photoactivated fluorescence from individual silver nanoclusters, *Science* **291**, 103–106 (2001).
32. O. Varnavski, R. G. Ispasoiu, L. Balogh, D. Tomalia, and T. Goodson, Ultrafast time-resolved photoluminescence from novel metal-dendrimer nanocomposites, *J. Chem. Phys.* **114**, 1962–1965 (2001).
33. A. Mooradian, Photoluminescence of metals, *Phys. Rev. Lett.* **22**, 185–187 (1969).
34. S. Link, M. B. Mohamed, and M. A. El-Sayed, Simulation of the optical absorption spectra of gold nanorods as a function of their aspect ratio and the effect of the medium dielectric constant, *J. Phys. Chem. B* **103**, 3073–3077 (1999).
35. T. P. Bigioni, R. L. Whetten, and O. Dag, Near-infrared luminescence from small gold nanocrystals, *J. Phys. Chem. B* **104**, 6983–6986 (2000).
36. T. Huang and R. W. Murray, Visible luminescence of water-soluble monolayer-protected gold clusters, *J. Phys. Chem. B* **105**, 12498–12502 (2001).
37. M. Moskovits. Surface-enhanced spectroscopy, *Rev. Mod. Phys.* **57**, 783–826 (1985).
38. R. G. Freeman, M. B. Hommer, K. C. Grabar, M. A. Jackson, and M. J. Natan, Ag-clad Au nanoparticles—novel aggregation, optical, and surface-enhanced Raman scattering properties, *J. Phys. Chem.* **100**, 718–724 (1996).
39. A. M. Ahern and R. L. Garrell, Protein–metal interactions in protein-colloid conjugates probed by surface-enhanced Raman spectroscopy, *Langmuir* **7**, 254–261 (1991).
40. M. Hidalgo, R. Montes, J. J. Laserna, and A. Ruperez, Surface-enhanced resonance Raman spectroscopy of 2-pyridylhydrazone and 1,10-phenanthroline chelate complexes with metal ions on colloidal silver, *Anal. Chim. Acta* **318**, 229–237 (1996).
41. S. M. Barnett, B. Vlckova, I. S. Butler, and T. S. Kanigan, Surface-enhanced Raman scattering spectroscopic study of 17-alpha-ethinylestradiol on silver colloid and in glass-deposited Ag-17-alpha-ethinylestradiol film, *Anal. Chem.* **66**, 1762–1765 (1994).
42. P. Matejka, B. Vlckova, J. Vohlidal, P. Pancoska, and V. Baumruk, The role of Triton X-100 as an adsorbate and a molecular spacer on the surface of silver colloid—a surface-enhanced Raman scattering study, *J. Phys. Chem.* **96**, 1361–1366 (1992).

43. K. Cermakova, O. Sestak, P. Matejka, V. Baumruk, and B. Vlckova, Surface-enhanced Raman scattering (SERS) spectroscopy with borohydride-reduced silver colloids—controlling adsorption of the scattering species by surface potential of silver colloid, *Collec. Czech. Chem. Commun.* **58**, 2682–2694 (1993).
44. S. Schneider, P. Halbig, H. Grau, and U. Nickel, Reproducible preparation of silver sols with uniform particle size for application in surface-enhanced Raman spectroscopy, *Photochem. Photobiol.* **60**, 605–610 (1994).
45. M. Fleischmann, P. J. Hendra, and A. J. McQuillan, Raman spectra of pyridine adsorbed at a silver electrode, *Chem. Phys. Lett.* **26**, 163–166 (1974).
46. C. H. Munro, W. E. Smith, M. Garner, J. Clarkson, and P. C. White, Characterization of the surface of a citrate-reduced colloid optimized for use as a substrate for surface-enhanced resonance Raman scattering, *Langmuir* **11**, 3712–3720 (1995).
47. T. M. Cotton, J. H. Kim, and G. D. Chumanov, Application of surface-enhanced Raman spectroscopy to biological systems, *J. Raman Spectrosc.* **22**, 729–742 (1991).
48. T. Klar, M. Perner, S. Grosse, G. vonPlessen, W. Spirkl, and J. Feldmann, Surface-plasmon resonances in single metallic nanoparticles, *Phys. Rev. Lett.* **80**, 4249–4252 (1998).
49. D. Steinmuller-Nethl, R. A. Hopfel, E. Gornik, A. Leitner, and F. R. Aussenegg, Femtosecond relaxation of localized plasma excitations in Ag islands, *Phys. Rev. Lett.* **68**, 389–392 (1992).
50. F. Stietz, J. Bosbach, T. Wenzel, T. Vartanyan, A. Goldmann, and F. Trager, Decay times of surface plasmon excitation in metal nanoparticles by persistent spectral hole burning, *Phys. Rev. Lett.* **84**, 5644–5647 (2000).
51. Y. H. Liau, A. N. Unterreiner, Q. Chang, and N. F. Scherer, Ultrafast dephasing of single nanoparticles studied by two-pulse second-order interferometry, *J. Phys. Chem. B* **105**, 2135–2142 (2001).
52. T. W. Roberti, B. A. Smith, and J. Z. Zhang, Ultrafast electron dynamics at the liquid-metal interface—femtosecond studies using surface plasmons in aqueous silver colloid, *J. Chem. Phys.* **102**, 3860–3866 (1995).
53. A. E. Faulhaber, B. A. Smith, J. K. Andersen, and J. Z. Zhang, Femtosecond electronic relaxation dynamics in metal nano-particles—effects of surface and size confinement, *Mol. Cryst. Liq. Cryst. Sci. Technol. Sec. A—Mol. Cryst. Liq. Cryst.* **283**, 25–30 (1996).
54. J. Y. Bigot, J. C. Merle, O. Cregut, and A. Daunois, Electron dynamics in copper metallic nanoparticles probed with femtosecond optical pulses, *Phys. Rev. Lett.* **75**, 4702–4705 (1995).
55. T. S. Ahmadi, S. L. Logunov, and M. A. El-Sayed, Picosecond dynamics of colloidal gold nanoparticles, *J. Phys. Chem.* **100**, 8053–8056 (1996).
56. B. A. Smith, J. Z. Zhang, U. Giebel, and G. Schmid, Direct probe of size-dependent electronic relaxation in single-sized Au and nearly monodisperse Pt colloidal nano-particles, *Chem. Phys. Lett.* **270**, 139–144 (1997).
57. S. L. Logunov, T. S. Ahmadi, M. A. El-Sayed, J. T. Khoury, and R. L. Whetten, Electron dynamics of passivated gold nanocrystals probed by subpicosecond transient absorption spectroscopy, *J. Phys. Chem. B* **101**, 3713–3719 (1997).
58. M. Perner, P. Bost, U. Lemmer, G. von Plessen, J. Feldmann, U. Becker, M. Mennig, M. Schmitt, and H. Schmidt, Optically induced damping of the surface plasmon resonance in gold colloids, *Phys. Rev. Lett.* **78**, 2192–2195 (1997).
59. J. Hodak, I. Martini, and G. V. Hartland, Ultrafast study of electron–phonon coupling in colloidal gold particles, *Chem. Phys. Lett.* **284**, 135–141 (1998).
60. J. H. Hodak, I. Martini, and G. V. Hartland, Spectroscopy and dynamics of nanometer-sized noble metal particles, *J. Phys. Chem. B* **102**, 6958–6967 (1998).
61. S. Link and M. A. El-Sayed, Spectral properties and relaxation dynamics of surface plasmon electronic oscillations in gold and silver nanodots and nanorods, *J. Phys. Chem. B* **103**, 8410–8426 (1999).
62. A. Stella, M. Nisoli, S. Desilvestri, O. Svelto, G. Lanzani, P. Cheyssac, and R. Kofman, Size effects in the ultrafast electronic dynamics of metallic tin nanoparticles, *Phys. Rev. B: Condens. Matter* **53**, 15497–15500 (1996).
63. J. Z. Zhang, B. A. Smith, A. E. Faulhaber, J. K. Andersen, and T. J. Rosales, in: *Ultrafast Processes in Spectroscopy*, edited by O. Svelto, S. De Silvestri and G. Denardo (Plenum Press, Trieste, Italy, 1995), p. 668.

64. M. Nisoli, S. Stagira, S. DeSilvestri, A. Stella, P. Tognini, P. Cheyssac, and R. Kofman, Ultrafast electronic dynamics in solid and liquid gallium nanoparticles, *Phys. Rev. Lett.* **78**, 3575–3578 (1997).
65. R. D. Averitt, S. L. Westcott, and N. J. Halas, Ultrafast electron dynamics in gold nanoshells, *Phys. Rev. B: Condens. Matter* **58**, 10203–10206 (1998).
66. R. H. M. Groeneveld, R. Sprik, and A. Lagendijk, Femtosecond spectroscopy of electron–electron and electron–phonon energy relaxation in Ag and Au, *Phys. Rev. B: Condens. Matter* **51**, 11433–11445 (1995).
67. H. E. Elsayed-Ali, T. B. Norris, M. A. Pessot, and G. A. Mourou, Time-resolved observation of electron–phonon relaxation in copper, *Phys. Rev. Lett.* **58**, 1212–1215 (1987).
68. R. W. Schoenlein, W. Z. Lin, J. G. Fujimoto, and G. L. Easley, Femtosecond studies of nonequilibrium electronic processes in metals, *Phys. Rev. Lett.* **58**, 1680–1683 (1987).
69. C. K. Sun, F. Vallee, L. H. Acioli, E. P. Ippen, and J. G. Fujimoto, Femtosecond-tunable measurement of electron thermalization in gold, *Phys. Rev. B: Condens. Matter* **50**, 15337–15348 (1994).
70. M. Nisoli, S. De Silvestri, A. Cavalleri, A. M. Malvezzi, A. Stella, G. Lanzani, P. Cheyssac, and R. Kofman, Coherent acoustic oscillations in metallic nanoparticles generated with femtosecond optical pulses, *Phys. Rev. B: Condens. Matter* **55**, 13424–13427 (1997).
71. J. H. Hodak, I. Martini, and G. V. Hartland, Observation of acoustic quantum beats in nanometer sized Au particles, *J. Chem. Phys.* **108**, 9210–9213 (1998).
72. E. D. Belotskii and P. M. Tomchuk, Surface electron phonon energy exchange in small metallic particles, *Int. J. Electron.* **73**, 955–957 (1992).
73. E. D. Belotskii and P. M. Tomchuk, Electron–phonon interaction and hot electrons in small metal islands, *Surf. Sci.* **239**, 143–155 (1990).
74. S. A. Gorban, S. A. Nepijko, and P. M. Tomchuk, Electron phonon interaction in small metal islands deposited on an insulating substrate, *Int. J. Electron.* **70**, 485–490 (1991).
75. S. M. Nie and S. R. Emery, Probing single molecules and single nanoparticles by surface-enhanced Raman scattering, *Science* **275**, 1102–1106 (1997).
76. S. R. Emory and S. M. Nie, Near-field surface-enhanced Raman spectroscopy on single silver nanoparticles, *Anal. Chem.* **69**, 2631–2635 (1997).
77. S. R. Emory, W. E. Haskins, and S. M. Nie, Direct observation of size-dependent optical enhancement in single metal nanoparticles, *J. Am. Chem. Soc.* **120**, 8009–8010 (1998).
78. M. Perner, T. Klar, S. Grosse, U. Lemmer, G. von Plessen, W. Spirkl, and J. Feldmann, Homogeneous line widths of surface plasmons in gold nanoparticles measured by femtosecond pump-and-probe and near-field optical spectroscopy, *J. Lumin.* **76–77**, 181–184 (1998).
79. T. Klaus, R. Joerger, E. Olsson, and C. G. Granqvist, Silver-based crystalline nanoparticles, microbially fabricated, *Proc. Nat. Acad. Sci. USA* **96**, 13611–13614 (1999).
80. G. Peleg, A. Lweis, O. Bouevitch, L. Loew, D. Parnas, and M. Linial, Gigantic optical nonlinearities from nanoparticle-enhanced molecule probes with potential for selectively imaging the structure and physiology of nanometric regions in cellular systems, *Bioimaging* **4**, 215–224 (1996).
81. C. P. Gibson, Synthesis and characterization of anisometric cobalt nanoclusters, *Science* **267**, 1338–1340 (1995).
82. A. T. Ngo, P. Bonville, and M. P. Pileni, Nanoparticles of CoxFey Square O-z(4): Synthesis and superparamagnetic properties, *Eur. Phys. J. B* **9**, 583–592 (1999).
83. X. X. Zhang, G. H. Wen, S. M. Huang, L. M. Dai, R. P. Gao, and Z. L. Wang, Magnetic properties of Fe nanoparticles trapped at the tips of the aligned carbon nanotubes, *J. Magn. Magn. Mater.* **231**, L9–L12 (2001).
84. Y. K. Gunko, S. C. Pillai, and D. McInerney, Magnetic nanoparticles and nanoparticle assemblies from metallorganic precursors, *J. Mater. Sci.—Mater. Electron.* **12**, 299–302 (2001).
85. C. D. Fernandez, C. Sangregorio, G. Mattei, C. Maurizio, G. Battaglin, F. Gonella, A. Lascialfari, S. Lo Russo, D. Gatteschi, P. Mazzoldi, J. M. Gonzalez, and F. D'Acapito, Magnetic properties of Co and Ni based alloy nanoparticles dispersed in a silica matrix, *Nucl. Instrum. Methods Phys. Res. Sect. B—Beam Interactions with Materials and Atoms* **175**, 479–484 (2001).
86. V. F. Puntes, K. M. Krishnan, and P. Alivisatos, Synthesis, self-assembly, and magnetic behavior of a two-dimensional superlattice of single-crystal epsilon-Co nanoparticles, *Appl. Phys. Lett.* **78**, 2187–2189 (2001).

10

Electrochemical Properties of Nanoparticle Assemblies

10.1. INTRODUCTION

Organized architectures of nanometer-sized electrodes and particles have been attracting extensive attention in diverse fields recently, in part, because of the fundamental importance and technological implications involved in these ordered arrays of quantum dots (QDs).[1–8] The great application potentialities of metal nanoparticles as building blocks for electronic nanodevices/nanocircuits have been largely motivated by the unique electronic/electrical properties associated with these nanoscale molecular entities. In particular, for metal nanoparticles that are passivated by a dielectric organic layer (or, *m*onolayer-*p*rotected *c*lusters, MPCs),[9] the particles exhibit a (sub)attofarad (aF, 10^{-18} F) molecular capacitance.[10] Upon the charging of a single electron, these nanoelectrodes exhibit a rather substantial potential change (and vice versa), the so-called electrochemical Coulomb staircase charging.[5,7,9,10]

These quantized charging behaviors are very sensitive to the MPC molecular structure, as the electrochemical resolution (i.e., the potential spacing between two neighboring charging peaks, ΔV) of these solution-phase single-electron-transfer processes is directly related to the particle molecular capacitance (C_{MPC}). Here,

$$\Delta V = e/C_{MPC} \tag{10.1}$$

$$C_{MPC} = 4\pi\varepsilon\varepsilon_0(r/d)(r+d) \tag{10.2}$$

where e is the electronic charge, ε_0 the vacuum permittivity, ε the effective dielectric constant of the MPC protecting monolayer, r the core radius and d the monolayer thickness. For MPCs with $r < 1.0$ nm, a well-resolved electrochemical discrete charging feature is generally observable.[10d] In addition, within this small size range, when MPC dimensions decrease, the charging features exhibit a transition from bulk double-layer charging to molecule-like redox behaviors, due to the evolution of a core-size-dependent HOMO–LUMO energy gap.[10b]

It is these *molecular-capacitor* characters that attract a great deal of attention lately in that these novel nanomaterials might be of great importance in the development of single-electron transistors (SETs).[7] Here, one major technological challenge lies in the development of efficient methods to assemble particles into 2D ordered and stable arrays. A variety of approaches have been described;[1,8] however, much remains to be done, especially, for electrochemical studies, where various fundamental questions remain unanswered. For example, what is the electron-transfer rate constant? What dictates the electron-transfer kinetics involved? How can one use these surface nanostructures to achieve more complicated manipulation of interfacial charging events? Surface-immobilized assemblies of electroactive species represent a simplified and yet well-defined system where various approaches have been developed to study the electron-transfer chemistry. Thus, the first and essential step is to develop effective routes to assemble the MPC molecules into organized ensembles on electrode surfaces.

In the following sections, we will first present an introduction of classical Coulomb staircase, then the synthesis and general properties of MPC molecules, and their quantized charging properties in solutions at ambient conditions, followed by the single-electron-transfer processes with MPC organized assemblies where effects of solvent media, as well as electrolyte ions on the charging characteristics will be discussed. In particular, the ion-induced rectification of MPC quantized charging will be examined in detail. The solid-state conductivity properties, as well as electrochemistry of chemical functionalizations of these jumbo molecules will also be discussed.

10.2. CLASSICAL COULOMB STAIRCASE

Classical Coulomb staircase refers to a series of single electron tunneling events involving a miniaturized object or junction.[10,12] The fundamental physics behind this unique character is that the energetic barrier for a single electron transfer (e^2/C) is much larger than the thermal kinetic energy (k_BT), where C is the (tunneling) junction capacitance and k_B is the Boltzmann constant, such that a substantial potential change (voltage) is needed to provide the necessary driving force for the discrete charging events (and vice versa). Thus, one can see that the staircase charging can be achieved either at low temperature or at a junction with a very small capacitance. In a typical configuration, a two-junction (RC) system is coupled in series (Sch. 10.1), where the current–voltage (I–V) response is reflected in a staircase fashion.[10,12] Figure 10.1 (A) depicts the experimental I–V characteristics of a two-junction system obtained by probing a small indium droplet (ca. 10 nm in diameter) at $T = 4.2$ K. Here, one junction is formed between the STM tip and the droplet and the other between the droplet and the substrate. The experimental data are then fitted by this simple equivalent circuit with the parameter values being $C_1 = 4.14$ aF, $C_2 = 2$ aF, $R_1 = 132$ MΩ, and $R_2 = 34.9$ MΩ (B).

To initiate ambient-temperature Coulomb staircase, much smaller objects/junctions are required.[5] This is made possible only recently with the latest advances in nanofabrication and synthesis of nanoscale materials. Here, again, in

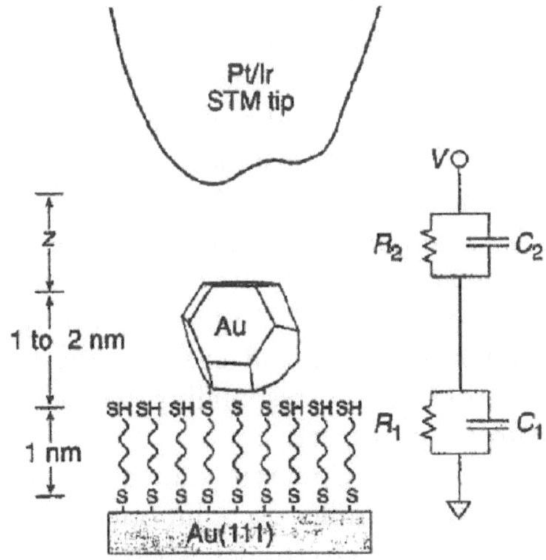

SCHEME 10.1. Double-junction model in an STM setup.[5]

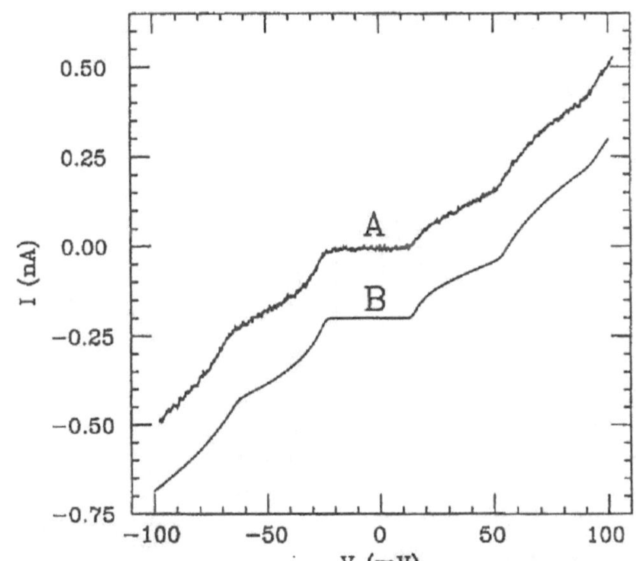

FIGURE 10.1. Curve A: Experimental I–V characteristic of a two-junction system obtained by probing a small indium droplet (approximately 10 nm in diameter) with a cryogenic STM. Curve B: Numerical fit of the experimental data of curve A, shifted by 0.2 nA for clarity. The parameters of the fitted curve are $C_1 = 4.14$ aF, $C_2 = 2$ aF, $R_1 = 132$ MΩ, $R_2 = 34.9$ MΩ, $V_p = 3.26$ mV, $T = 4.2$ K and $\alpha = 24\ V^{-2}$ (where $\alpha V^3/R$ is a nonlinear background term added to the tunneling rates). (Ref. 11b ©1991 American Physical Society).

a configuration similar to the aforementioned double-junction model, a nanoparticle is anchored to a substrate surface through a chemical bridge (e.g., rigid arenedithiols). When an STM tip is placed above the nanoparticle, the tunneling current generated by varying the potential bias between the STM tip and the substrate is the combined results of the tip/nanoparticle/substrate two-junction system (Sch. 10.1).[5] For instance, when a gold colloidal particle (~1.8 nm in diameter consisting of ~500 Au atoms) is anchored to a Au (111) single crystal surface by a self-assembled monolayer of α,α′-xylyldithiol (XYL), the corresponding tunneling current at ambient temperature exhibits the Coulomb staircase behavior (Fig. 10.2). Again, by fitting the experimental data with the double-unction model, one can evaluate the associated capacitances: $C_1 = 0.08$ aF, and $C_2 = 0.13$ aF, corresponding to the junctions of STM tip/particle and particle/substrate, respectively.

A comprehensive review of Coulomb staircase is beyond the scope of this chapter. However, the above examples do point out the technological importance of "nanoaturition" in molecular electronics where the applications of quantum effects can now be extended to ambient conditions.

It should be noted that the Coulomb staircase studies mentioned above are focused on a single particle. Analogous discrete charging phenomena have also been observed recently with an ensemble of (relatively) monodisperse nanoparticles in solutions at ambient temperature where the current responses are much larger such that a sophisticated STM setup is not required. This is dubbed the *Electrochemical Coulomb Staircase*, or quantized capacitance charging, an unprecedented electrochemical charge-transfer phenomenon. Detailed discussions are presented below.

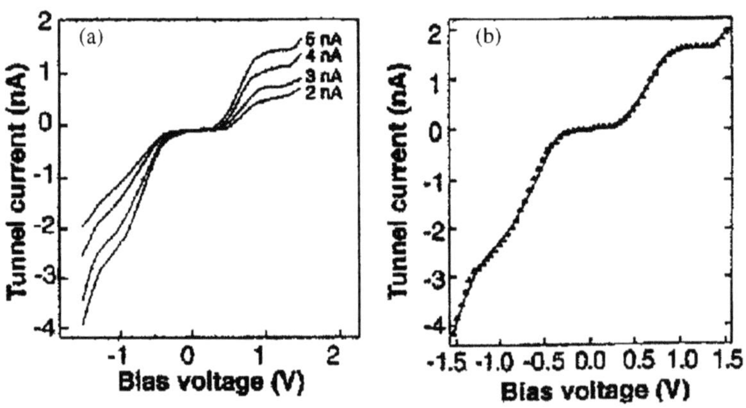

FIGURE 10.2. (a) Plot of $I(V, Z)$ data for a tethered 1.8 nm Au cluster on XYL, exhibiting Coulomb staircase behavior. The $I(V)$ curves were taken with a bias voltage set point of -1.75 V and tunneling current set points of 2–5 nA. Each of the $I(V)$ curves is an average of 100 individual voltage sweeps. (b) A typical comparison between the least square fit from the semiclassical model of Coulomb blockade and $I(V)$ data from an XYL-tethered Au cluster (see A). The data were obtained at room temperature using a Pt/Ir tip in a UHV STM. (Reprinted with permission Ref. 5 ©1996 American Association for the Advancement of Science)

10.3. NANOPARTICLE QUANTIZED CAPACITANCE CHARGING

10.3.1. Synthesis and Separation of Monolayer-Protected Nanoparticles

Monolayer-protected nanoparticles (MPCs) were first synthesized by Schiffrin *et al.*[9a] using a biphasic system. In a typical reaction, tetrachloroaurate ions were first dissolved in water and subsequently transferred to an organic (toluene) phase by a phase transfer catalyst (tetra-*n*-octylammonium bromide). The completion of the phase transfer was evidenced by the appearance of an orange organic layer and a colorless aqueous layer. After phase separation, a calculated amount of alkanethiols was added into the toluene phase which reacted with Au(III) to form a Au(I)-SR polymer intermediate. The subsequent addition of an aqueous solution of $NaBH_4$ into the solution initiated the reduction of Au(I) to Au(0), where the solution turned dark brown immediately, signaling the formation of nanosized gold particles. There are several experimental parameters that can be used to tailor the resulting nanoparticle dimensions, which include the initial feed ratio of gold and thiol, reaction temperature, reducing agents, specific thiol ligands, etc. For instance, Hostetler *et al.*[13] carried out a detailed study and found that the nanoparticle core diameter increases with deficient thiols, at higher temperatures, and with less potent reducing agents.[9] Typically, nanoparticles of size within the range of 1–5 nm in diameter could be readily accessible by manipulating the reaction temperature (from 0°C to room temperature, RT) and varying the S : Au ratio from 3 to 1/6. In addition, it is found that nanoparticles protected by arenethiols tend to be somewhat larger than those with alkanethiol protecting monolayers.[14] These effects can be understood in the context of nanoparticle formation mechanism which involves at least two competing processes, nucleation, and passivation.[15] By assuming a truncated octahedron core configuration, the number of gold atoms in the core can then be estimated (Table 10.1). In combination with other characterization techniques such as thermogravimetric analysis (TGA), nuclear magnetic resonance (NMR), and X-ray photoelectron spectroscopy (XPS), the number of organic protecting ligands per particle can be evaluated as well. Overall, the nanoparticle stoichiometric composition can then be obtained. Thus, the particles are referred to as jumbo molecules.

The above synthetic route can also be modified and extended to the fabrication of other metal nanoparticles, even alloy particles.[16-20] For instance, monolayer-protected platinum,[16] palladium,[17] silver,[18] and copper[19] particles within the similar size range have also been synthesized. In general, the particle sizes increase while the size dispersity decreases with a decrease in density of the metal elements under otherwise identical synthetic conditions. The former might be ascribed to the weaker interactions between the metal and thiols, which compromise somewhat the passivation process during particle growth. The latter can be understood in terms of the difference in metal–metal bonding interactions where a stronger interaction in the metal–metal bond leads to more tolerance in nanocrystal defects and hence more disperse distribution of the final particle size.[17]

In the synthesis of monolayer-protected alloy nanoparticles,[20] two fundamental questions arise. First, will the core composition reflect the initial feed ratio

TABLE 10.1. Results from Modeling of Gold Core Sizes, Shapes, and Alkanethiolate Coverages Based on SAXS, TGA and HRTEM-Determined Core Radii and Presumed Polyhedral Shapes (Compiled from Ref. 13)

Preparation, conditions[a]	Core radius R_{AVG}, (nm)	Number of atoms in avg. cluster (shape[b])	Number of surface atoms/%defect/area (nm^2)	Calc. TGA %organic/%coverage/number of chains
−78°, 2×, sd	0.76	116 (TO)	78/61%/11.36	31.8/68%/53
0°, 2×, fd	0.96	116 (TO)	78/61%/11.36	31.8/68%/53
		140 (TO^{+})	96/50%/11.43	27.9/55%/53
0°, 2×, md	0.98	201 (TO^{+})	128/47%/15.22	26.5/55%/71
0°, 2×, sd	0.98	201 (TO^{+})	128/47%/15.22	26.5/55%/71
		225 (TO^{+})	140/43%/15.19	24.4/51%/71
RT, 1×, fd	1.0	225 (TO^{+})	140/43%/15.19	24.4/51%/71
		314 (TO^{-})	174/41%/19.46	22.9/52%/91
RT, 4×, fd	1.0	225 (TO^{+})	140/43%/15.19	24.4/51%/71
		314 (TO^{+})	174/41%/19.46	22.9/52%/91
RT, 2×, sd	1.0	225 (TO^{-})	140/43%/15.19	24.4/51%/71
		314 (TO^{-})	174/41%/19.46	22.9/52%/91
RT, 2×, fd	1.0	225 (TO^{+})	140/43%/15.19	24.4/51%/71
		314 (TO^{+})	174/41%/19.46	22.9/52%/91
60°, 2×, sd	1.0	225 (TO^{+})	140/43%/15.19	24.4/51%/71
		314 (TO^{+})	174/41%/19.46	22.9/52%/91
90°, 2×, sd	0.96	201 (TO^{+})	128/47%/15.22	26.5/55%/71
RT, 1 2×, fd	1.4	586 (TO)	272/35%/28.94	19.1/50%/135
RT, 1 3×, fd	1.5	976 (TO)	390/31%/40.02	16.4/48%/187
RT, 1 4×, fd	2.2	2406 (TO)	752/22%/69.86	12.2/43%/326
RT, 1 6×, fd	2.2	2951 (TO^{+})	876/21%/79.44	11.4/42%/371
RT, 1 8×, fd	2.2	2951 (TO^{+})	876/21%/79.44	11.4/42%/371
RT, 1 10×, fd	2.4	4033 (TO)	1082/19%/97.00	10.3/42%/453
RT, 1 12×, fd	2.6	4794 (TO^{+})	1230/18%/108.28	9.7/41%/506

[a] sd, md, and fd, respectively denote slow, medium, and fast delivery of the reducing agent ($NaBH_4$) into the solution during particle synthesis, where the corresponding particle size dispersity is decreasing.
[b] TO = ideal truncoctahedron (all sides equal); TO^{+} = truncoctahedron in which ($0 < n - m \leq 4$), where n is the number of atoms between (111) facets and m is the number of atoms between (111) and (100) facets; TO^{-} = truncoctahedron in which ($-4 \leq n - m < 0$, $m > 1$).

of the varied metal elements? Second, what is the spatial distribution of these different metal elements within the nanosized cores? Using elemental analysis and XPS, Hostetler *et al.*[20] probed a series of binary, tertiary, and quaternary alloy nanoparticles and found that overall metal composition of the particle core vary with, but differ from, the original metal salt feed ratio with the more noble metal more favorably incorporated into the core. They ascribed this observation of differential incorporation to galvanic effects. For example, in Ag/Cu mixtures the Cu core content could be depressed by Cu^0 acting as an ancillary reductant of Ag^+, which leaches Cu^0 from the nascent core. As to the spatial distribution of different metals in the nanoparticle cores, the main issue is related to whether they form core-shell structures, partially segregated alloys, or pure alloys, and whether the ligand monolayer influences the core structure. By exploiting the difference in the metal–thiol bonding energy as examined by XPS and in combination with results of elemental analyses, they found that the more noble metals tend to prefer non-surface sites, but pure alloy structures have also been observed for AgPd nanoparticles.

Certainly, the nanoparticle obtained above are somewhat polydisperse and the exact dispersity is a function of a variety of parameters as well. A typical size histogram is shown in Fig. 10.3. For instance, one can see that a slow delivery of reducing agents into the gold–thiol intermediate solution leads to a disperse distribution of the nanoparticle core size. Thus, a challenging task in nanoparticle research is to separate these nanomaterials into monodisperse fractions.[21–23] One common approach is taking advantage of the size dependence of nanoparticle solubility.[22] For instance, the alkanethiolate-protected nanoparticles are only soluble in apolar solvents (e.g., alkanes, toluene, benzene, etc.) but insoluble in polar solvents (e.g., alcohols, acetone, and water). By using a binary mixture of solvents and non-solvents in varied mixing ratios, the nanoparticles of varied size fractions can be separated, as characterized by matrix-assisted laser-desorption ionization (MALDI) mass spectrometry (Fig. 10.4). In another approach,[23] supercritical ethane was used to separate gold nanoparticles into varied size fractions. This is understood in terms of the nanoparticle polarizability that is size sensitive.[23] For instance, for a conducting sphere of radius R, the polarizability (α_s) is proportional to the volume of the sphere, $\alpha_s \propto R^3$, whereas the total polarizability of an alkanethiolate-protected nanoparticle is approximately proportional to R^2, which takes into account the contributions of the alkane chains. Thus, for small particles, the relative contributions of these two components (protecting layers and cores) to the nanoparticle total polarizability are complicated. With nanoparticles of decreasing core sizes, the alkane layers will become the significant contributing component and hence dictate the particle solubility.

10.3.2. Nanoparticle Quantized Capacitance Charging

The alkanethiolate monolayers of the nanoparticles serve at least three purposes.[9] First, they act as the protecting layers against coalescence and aggregation of the particles so that the particles can be stable in both solution and air forms. This is one of the striking differences from conventional "naked" colloids. Second, the protecting ligands can be replaced by other thiol molecules by exchange reactions and/or further chemically functionalized by surface coupling reactions, thus providing a nanoscale platform for chemical manipulation. Third, because of the dielectric nature of the alkyl layers, the nanoparticles exhibit a molecular capacitance of the order of attofarad (aF, 10^{-18} F). It is these molecular capacitor characters that give rise to the Coulomb staircase observed in scanning electron microscopic (STM) measurements.[10a]

In 1997, Murray and coworkers[10a] first discovered that alkanethiolate-protected gold nanoparticles in solution exhibited electrochemistry analogous to STM Coulomb staircase. Here, the particles are protected by a monolayer of hexanethiolates and the core mass is 28 kDa, as determined by MALDI mass spectrometry, corresponding to 140 gold atoms and a core diameter of 1.6 nm.

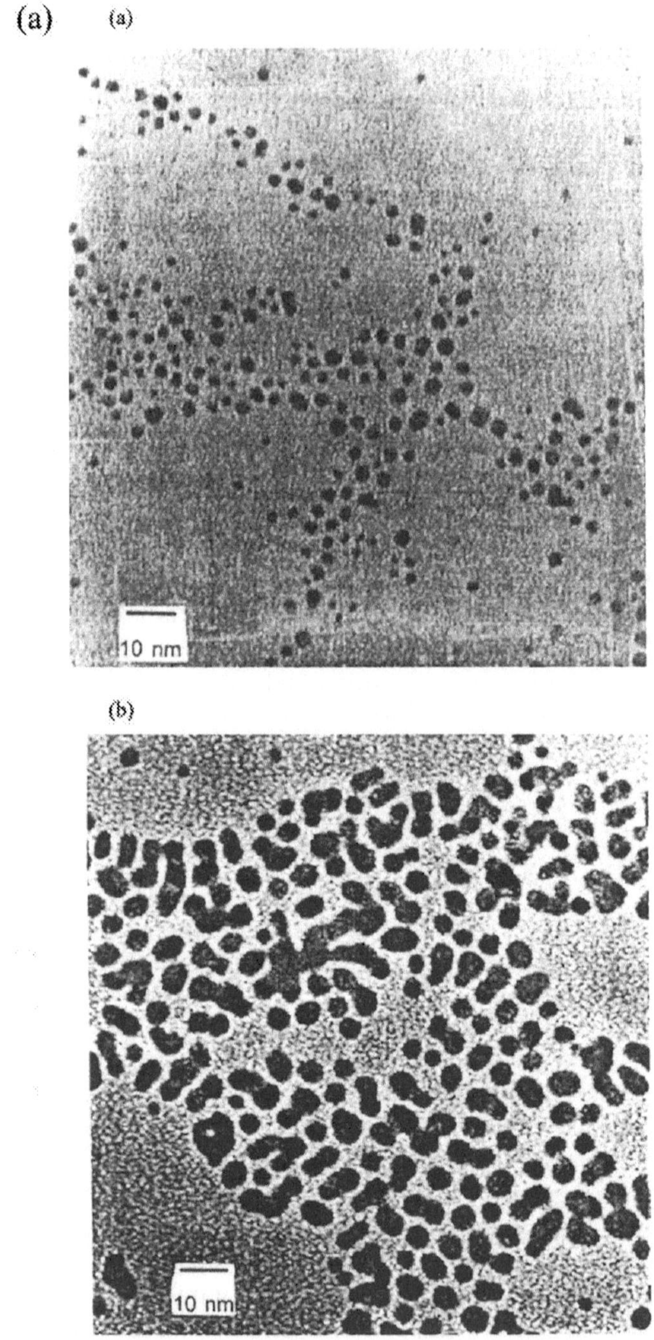

FIGURE 10.3. (a) Transmission electron micrographs of dropcast films of dodecanethiolate-protected Au clusters: (a) (0°, 2×, fd), (b) (RT, 1/6×, fd). (b) Size histograms: (a) (0°, 2×, fd); (b) (0°, 2×, sd); (c) (RT, 1/4×, fd); (d) (RT, 1/6×, fd). (Reprinted with permission from Ref. 13 ©1998 American Chemical Society)

(b)

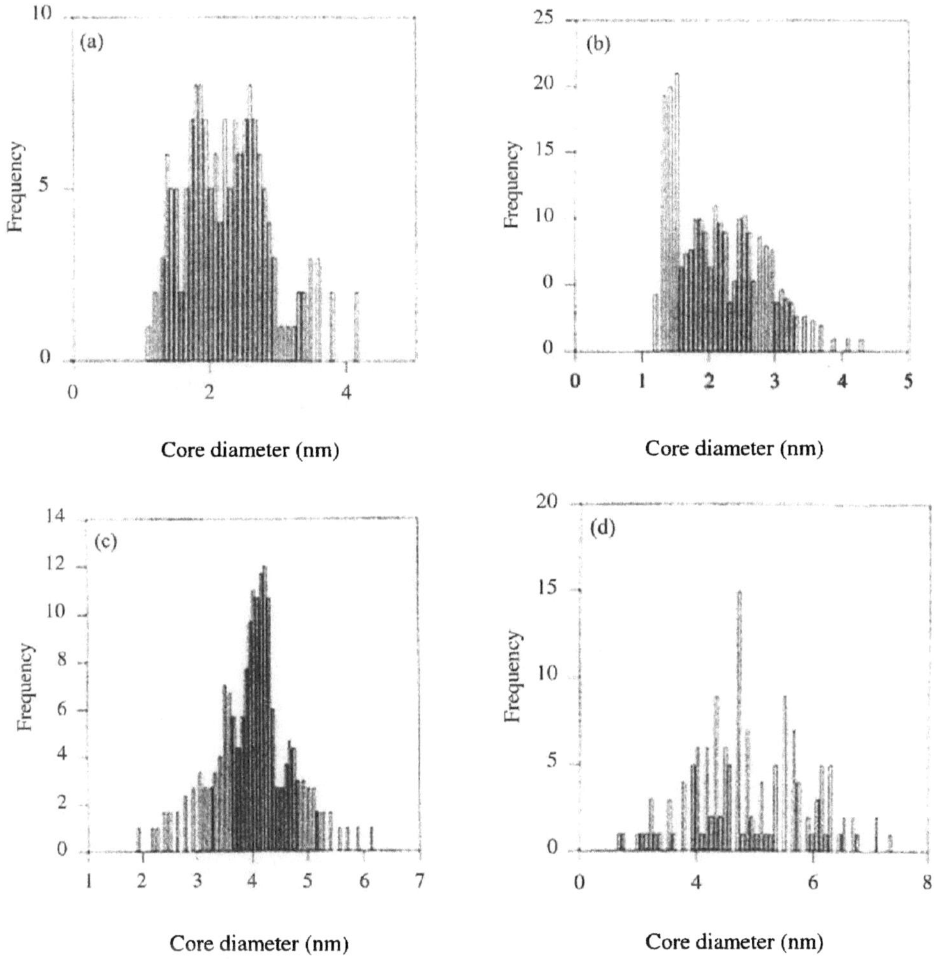

FIGURE 10.3. (*Continued*).

Figure 10.5 shows the STM I–V curve as well as the electrochemical responses (cyclic voltammograms (CVs) and differential pulse voltammograms (DPVs)). The main difference here is that in STM measurements, a single particle is probed whereas in electrochemical experiments, an ensemble of nanoparticles are charged and discharged at the electrode–electrolyte interface. Additionally, in contrast to the staircase features, electrochemical responses are a series of voltammetric peaks that appear to be diffusion-controlled. These voltammetric responses are ascribed to the quantized charging of the nanoscale double-layer of the particle molecules, that is, the nanoparticles behave as diffusive nanoelectrodes. From the potential spacing between neighboring charging peaks (ΔV), the nanoparticle molecular capacitance (C_{MPC}) can be evaluated, $C_{MPC} = e/\Delta V$ with e being the electronic charge. For instance, in the present study (Fig. 10.5), $C_{MPC} = 0.55$ aF.

Similar responses are also observed with nanoparticles of varied sizes and protected by monolayers of alkanethiolates of varied chain lengths (Fig. 10.6).[23]

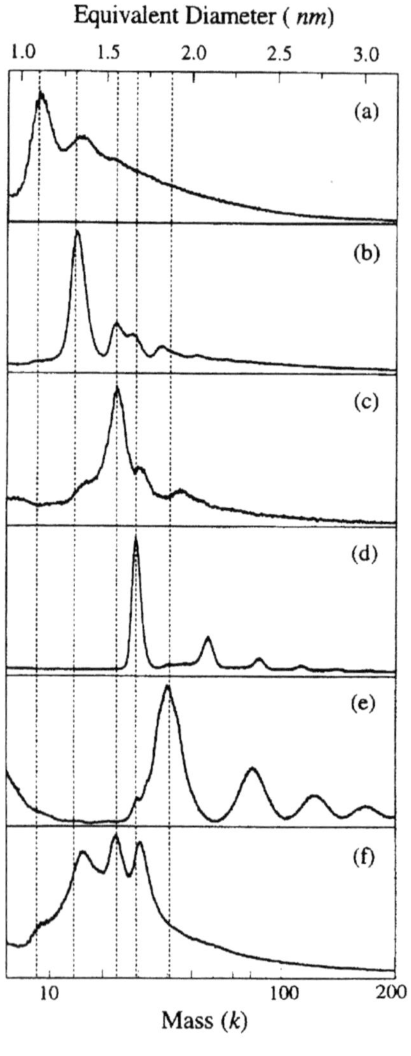

FIGURE 10.4. Low-resolution mass spectra, plotted as abundance vs. $(\text{mass})^{1/3}$, of size-separated clusters samples (b, c) and an unseparated raw mixture (a,f), obtained from growth of Au_N clusters from polymeric AuSR, where R is denoted by the n-alkyl chain length, C_n. The equivalent diameter of the cluster core (top axis) is calculated from the mass, assuming the density is that of bulk gold, and neglecting any contribution from bound SR groups retained, the dispersion in which is a principal source of peak broadening. The primary peaks lie in the 5–44 k range, corresponding to $N = \sim 30$–220 atoms. The spectra are obtained by laser desorption time-of-flight mass spectrometry, with negative ion detection, performed on neat (matrix-free) films of the molecular solid; this results in a series of secondary or "coalescence" peaks at multiples and combinations of the primary masses. All mass spectra are calibrated against protein standards and are corrected for ejection velocity effects. The vertical dashed lines are provided to indicate the correspondence of the various principal peaks of each separated sample to the species detected in raw mixtures, as well as to the impurities remaining as molecularly mixed (not phase separated) substances. Estimates of the (percentage) abundances of the main species present in the samples are (a) 8k (>75%), 14k (<25%); (b) 8k (~5%), 14k (80%), 22k (15%); (c) 14k (15%), 22k (60%), 28k (25%); (d) 28k (95%), 38k (<5%); (e) 28k (~10%); 34k (~30%), 38k (~60%); (f) example crude mixture includes 8k (10%), 14k (~30%), 22k (~30%), 28k (~30%), >30k (<10%). The R groups used in the preparation of the samples shown are (a) C_{12},(b) C_{18}, (c–e) C_6, (f) C_{12}. (Reprinted with permission from Ref. 22 ©1997 American Chemical Society)

The corresponding nanoparticle capacitances can also be evaluated. By assuming a concentric configuration of the nanoparticles, the nanoparticle capacitance is expressed as $C_{MPC} = 4\pi\varepsilon_0 \frac{r}{d}(r + d)$ based on electrostatic interactions (Eq. 10.2). The calculated values are consistent with the experimental data where the monolayer thickness is approximated to be the fully-extended chain lengths as calculated by Hyperchem® (Table 10.2).[10d] Overall, it is found that the nanoparticle capacitance increases with increasing core size but decreases with increasing alkanethiolate chain lengths. In addition, the capacitance is found to be sensitive to the peripheral charged groups as well. For instance, several-fold increase has been found in the nanoparticle capacitance when the peripheral ferrocene moieties are oxidized into ferroceneium (*vide infra*).

Similar quantized charging responses have also been observed with nanometer-sized alkanethiolate-protected particles of other metal elements, including palladium,[17] copper,[19] etc. These unique properties of single electron transfers,

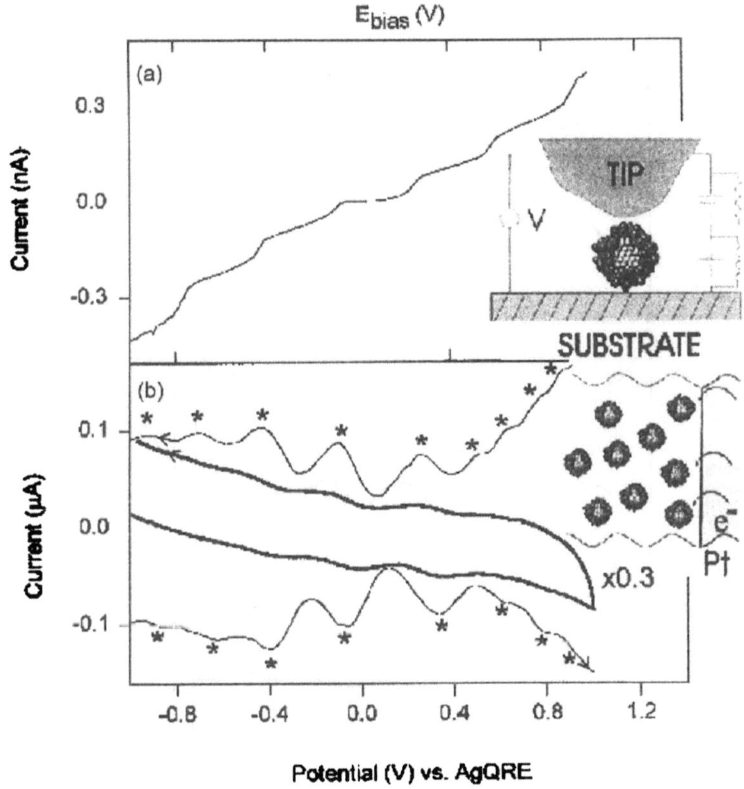

FIGURE 10.5. Panel A: Au STM tip addressing a single cluster adsorbed on an Au-on-mica substrate (inset) and Coulomb staircase *I–V* curve at 83 K; potential is tip–substrate bias; equivalent circuit of the double tunnel junction gives capacitances $C_{upper} = 0.59$ aF and $C_{lower} = 0.48$ aF. Panel B: voltammetry (CV —, 100 mV/s; DPV —. current peaks are indicated by asterisk, 20 mV/s, 25 mV pulse, top and bottom are negative and positive scans, respectively) of a 0.1 mM, 28 kDa cluster solution in 2:1 toluene:acetonitrile/0.05 M Hx_4NClO_4 at a 7.9×10^{-3} cm^2 Pt electrode, 298 K, Ag wire pseudoreference electrode. (Reprinted with permission from Ref. 10a ©1997 American Chemical Society)

along with their distinct optical characteristics, might provide a molecular platform for the manipulation of nanoscale electronic structures and devices.[1,2]

More importantly, when the nanoparticle core size decreases, a transition into molecule-like redox behaviors is found.[10b] Figure 10.7 depicts a series of DPVs of gold nanoparticles of varied sizes. One can see that with particles larger than 28 kDa (diameter 1.6 nm), the voltammetric charging peaks are quite evenly separated at potentials near 0 V (with a spacing of about 0.3 V), whereas with smaller particles, a central gap starts to emerge between the first positive and negative charging peaks. For instance, for particle core size of 8 kDa, the central gap is about 1.2 V. This is attributed to the evolution of a HOMO–LUMO energy gap due to shrinking size of the nanoparticle core, namely, quantum size effect. This is akin to the voltammetric responses of transition–metal complexes where

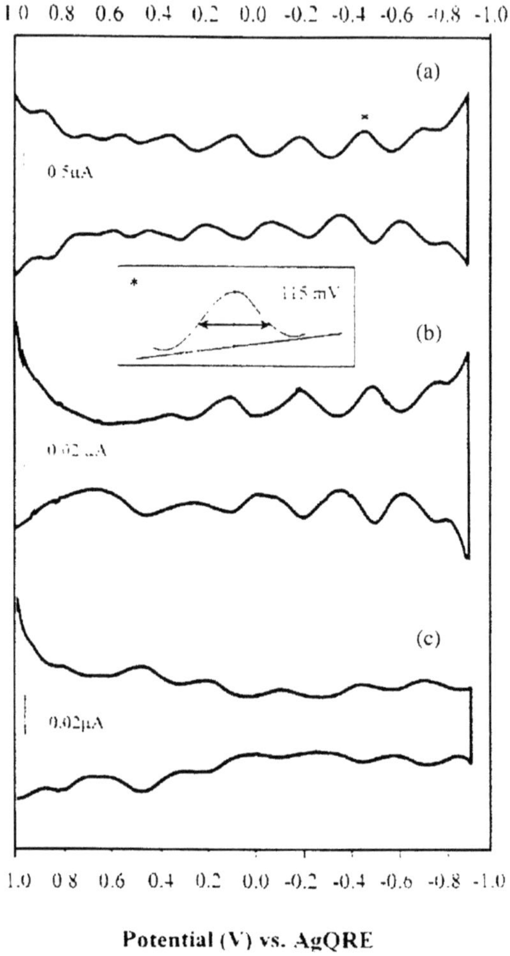

FIGURE 10.6. DPVs showing charging events for ca. 0.1 mM fractionated Au MPCs in dichloromethane (DCM) at a 1.6 mm diameter Au working electrode (0.05 M Bu_4NClO_4, potential vs. Ag QRE reference Pt coil counter electrode). All charging events shown are above background. (a) C6 MPC; (b) C8 MPC; (c) C10 MPC. Inset shows that peak* on expanded scale has FWHM 115 mV. (Reprinted with permission from Ref. 10c © 1999 American Chemical Society)

successive electron transfers are reflected by a series of voltammetric peaks with unequal potential spacings (the so-called odd–even effect). The growth of this energy gap with decreasing particle dimension is further manifested by NIR spectroscopic measurements. Figure 10.8 shows the spectra of gold nanoparticles of varied sizes. One can see that the threshold energy (HOMO–LUMO gap) grows with decreasing particle size, from <0.4 eV (limited by instrumental sensitivity) for large gold colloids (bulk-like) to 0.9 eV for 8 kDa size gold particles. Taking into account the electrostatic energetic barrier of 0.3 eV, the NIR result is very consistent with the electrochemical measurements. These observations clearly indicate a transition of the nanoparticle quantized charging from classical double layer charging to molecule-like redox behaviors with decreasing nanoparticle core size.

TABLE 10.2. Molecular Capacitance of Gold Nanoparticles with Varied Protecting Monolayers (Compiled from Ref. 10d)

Au MPCs	$C_{MPC,1}$ (aF)[a]	$C_{MPC,2}$ (aF)[b]	$C_{MPC,3}$ (aF)[c]
C4Au (14 kD[a])	0.22	0.49	0.35
C4Au (22 kD[a])	0.40	0.56	0.44
C4Au (28 kD[a])	0.50	0.69	0.59
C6Au (8 kD[a])	0.13	0.31	0.25
C6Au (22 kD[a])	0.41	0.45	0.51
C6Au (28 kD[a])	0.41	0.55	0.56
C6Au (38 kD[a])	0.53	0.71	0.62
PhC4SAu	0.66	0.74	0.49
PhC2SAu	0.89	1.22	0.93
4-CresolSAu	1.11	2.79	0.97

[a] From Eq. (10.1), $C_{MPC,1} = e/\Delta V$, where ΔV is the difference between peaks for $z = \pm 1/0$ (from Refs. 10b and 14).
[b] From Eq. (10.2), for C4 and C6 monolayers, $\varepsilon = 3$, $d_{C4} = 0.52$ nm, and $d_{C6} = 0.77$ nm; for PhC4S monolayer $\varepsilon = 3$, $d = 0.94$ nm, $r = 1.1$ nm; for PhC2S monolayer $\varepsilon = 4$, $d = 0.67$ nm, $r = 1.1$ nm; for 4-CresolS monolayer $\varepsilon = 5$, $d = 0.64$ nm, $r = 1.5$ nm; monolayer thicknesses (d) are approximated as fully extended chainlengths of the corresponding ligands (Hyperchem®).
[c] From plots of $E^{o\prime}_{Z,Z-1}$ vs. Z, where $E^{o\prime}_{Z,Z-1} = E_{PZC} + \frac{(Z-1/2)e}{(C_{MPC})}$.

Additionally, these studies clearly demonstrate that electrochemical methods are very effective in estimating the band gap energy of nanoparticles, which is also particularly useful in the investigations of semiconductor nanoparticles.[24] The band gap of semiconductor nanoparticles is generally much larger than that of the metal counterparts, and also increases with decreasing particle size (which typically reaches a few eV with the size in the nanometer regime). For instance, monolayer-protected PbS[24a] and CdS[24b] nanoparticles exhibit a series of voltammetric peaks at extreme negative and positive potentials, which are ascribed to the cathodic and anodic decomposition reactions of the nanoparticle molecules. Again, there is a large featureless region between the first positive and negative peaks, and the potential difference between them represents the HOMO–LUMO energy gap. Thus, electrochemical methods provide a convenient way to evaluate the nanosized particle band gap structure, which might serve as a complementary approach to optical methods (e.g., NIR spectroscopy[10b]).

10.3.3. Self-Assembled Monolayers of Nanoparticles

The unique properties of quantized charging of monolayer-protected nanoparticles demonstrate great application potentialities of these nanoscale building blocks in the development of novel electronic nanodevices/nanocircuits. One of the technological challenges is the fabrication of organized and robust assemblies of nanoparticles in a controllable manner. A variety of routes have been described,[1–8] among which a common approach is to utilize bifunctional ligands where one end is anchored to the substrate surface and the other attached to the nanoparticles. Typically, a monolayer of these bifunctional ligands is preformed which is then incubated into a nanoparticle solution to anchor the particles. For instance, Kubiak *et al.*[5] used XYL to assemble nanoparticles in

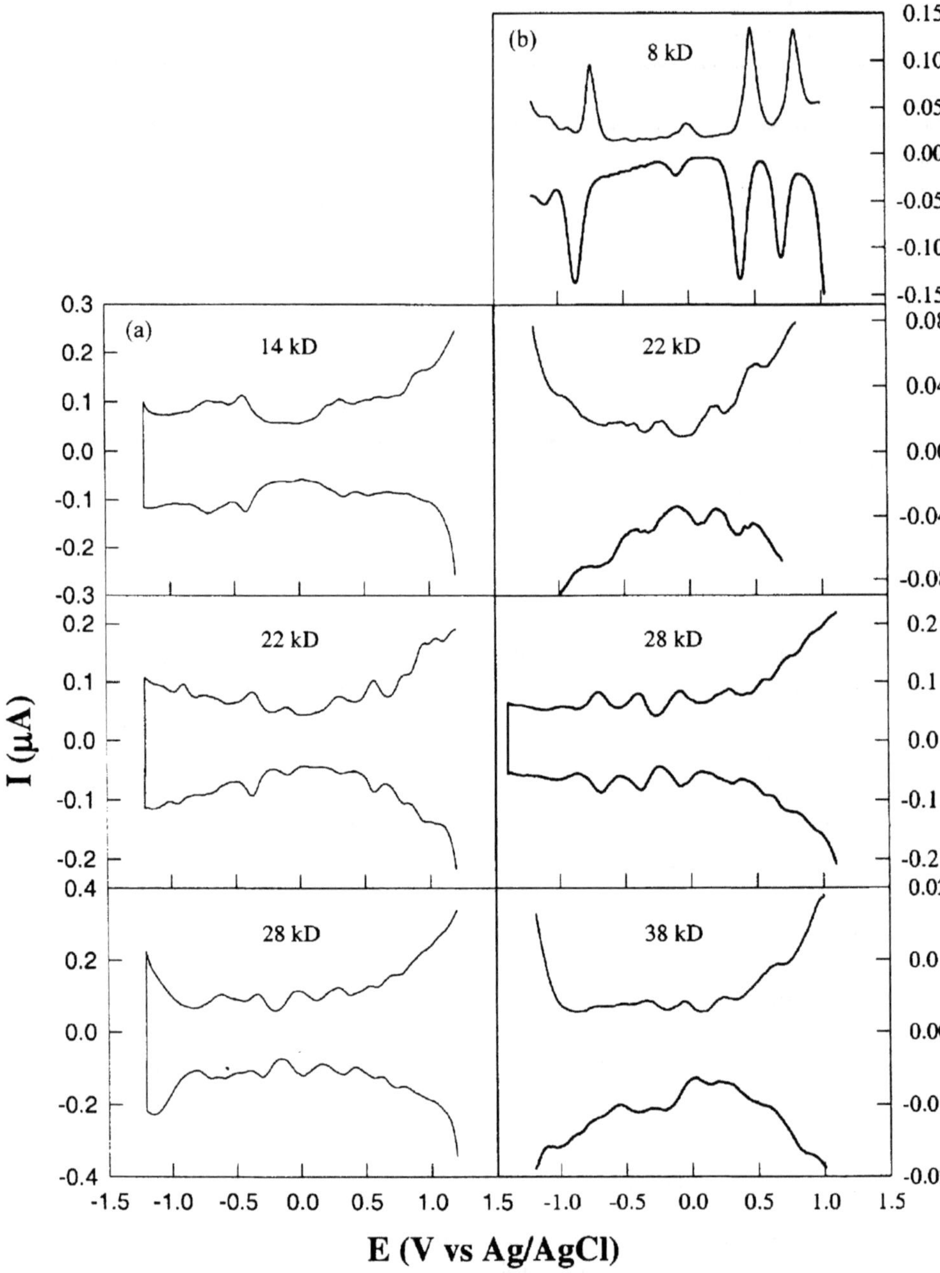

FIGURE 10.7. DPVs of (a) Butanethiolate (C4) and (b) hexanethiolate (C6) Au MPCs as a function of uniform core size, in 0.05 M Hex_4NClO_4/toluene/acetonitrile (2/1 v:v), at 9.5×10^{-3} cm^2 Pt electrode; DC potential scan 10 mV/s, pulse amplitude 50 mV. Concentrations are (a) 14 kDa, 0.086 mM; 22 kDa, 0.032 mM; 28 kDa, 0.10 mM; (b) 8 kDa, 0.30 mM; 22 kDa, 0.10 mM; 28 kDa, 0.10 mM; 38 kDa, 0.10 mM. (Reprinted with permission from Ref. 10b ©1998 American Association for the Advancement of Science)

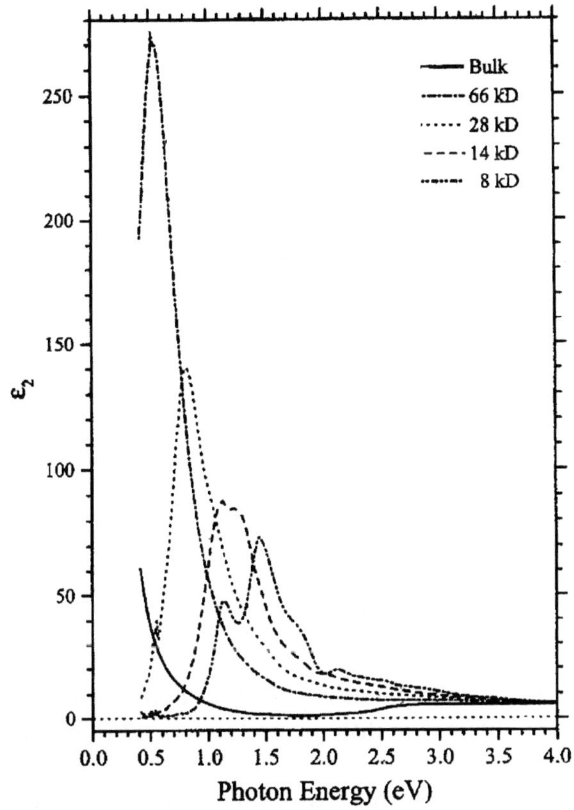

FIGURE 10.8. Optical response (ε_2) of monodisperse Au MPCs with various core sizes. The HOMO–LUMO gaps vary with core size: bulk Au, 0 eV; 66 kDa, $\ll$ 0.4 eV; 28 kDa, $\ll$ 0.4 eV; 14 kDa, ~0.6 eV; 8 kDa, ~0.9 eV. These curves were derived from a dispersion analysis (Kramers–Kronig relation) of the optical absorbance spectra, across the 0.2–6.4 eV range, of dilute solutions at room temperature. Absorbances at lower energies were obtained by extrapolating measured spectra and those in the far UV by assuming convergence to bulk gold values. Dielectric functions (ε_1 and ε_2) of the (assumed spherical) nanoparticles were computed from absorption coefficients and calculated index of refraction using a model of the effective dielectric constants for the nanoparticle–medium composite (Mie theory). (Reprinted with permission from Ref. 10b ©1998 American Association for the Advancement of Science)

their STM Coulomb staircase study. Schiffrin *et al.*[25] used viologen dithiol to build nanoparticle multilayer structures. Schmid *et al.*[2] used a similar approach to anchor gold nanoparticles onto a variety of substrates including glass, mica, and gold by first functionalizing the substrate surfaces with thiol-terminated groups. For large and "naked" colloids, this sequential deposition procedure is very effective for particle assembling. However, for small and monolayer-protected nanoparticles, the process could be quite tedious, presumably due to steric hindrance.[26] Thus, Chen *et al.* developed a new two-step procedure involving place-exchange reactions and self-assembling for the efficient surface-immobilization of nanoparticles.[27] In this approach (Sch. 10.2), alkanethiolate-protected nanoparticles first undergo surface place-exchange reactions with alkanedithiols of similar

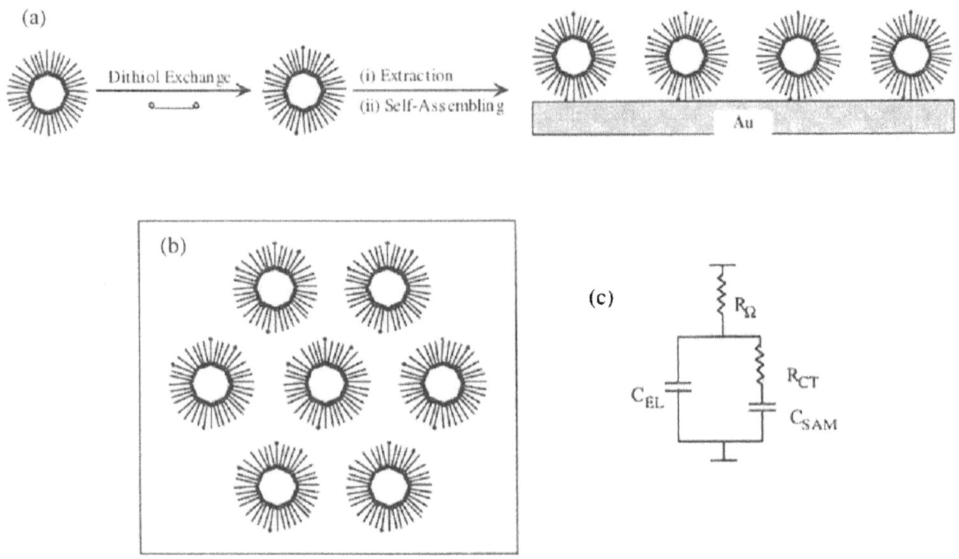

SCHEME 10.2. Self-assembling of surface-active nanoparticles.[27]

chain lengths, rendering the particles surface active with varied copies of peripheral thiol groups. Excessive dithiols and displaced thiolates are removed by phase extraction using a hexane–methanol system. Self-assembling of these surface-active nanoparticles is effected by incubating the electrode into the solution, just like monomeric alkanethiols. This approach can also be extended for the fabrication of other particle molecules, such as palladium nanoparticles.[28] The surface coverage can be readily assessed by a variety of techniques, including quartz crystal microbalance (QCM), UV–vis spectroscopy, etc.

Nanoparticle anchoring based on other specific interactions has also been reported. For instance, using the chelating interactions between transition-metal ions and carboxylic groups, monolayers and multilayers of nanoparticle have been constructed.[29] Nanoparticle organized assemblies have also been fabricated based on the specific chelation of transition metal ions to pyridine moiety.[30] In fact, one can envision that by exploiting the specific interactions between particle peripheral groups and surface terminated moieties, there will be almost countless ways for nanoparticle surface anchoring. Some interests have been focused on those that exhibit discrete electron transfers similar to those dissolved in solutions. Certainly, the primary criterion is that these particles are small (preferably ≤ 2 nm in diameter) and (relatively) monodisperse, where the difference involved in their chemistry is anticipated to give rise to variations of their charge-transfer properties (details given below).

10.3.4. Solvent Effects on Nanoparticle Quantized Capacitance Charging

Electrochemistry of these surface ensembles of nanoparticles also exhibit well-defined quantized capacitance charging features.[27,29,30] Figure 10.9 depicts the CVs and DPVs of a C6Au MPC self-assembled monolayer (Sch. 10.2) in 0.10 M tetra-*n*-butylammonium perchlorate (TBAP) in CH_2Cl_2.[27a] One can see that there are various voltammetric peaks within the wide potential range of −1.4 to + 1.0 V. The potential spacings between neighboring peaks are slightly smaller than those when the particles are dissolved in solutions (*vide infra*). Impedance spectroscopic measurements show a similar modulation of the interfacial

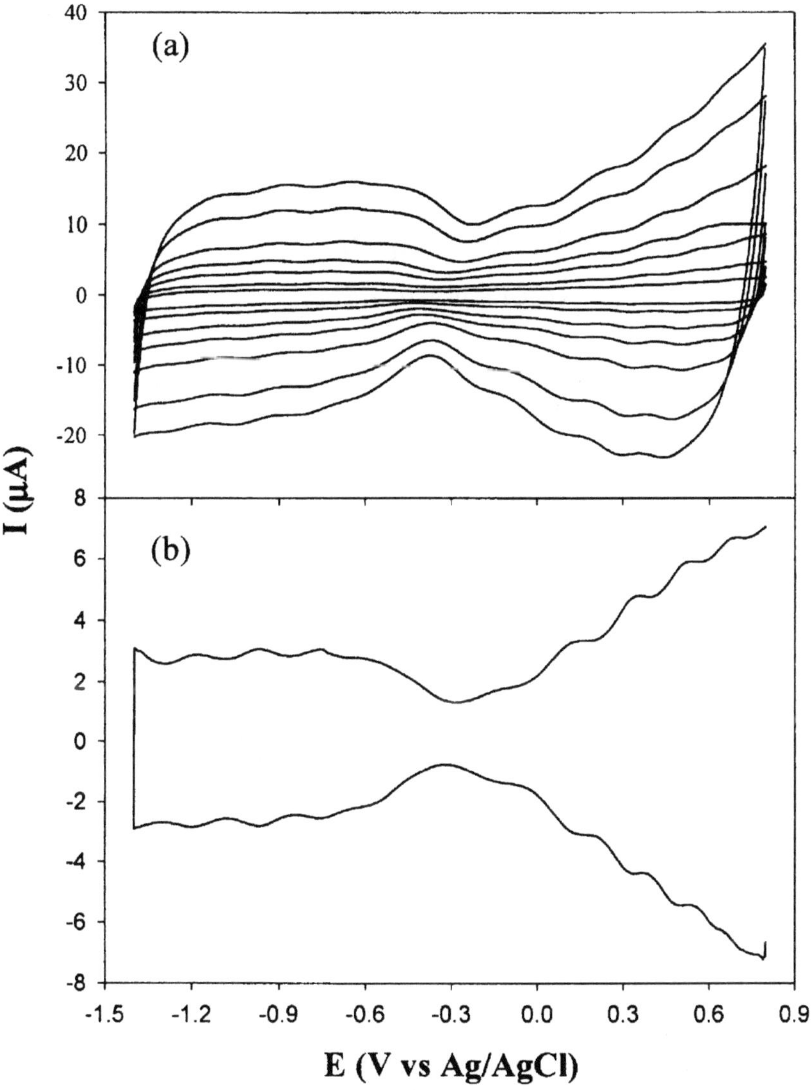

FIGURE 10.9. Cyclic (a, CV) and differential pulse (b, DPV) voltammograms of a self-assembled monolayer of C6Au particles onto a gold electrode surface in CH_2Cl_2 containing 0.1 M TBAP. CV sweep rates increase from 100 to 200, 400, 600, 902, 1505, and 2000 mV/s; and in DPV, the DC potential sweep rate 10 mV/s, pulse amplitude 50 mV. Electrode area 0.116 cm^2. (c) Plot of variation of charging formal potentials with MPC charge state (*z*-plot). Data obtained from DPV measurements. (Reprinted with permission from Ref. 27a ©2000 American Chemical Society)

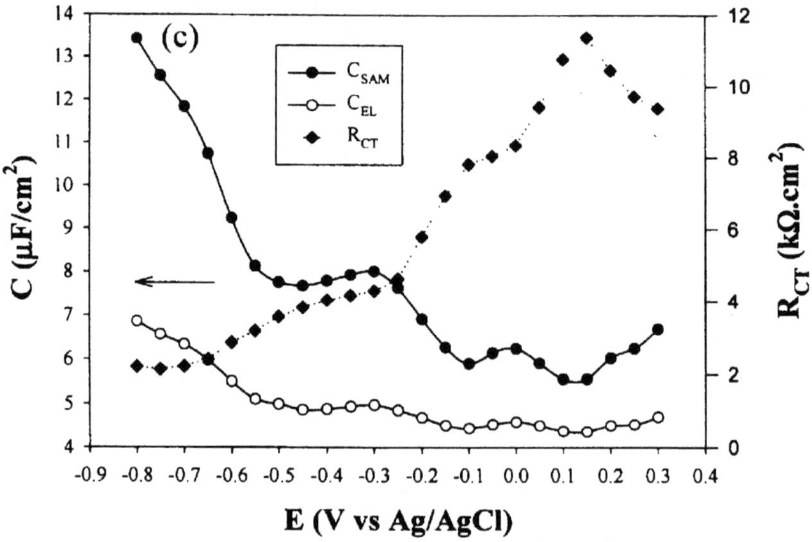

FIGURE 10.9. (*Continued*).

capacitance with electrode potentials (Fig. 10.9c), where the parameters are evaluated by using the Randles equivalent circuit (Sch. 10.2 inset) to fit the experimental data. Here, R_{Ω} reflects the solution uncompensated resistance, R_{CT} the charge-transfer resistance of the nanoparticle molecules, and C_{EL} and C_{SAM} the electrode interfacial capacitance from the "naked" electrode surface (interparticle void space) and the collective contributions of all surface-confined nanoparticles, respectively. Of note is that the interfacial capacitance exhibits a minimum at around –0.2 V. This is defined as the potential of zero charge (PZC) of the nanoparticle self-assembled monolayers. The technological significance is that based on this, one can further define the charge states of the nanoparticle molecules depending on the electrode potentials. For instance, the valleys will dictate the particles at a specific charge state while at the peak potentials, a mixed-valence state is anticipated. One can envision that by manipulating the electrode potentials, particles of different charge states can be prepared by using bulk electrolysis. These can then be used as potent reducing or oxidizing reagents. In fact, Murray *et al.*[31] demonstrated that this is indeed a very feasible route to use nanoparticles as the nanoscale charge storage devices.

Nanoparticle assemblies fabricated on the basis of other specific interactions also exhibit well-defined quantized charging features.[29,30] Figure 10.10 shows the CVs and DPVs of nanoparticles surface ensembles of varied layers by divalent metal ion and carboxylic interactions. Again, there are various voltammetric peaks corresponding to the successive single-electron transfer events; and the voltammetric currents increase with the number of layers of nanoparticles. Of note is that with the long chemical bridge between the electrode and the nanoparticles,[29] the peak splitting (ΔE_P) in these cases is somewhat larger than that with nanoparticles linked by alkanedithiols of similar chainlengths,[27] indicating a sluggish electron-ransfer process due to the long tunneling distance.

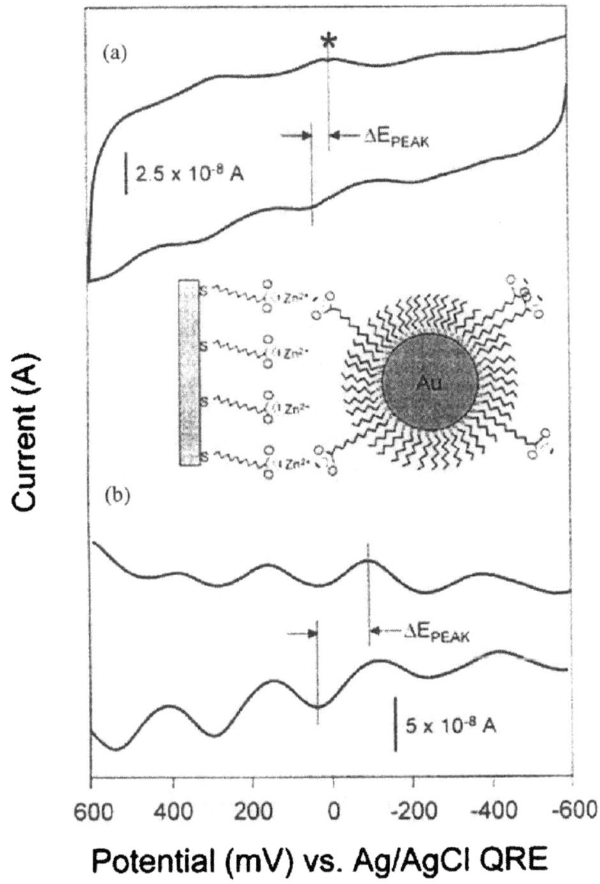

FIGURE 10.10. (a) Cyclic and (b) differential pulse voltammetry of Au C6/C10COOH MPCs (~48/5) attached with one dipping cycle to a Au (0.02 cm^2)/MUA substrate as illustrated. DC potential scan rate 50 mV/s, 0.1 M Bu_4NPF_6/CH_2Cl_2. (Reprinted with permission from Ref. 29 ©2000 American Chemical Society)

It has to be noted that these studies are carried out in (low-dielectric) organic media. In aqueous solutions, the electrochemistry of these nanoparticle assemblies is drastically different.[27b,c] Figure 10.11 shows the (a) CVs and (b) DPVs of a C6Au self-assembled monolayer in an aqueous solution containing 0.1 M NH_4PF_6. There are at least three aspects that warrant attention. First, several well-defined voltammetric peaks are found in the positive potential regime, which are ascribed to the quantized charging of nanoparticles. This is the first observation of nanoparticle electrochemical quantized charging in *aqueous* solutions. Second, the quantized charging features are sensitive to electrolyte composition. In the present study, the quantized charging features are only observed in the presence of PF_6^- ions but not NO_3^-. Also, the voltammetric responses do not appear to be dependent on the cationic species. Third, the voltammetric current is much larger in the positive potential regime (compared to that at the same bare electrode) but somewhat suppressed in

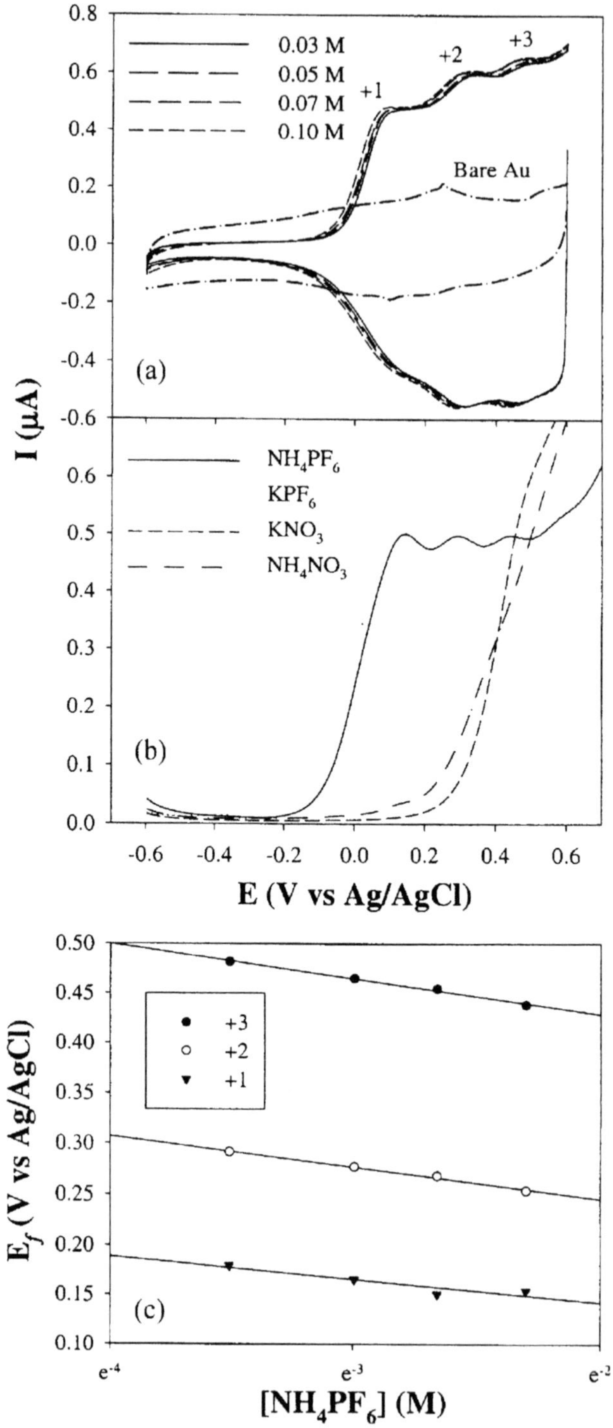

FIGURE 10.11. (a) CVs of a C6Au MPC-modified Au electrode in aqueous NH_4PF_6 solutions of various concentrations. Also shown is the CV of the same bare electrode in 0.1 M NH_4PF_6. Electrode area 1.1 mm^2. Sweep rate 100 mV/s. (b) DPVs of the same MPC-modified electrode in various electrolyte solutions (0.1 M). Pulse amplitude 50 mV, dc ramp 4 mV/s. (c) Variation of the formal potentials with the concentration of NH_4PF_6. Lines are linear regressions. +1, +2, and +3 refer to the three voltammetric peaks observed in (a), respectively. (Reprinted with permission from Ref. 27b © 2000 American Chemical Society)

the negative region. In essence, this behaves like a molecular diode (current rectifier). This is interpreted on the basis of the effects of the binding of hydrophobic anions like PF_6^- to the surface-confined nanoparticle molecules on interfacial double layer capacitance. As in (low-dielectric) organic media, C_{EL} is generally smaller than C_{SAM}, the double-layer charging is mainly through the surface-immobilized particle molecules, which is observed at both positive and negative potentials; whereas in aqueous solutions, $C_{EL} > C_{SAM}$, thus generally the double-layer charging is through the naked electrode part, resulting in featureless responses only. However, in the presence of hydrophobic anions, they bind to positively charged particle molecules (at positive electrode potentials), repelling water molecules from the interface and hence rendering $C_{SAM} > C_{EL}$. This ion-pairing hypothesis is further evidenced by the cathodic shift of the voltammetric profile with increasing PF_6^- concentration, as

$$E_f = E^{o\prime} + (RT/n_aF)\ln(K_2/K_1) - [(p-q)RT/n_aF]\ln[PF_6^-] \tag{10.3}$$

where E_f and $E^{o\prime}$ are the formal potentials in the presence and absence of ion-pairing, respectively, K_2 and q (K_1 and p) are the equilibrium constant and number of ions bound to the reduced (oxidized) form of particle molecules, respectively; other parameters have their usual significance. One can see from Fig. 10.11c that all peak potentials shift cathodically with increasing PF_6^- concentration, and the slopes evaluated from linear regressions are −23, −30, and −35 mV for the first, second, and third charging peaks, respectively, effectively indicating a 1:1 ratio of the number of PF_6^- ions and the MPC charge states ($n_a = 1$). It is also anticipated that the binding of anionic PF_6^- to surface MPCs increases the *effective* dielectric of the MPC protecting monolayers, leading to the larger value of the MPC capacitance (1.05 aF).

More detailed studies involving other hydrophobic (soft) anions, such as ClO_4^-, BF_4^-, exhibit similar rectified charging features, further supporting the notion that the nanoparticle quantized charging can be manipulated by simple ion-binding chemistry. In Fig. 10.11, one might also note that in the presence of "hard" anions (e.g., NO_3^-) the charging features are similar to those of a conventional molecular diode. At potentials more negative than the threshold value, the current is significantly suppressed; whereas at potentials more positive than the threshold potential, the current starts to increase rapidly (without the discrete charging characters). This also strongly suggests that increasing the "softness" of the electrolyte ions leads to a transition from conventional molecular diodes to single-electron rectifiers of the nanoparticle assemblies.[27c]

10.3.5. Potential Control of Rectification

As stipulated above,[27] the onset of the rectified quantized charging is ascribed to the binding of electrolyte anions to positively charged MPC molecules at positive electrode potentials. This is anticipated to be closely related to the MPC PZC (E_{PZC}) in the specific electrolyte solutions, akin to the effect of specific adsorption on the electrode interfacial double-layer structures.[32] From Fig. 10.11, one can see that the onset of the rectified charging current varies with different ions in the solutions, and the onset potential (E_{on}, which is close to E_{PZC}) shows a negative shift with anions of increasing hydrophobicity, for instance, from NO_6^- to PF_6^-. This might be accounted for by the binding of "soft" anions to the electrode interface which shifts the E_{PZC} to a more negative potential position and hence the negative onset of the quantized electron-transfer reactions of the particles. Table 10.3 gives the onset potentials for the MPC monolayers in a series of aqueous electrolyte solutions, where one can see that the anion effects can be grouped into the following sequence: (NH_4PF_6, NH_4PF_6, $TEAPF_6$) < (NH_4ClO_4, NH_4BF_4, $TEAClO_4$, $TMABF_4$) < (NH_4NO_3, KNO_3) < ($TEANO_3$, $TMANO_3$). In other words, the effect of anion-binding is decreasing in the order of $PF_6^- > ClO_4^- \approx BF_4^- > NO_6^-$, which is in the same order of the ionic radii (and hydrophobicity), PF_6^- (0.255 nm) > ClO_4^- (0.226 nm) > BF_6^- (0.218 nm) > NO_6^- (0.165 nm).[33] Furthermore, the effects of cations on the onset potentials appear to be consistent with the ion hydrophobicity as well, where one can see that in the presence of identical anions, the onset potential shifts anodically with increasing cation hydrophobicity, K^+ (0.151 nm) ≈ NH_4^+ (0.151 nm) < TEA^+ (0.265 nm) < TMA^+ (0.215 nm).[33]

These observations provide a mechanistic basis on which the potential regulation of MPC rectified quantized charging can be manipulated by electrolyte compositions. As mentioned above, the rectification of MPC quantized charging is more sensitive to electrolyte anions than to cations, in the context of the present experimental conditions. However, at the moment, well-defined rectification can only be initiated at positive electrode potentials. In order to achieve rectification in the negative potential regime, one will have to extend the system to other bulky cationic species as well as to exploit structural manipulation of the MPC surface structures to enhance the MPC and electrolyte ion interactions that might lead to the chemical regulation of the interfacial double-layer capacitance. Further studies are currently underway.

TABLE 10.3. Variation of Onset Potentials (E_{on}) of MPC rectification with Electrolyte Compositions[27c]

E_{on} (V)	C4Au	C6Au	C8Au	C10Au
NH_4PF_6	−0.26	−0.20	−0.28	−0.32
KPF_6	−0.26	−0.20	−0.28	−0.32
$TEAPF_6$	−0.18	−0.16	−0.20	−0.28
NH_4ClO_4	−0.12	−0.07	−0.16	−0.24
NH_4BF_4	−0.08	0	−0.14	−0.18
TEAP	−0.10	−0.05	−0.15	−0.20
$TMABF_4$	−0.02	0.01	−0.12	−0.16
NH_4NO_3	0.06	0.14	0	0
KNO_3	0.12	0.16	0.12	0.08
$TEANO_3$	0.15	0.28	0.16	0.20
$TMANO_3$	0.26	0.30	0.24	0.22

Note: (i) All electrolyte concentrations are 0.10 M. (ii) For C_6Au particles, the onset potentials in other electrolyte solutions are: TMAF, 0.28 V; TMAOH, 0.33 V; $TMACH_3SO_4$, 0.27 V; $TBAH_2PO_4$, 0.35 V; and $TBAHSO_4$, 0.35 V.

10.3.6. Electron Transfer Kinetics

The MPC quantized charging represents a novel electrochemical redox phenomenon, whose electron-transfer chemistry remains largely unexplored. Previously, two electrochemical techniques have been used to probe the kinetic aspects of these unique electron-transfer processes, one based on the Laviron method and the other based on AC voltammetry (or impedance spectroscopy).[27a] Both methods yielded a rather consistent rate constant. Some further studies are described here using MPCs with varied protecting alkanethiolate chainlengths (and the corresponding alkanedithiol linkers) to examine the electron-transfer mechanism. Figure 10.12a and b respectively shows a series of AC voltammograms measured with a C6Au MPC monolayer in CH_2Cl_2 containing 0.10 M TBAP and in aqueous 0.10 M NH_4PF_6. One can see that at low frequencies (e.g., 1 and 10 Hz), the quantized charging features of the surface MPC molecules are very visible; whereas at high frequencies (e.g., 200 and 1000 Hz), the discrete charge currents are only slightly larger than that of the background and become less well-defined. Figure 10.12c and d show the corresponding plots of the variation of the ratio of the peak currents (I_P) vs. background currents (I_B) with AC frequency. It can be seen that the overall behaviors are very similar to those observed with ω-ferrocenated alkanethiol monolayers on gold surfaces,[34,35] where at low frequencies, the ratio I_P/I_B is much greater than one whereas at high frequencies, the ratio I_P/I_B approaches unity. This can be understood in terms of the relative amplitudes of the electron-transfer kinetics between the surface-confined MPC molecules and the electrode electrons vs. the AC signal oscillation, and when the electron-transfer kinetics is not fast enough to keep up with the AC perturbation, the voltammetric responses start to level off. By fitting the curves with the Randles equivalent circuit (Sch. 10.2 inset), one can quantitatively estimate the corresponding electron-transfer rate constant for the MPC molecules,[35]

$$k_{et} = \frac{1}{2R_{CT}C_{SAM}} \tag{10.4}$$

Similar observations are also found with other MPCs with different protecting ligands, with the results summarized in Table 10.4. It can be seen that the electron-transfer rate constants evaluated here generally fall within the range of 10–100 s^{-1} which seem to vary only slightly with the charge states of the MPC molecules, and decrease with increasing chainlengths of the alkyl linkers. In addition, the kinetics appear to be somewhat faster in organic solutions than in aqueous media, which might be ascribed to the binding of electrolyte anions to the nanoparticles rendering the energetic barrier greater for further oxidation of the MPC molecules. Another contributing factor might be related to the solvation of the alkanethiolate protecting monolayers in organic media, where the particles were squashed down so that the gold cores more closely approach the electrode surface.

For electron tunneling through an organic barrier, the electron-transfer rate constant (k_{et}) has been found to decrease exponentially with the thickness of the

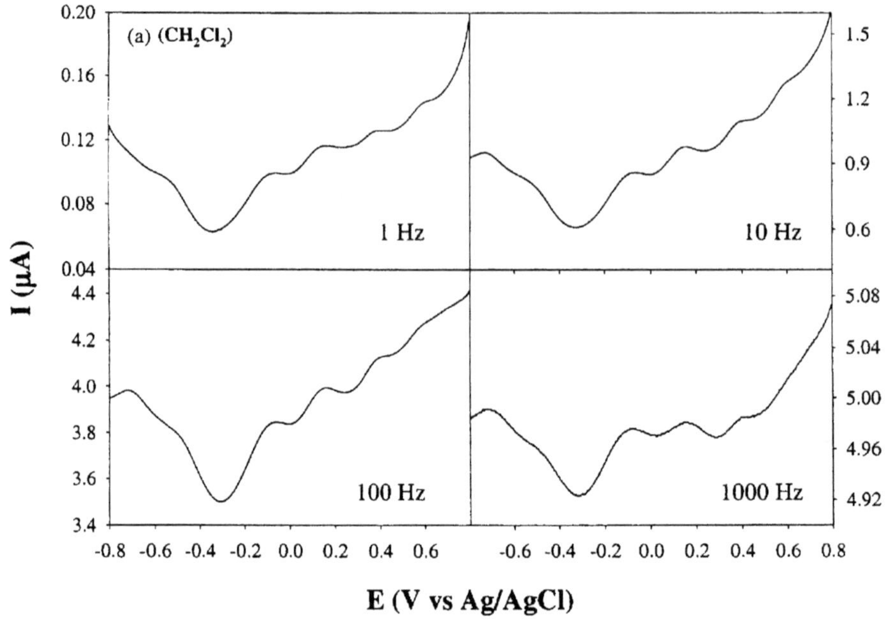

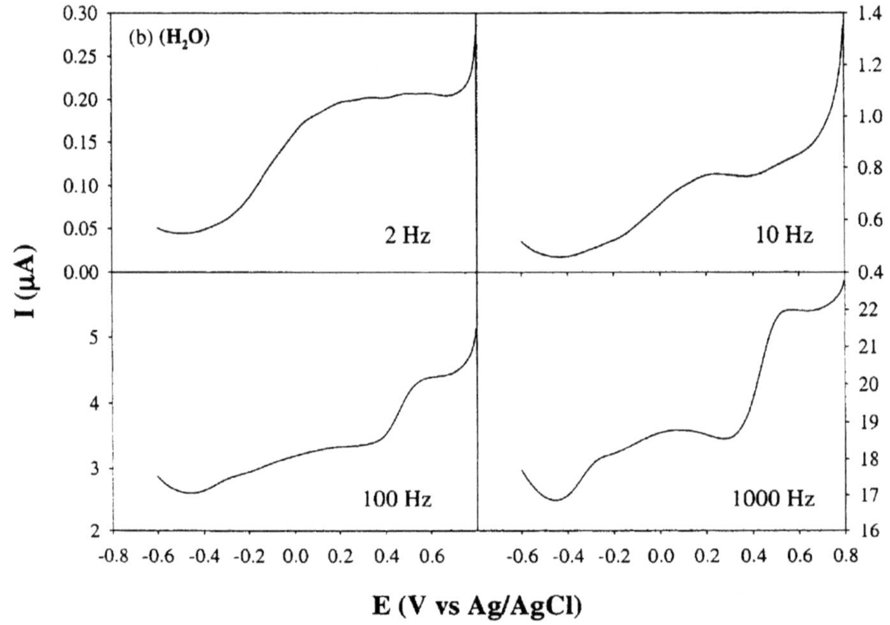

FIGURE 10.12. AC voltammograms of a C6Au MPC monolayer in (a) CH_2Cl_2 containing 0.10 M TBAP; and (b) aqueous 0.10 M NH_4PF_6 solution. Electrode area ca. 1 mm^2. The DC potential ramp 10 mV/s, AC amplitude 10 mV and the frequencies as shown. (c) and (d) show the variation of the ratios of the peak currents vs. background current with AC frequency at varied MPC charge states in CH_2Cl_2 and aqueous solutions, respectively. Symbols are experimental data and lines are for eye-guiding only. (Reprinted with permission from Ref. 27c ©2001 American Chemical Society)

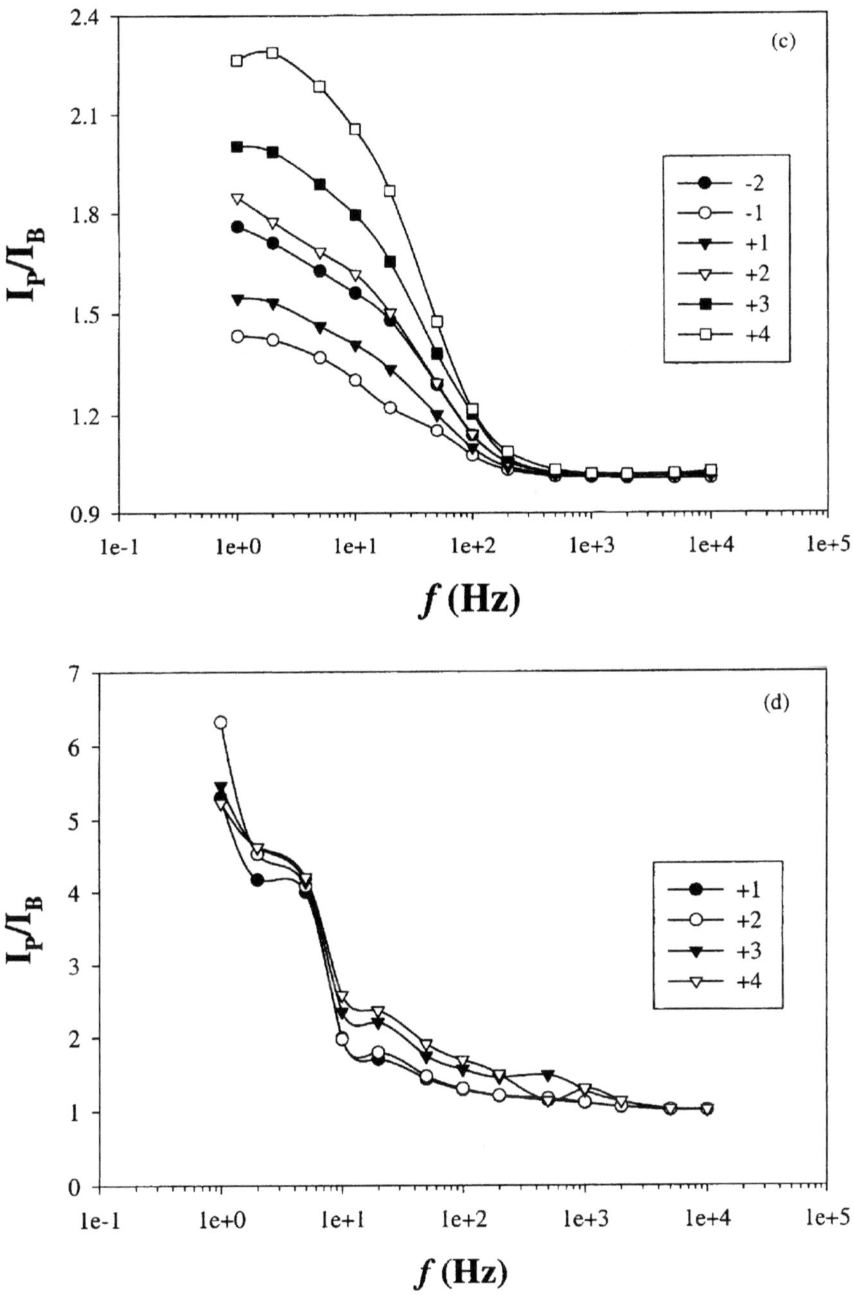

FIGURE 10.12. (*Continued*).

barrier (*d*), $k_{et} \propto e^{\beta d}$ with β being the coupling coefficient.[34] For straight alkyl spacers, the β values are typically found around 0.8–1.2 Å^{-1}.[34] Figure 10.13 shows the variation of the MPC electron-transfer rate constants in CH_2Cl_2 with the alkyl spacer length, where a relatively good linearity ($R^2 = 0.98$) can be seen. From

TABLE 10.4. Electron-transfer rate constants (k_{et}) of MPCs with varied protecting ligands and charge states (z) in different solution media[a]

	z						
k_{et} (s^{-1})	−2	−1	+1	+2	+3	+4	Average[b]
C4Au (CH_2Cl_2)	642.1	396.6	495.7	349	392.3	284.8	426.7 ± 126
C4Au (H_2O)			23.7	48.8	65.0	59.6	49.3 ± 18.3
C6Au (CH_2Cl_2)	113.4	83.7	108.5	100.4	106.7	98.3	101.8 ± 10.4
C6Au (H_2O)			15.2	11.7	18.2	22.9	17 ± 4.7
C8Au (CH_2Cl_2)		9.0	8.2	9.9	10.0		9.3 ± 0.8

[a]In H_2O, 0.10 M NH_4PF_6 was used as the supporting electrolyte while in CH_2Cl_2, 0.10 M TBAP was used instead.
[b]Averaged over all charge states.

the slope one can estimate the β value to be about 0.8 $Å^{-1}$. In the investigations of solid-state electronic conductivity of alkanethiolate-protected gold nanocluster molecules, Wuelfing *et al.*[36a] also obtained an electron-tunneling coefficient of $\beta = 0.8$ $Å^{-1}$. This value is close to that found previously with 2D self-assembled monolayers of alkanethiols with ω-functional groups (e.g., ferrocene).[34,35]

The above studies indicate that the electron-transfer kinetics of MPC molecules might not be as fast as one would think, which is a few orders of

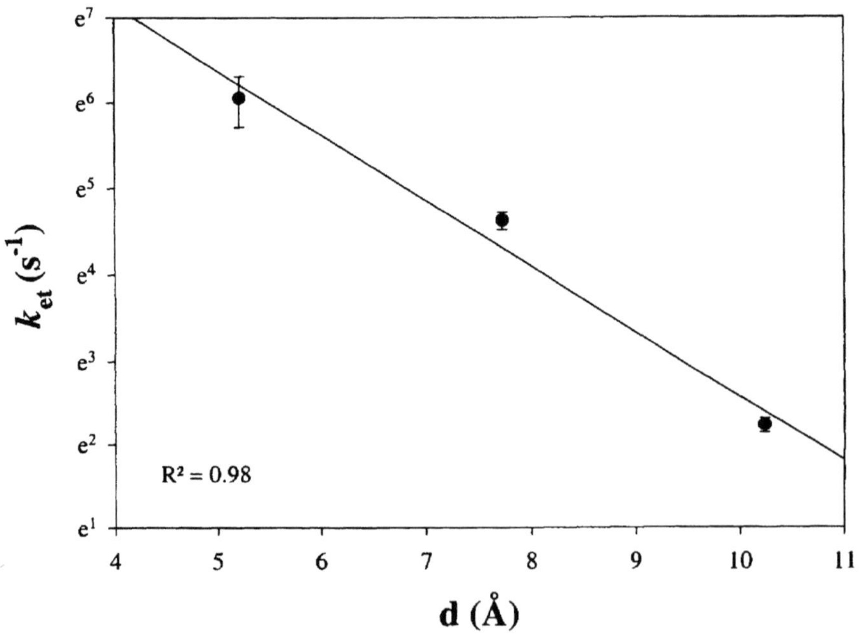

FIGURE 10.13. Variation of MPC electron-transfer rate constants with protecting ligand chainlengths in CH_2Cl_2 containing 0.10 M TBAP. Symbols are experimental data (Table 10.4) and line is the linear regression. (Reprinted with permission from Ref. 27c ©2001 American Chemical Society)

magnitude smaller than that of ferrocene moiety tethered to a similar alkyl spacer on the electrode surface,[34] or the electron self-exchange rate in nanoparticle multilayer thin films.[36b] Thus, one might argue that the rate constants measured above might actually reflect the ion-binding processes rather than the electron-transfer steps between MPCs and electrode. However, if this is the case, one would anticipate a rate constant (for a specific anion or cation) that is either invariant of the MPC structures, or more likely, (slightly) increasing with longer alkyl spacers due to increasing hydrophobicity of the particle protecting monolayers. This is contrary to the experimental observations shown above.

To provide further supporting evidence for the arguments presented above, Chen *et al.* [27c] carried out additional studies to examine the MPC electron-transfer chemistry by employing an electroactive species as the probe molecule, such as dopamine. Figure 10.14a shows the cyclic voltammograms of 1 mM dopamine in 0.10 M KPF_6 at a naked gold electrode surface, where a pair of voltammetric waves can be found at +0.08 and +0.21 V (at 100 mV/s) respectively. These are ascribed to the 2-electron, 2-proton redox reactions of dopamine (DA) into dopamine-o-quinone (DOQ),[37]

$$\mathrm{DA} \Leftrightarrow \mathrm{DOQ} + 2e^- + 2H^+ \tag{10.5}$$

Figure 10.14b depicts the variation of peak currents with the square roots of potential sweep rates where a good linearity is found, indicating diffusion-controlled reactions. However, the peak splitting ($\Delta E_p = 0.13$ V at 100 mV/s) is much greater than that expected for a reversible electron-transfer reaction (0.0285 V at $n = 2$, and 25°C),[32] suggesting that the reaction is kinetically rather sluggish. Thus, one might suspect that if the MPC electron-transfer is indeed very facile, they might be able to serve as electron-transfer mediators to facilitate the redox chemistry of dopamine. Or, if the MPC electron-transfer kinetics is not so fast, the redox processes of dopamine will be further impeded by the additional energetic barriers of MPC monolayers.

Figure 10.15a shows the cyclic voltammograms of the dopamine solution at a gold electrode with MPC monolayers with varied protecting monolayer ligands (the respective peak currents of dopamine redox reactions are again linearly proportional to the square root of potential sweep rates; not shown). One can see that the dopamine voltammetric currents are suppressed somewhat and the peak splitting becomes wider at MPC-modified electrodes than at the naked counterpart. These indicate that the MPC monolayers in fact serve as a blocking layer against the dopamine electron-transfer instead of being efficient electron-transfer mediators, as the overpotentials become even bigger with longer MPC protecting ligands. This implies that the electron-transfer kinetics of the MPC molecules cannot be very fast; in other words, the relatively sluggish electron-transfer kinetics of the MPC monolayers further slows down the relay of electron transfer for solution-phase dopamine to the electrode. Thus, the additional energetic barrier created by the MPC monolayers, $F\eta$, has to be taken into account in the evaluation of the dopamine electron-transfer rate constants,[32]

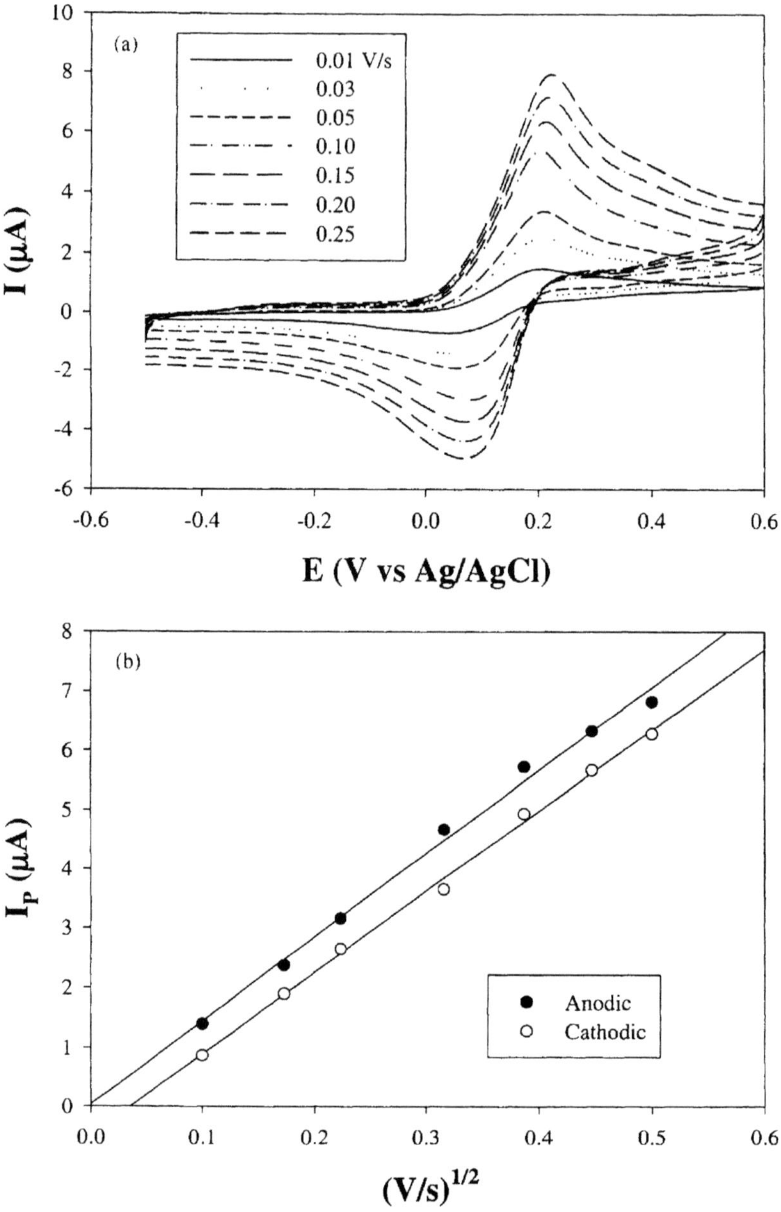

FIGURE 10.14. (a) Cyclic voltammograms of 1 mM dopamine in 0.10 M KPF_6 at a naked gold electrode surface (area 0.167 mm^2) at varied sweep rates. (b) Variation of peak currents with the square roots of the sweep rates. Symbols are experimental data and lines are linear regressions. (Reprinted with permission from Ref. 27c © 2001 American Chemical Society)

$$k_f = k_{f,o}e^{-\alpha f \eta} \quad \text{and} \quad k_b = k_{b,o}e^{(1-\alpha)f\eta} \tag{10.6}$$

where η is the measured overpotential, subscripts f and b refer to forward and back reactions, respectively; subscript o denotes the rate constants at the naked electrode; $f = nF/RT$ (n, F, R and T have their usual significance; and here $n = 2$);

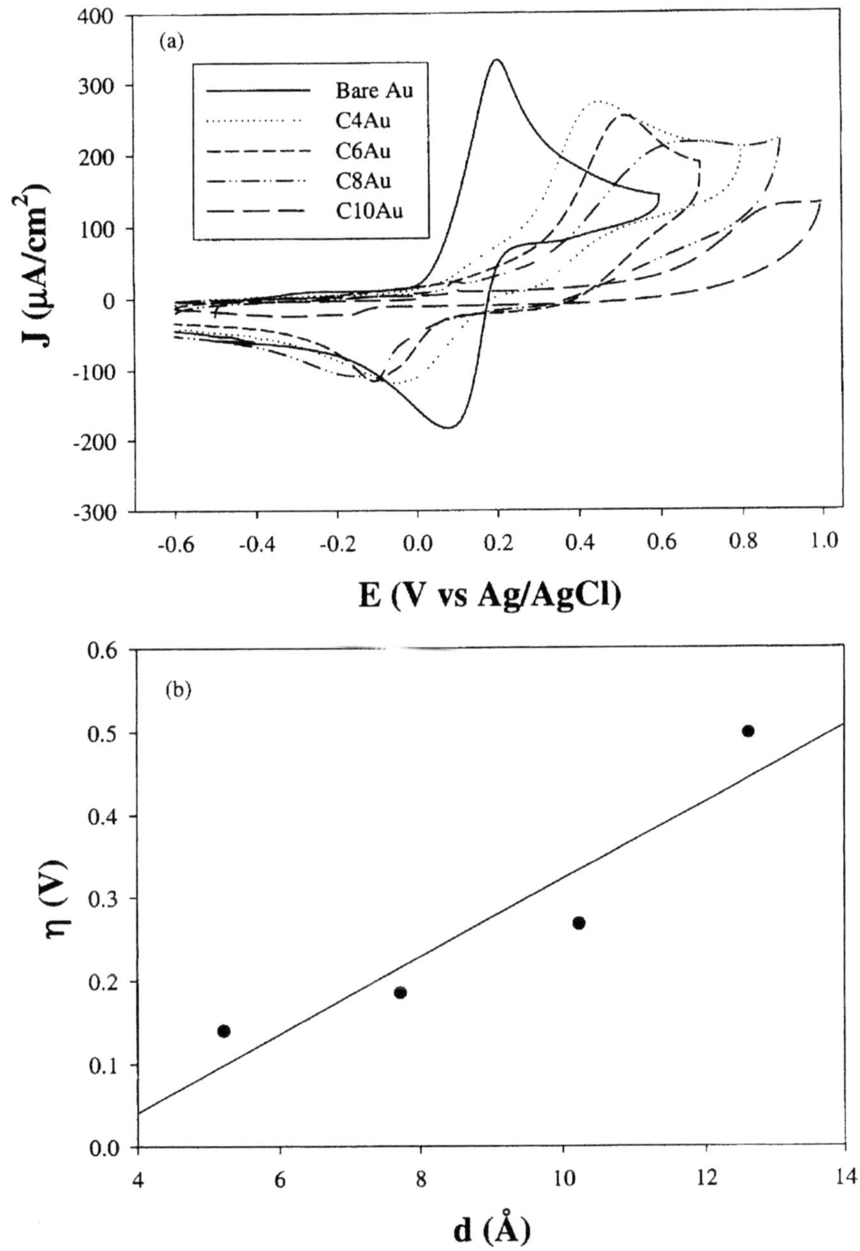

FIGURE 10.15. (a) Cyclic voltammograms of 1 mM dopamine in 0.10 M KPF_6 at a gold electrode with MPC monolayers of varied protecting alkanethiolates. Potential sweep rate 100 mV/s. (b) Variation of dopamine overpotentials with the MPC protecting ligand chainlengths. Symbols are experimental data and line is the linear regression. (Reprinted with permission from Ref. 27c ©2001 American Chemical Society)

and α is the transfer coefficient. As a first-order approximation, it is assumed that the additional (effective) electron-transfer pathway involves electron tunneling through two alkanethiolate ligands and a metal core (Sch. 10.3), since it has been

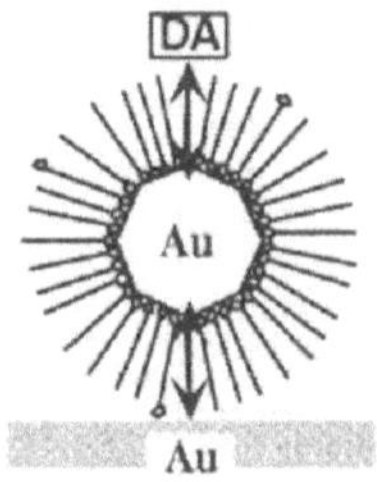

SCHEME 10.3. Schematic chart of electron tunneling through surface-confined nanoparticle layers.[27c]

found that through-bond electron-transfer is much more favorable than the chain-to-chain path.[34b] That is, the rate constant should bear a similar exponential dependence on the particle organic layer thickness (d), $k \propto e^{-2\beta d}$. Thus, in combination with Eq. (10.6), one can obtain

$$\eta \propto 2\beta d(RT/F) \tag{10.7}$$

by assuming $\alpha = 0.5$.[27a] Figure 10.15b depicts the variation of overpotential with the alkanethiolate chain lengths (relative to the peak potential positions at the naked Au electrode).[32] One can see that the data points are not in a good linearity as expected from Eq. (10.7), which might be ascribed to the over-simplification of modeling nanoparticle layers as a flat film with uniform thickness (equal to the particle diameter). However, for the sake of a qualitative comparison, from the slope of the linear regression, one can obtain a β value of about 0.9 Å^{-1}, which is somewhat greater than that calculated from the AC voltammetric measurements as shown above.

10.4. BULK-PHASE ELECTROCHEMISTRY OF NANOPARTICLES

10.4.1. Solid-State Conductivity

With the unique structure of (metal) core and (organic) shell, these nanoparticle materials possess interesting electrical properties. One can anticipate that these nanoscale composites behave as an intermediate of conductors and insulators, manifested readily by solid-state conductivity measurements.

Using dropcast thick films of nanoparticles onto an interdigitated array (IDA) electrode, Murray *et al.*[38] investigated the conductivity properties of these novel nanomaterials. Figure 10.16 shows the temperature-dependence of the nanoparticles with varied protecting monolayers. One can see that the materials resistivity increases with increasing chain length of the nanoparticle protecting monolayers and decreases with increasing temperature. These observations were ascribed to electron hopping between neighboring gold cores through the alkanethiolate layers. The electron transport rate is thought to be controlled by the combination of two mechanisms: (1) thermally activated electron transfer to create oppositely charged gold cores (cermet theory) and (2) distance-dependent tunneling through the oriented alkanethiolate layers separating them. From the fitting with an Arrhenius model, the activation energy was evaluated to be 9.6 $\pm$ 0.6, 16 $\pm$ 3, and 19 $\pm$ 2 kJ/mol for the C8, C12, and C16 particles,

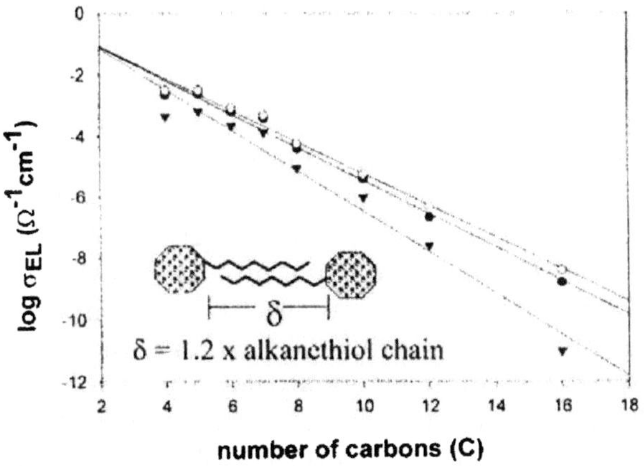

FIGURE 10.16. Plot of 70 (○), 30 (●), and −60°C (▼) conductivity vs. number of carbons in the alkanethiolate chains of $Au_{309}(C_n)_{92}$ MPCs. The inset is a schematic describing the interdigitation of monolayer chains in solid-state MPC films. (Reprinted with permission from Ref. 38 ©1995 American Chemical Society)

respectively. In addition, the tunneling coupling constant (β) is about 1.2 $Å^{-1}$. In another study,[39] using a similar approach, a transition from metal to insulator was observed as the particle size decreased from 7.2 to 1.7 nm in diameter (with an identical *n*-dodecanthiolate protecting monolayer).

The electrical properties of nanoparticle materials have also been probed using Langmuir monolayers at the air–water interface or Langmuir–Bldgett thin films on a substrate surface.[40,41] For instance, Majda *et al.*[40] used a line microband electrode that was brought into contact with the nanoparticle monolayer at the air–water interface and measured the current potential profiles at varied surface pressures. However, they found negligible conductivity. In contrast, Heath *et al.*[41] used a Langmuir–Bldgett trough to manipulated the interparticle distance of a monolayer of alkanethiolate-protected silver nanoparticles and measured the corresponding linear and nonlinear optical responses. They found that when the interparticle spacing was compressed to a mere 5 Å, the particle thin films were found to exhibit a transition from an insulator to a conductor. This transition was reversible by the simple compression and release of the mechanical barrier of the trough.

In summary, because of the composite nature of these monolayer-protected nanoparticles, their electrical properties can be tailored by the relative composition of the metal and organic components. Additionally, one can anticipate that the chemical structure of the organic layers (e.g., saturated vs. unsaturated chains) will provide another experimental parameter for the manipulation of these nanomaterials properties.

10.4.2. Electroactive Functional Groups

As mentioned earlier, the 3D monolayer structure of these nanoparticles also provides a unique platform for nanoscale functionalization. Here, new functional moieties can be incorporated into the particle surface by simple surface place exchange reactions.[28,42,43] For instance, by mixing gold nanoparticles protected by a monolayer of 8-octanethiolates (C8Au) with different calculated amounts of 8-mercaptooctylferrocene (C8FcSH),[43] one can introduce varied copies of ferrocene moieties into the particle. The overall electrochemistry of these ferrocene moieties is very similar to that of ferrocene monomers in terms of the formal potential position. However, the voltammetric responses at a steady-state rotating disk electrode (RDE) exhibit some unusual characters. In the solutions of monomeric ferrocene, the RDE profile depicts a (flat) limiting current plateau at potentials positive of the Fc formal potential, as dictated by the Levich equation.[32] In contrast, when the ferrocene moieties are immobilized onto the particle surface, substantially titled prewave and postwave are found (Fig. 10.17) and their slopes are found to increase with the square-root of rotation rate, indicating a diffusion-ontrolled process. These are interpreted based on the molecular capacitance characters of the nanoparticles, that is, the nanoparticles behave as diffusive nanoelectrodes in solutions, with a

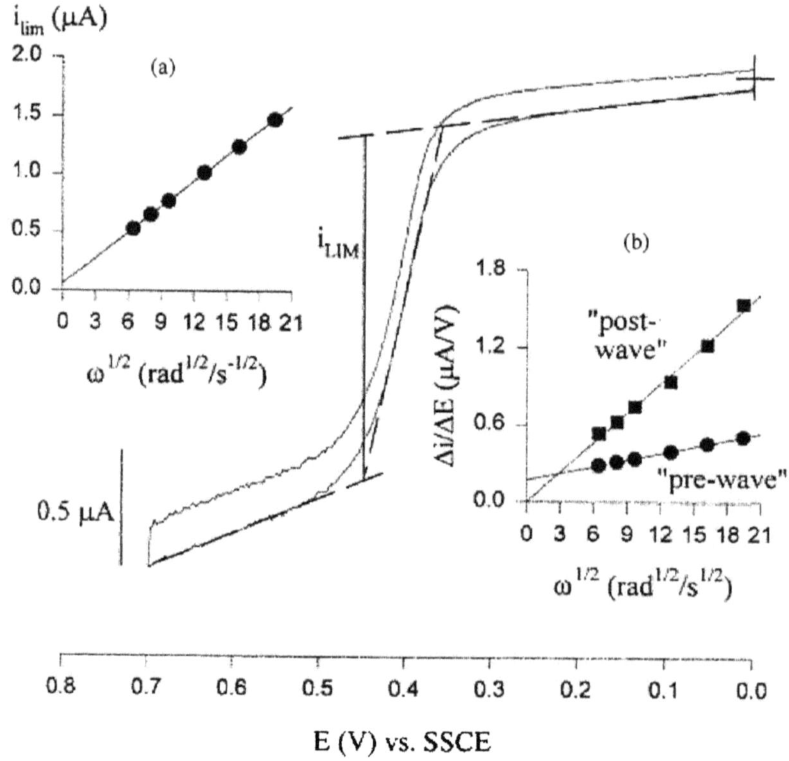

FIGURE 10.17. Voltammogram recorded at 10 mV/s at a 0.15 cm^2 glassy carbon rotating disk electrode (3600 rpm) in ca. 1 mM 1:5.5 C_8Fc/C_8 gold cluster compound in CH_2Cl_2/0.1 M Bu_4NClO_4. Inset A shows the dependence of the limiting current for this voltammetry on the square root of the electrode rotation rate ($\omega^{1/2}$), while inset B shows the dependence on $\omega^{1/2}$ of the prewave and postwave slopes of the voltammetry. (Reprinted with permission from Ref. 43a ©1997 American Chemical Society)

nanoscale double layer. The charging of these nanodouble-layers gives rise to the additional voltammetric currents measured. For instance, the prewave current is

$$I_{PRE} = 0.62AD^{2/3}\omega^{1/2}\nu^{-1/6}C^{*}[(\sigma_S - \sigma_{PRE})N_A A_{MPC} \quad (10.8)$$

where D and C^* are the MPC diffusion coefficient and bulk concentration, respectively; σ_S and σ_{PRE} are the charge densities on the MPC core surfaces at ambient solution and electrode (prewave) potentials, respectively; N_A is the Avogadro's constant, A_{MPC} and A are the MPC core and electrode surface areas, respectively. Thus, the MPC capacitance is

$$C_{MPC} = \left(\frac{\Delta i}{\Delta E}\right)[0.62AD^{2/3}\omega^{1/2}\nu^{1/6}C^{*}N_A]^{-1} \quad (10.9)$$

where $(\Delta i/\Delta E)$ is the slope of the (prewave or postwave) voltammetric profile. From the dependency on rotation rate (ω), the nanoparticle capacitance can then be evaluated. The MPC capacitance at postwave potentials is estimated in a similar manner. From Table 10.5, one can see that the particle capacitance increases several-fold with the peripheral ferrocene moieties oxidized to positively charged ferrocenium.

Of course, other chemical functional groups can be introduced into the particle monolayers as well. For instance, viologen moieties[28] have been incorporated into palladium nanoparticle protecting monolayers, which exhibit two well-defined voltammetric peaks corresponding to the successive $1e^-$ redox reactions of viologen, $V^{2+} \Longleftrightarrow V^{+\cdot} \Longleftrightarrow V^0$. Thiol-erivatives of anthroquinone[44] and phenothiazine[45] have also been exchanged into nanoparticle protecting monolayers. The electrocatalytic activities of particle-bound anthroquinone in the reduction of 1,1-dinitrocyclohexane are found to be better than the monomeric anthroquinone in identical concentrations.[44]

TABLE 10.5. Results of Diffusion Coefficients and Double Layer Capacitances for Ferrocenated Cluster Molecules (Compiled from Ref. 43)

		Ratio of postwave to prewave	C_{MPC} ($\mu F/cm^2$) 309 atom cuboctahedron model			C_{MPC} ($\mu F/cm^2$) 314 atom truncated octahedron model		
C_8Fc/C_8[a]	D_{MPC}[b] (10^{-6} cm^2/s)	C_{MPC}	θ_{Fc}[c]	Prewave	Postwave	θ_{Fc}[c]	Prewave	Postwave
1:2	3.6	7.9	34	3.1	24	35	2.4	19
1:5.5	2.8 (±11%)	5.7	16	2.5 (±25%)	14 (±36%)	16	1.9 (±25%)	11 (±36%)
1:9.5	2.7 (±4%)	3.4	10	2.8 (±3%)	9.6 (±10%)	10	2.2 (±3%)	7.5 (±10%)
1:24	2.6	3.4	4	3.5	12	4	2.7	9.2

[a] By ^{1}H-NMR; the relation numbers of terminal ferrocene to methyl groups, average, on a cluster molecule.
[b] Calculated from the limiting current using the Levich equation.
[c] Average ferrocene coverage on particle surface.

In combination with surface coupling reactions,[46] the chemistry of the nanoparticle molecules can be further decorated in seemingly endless ways, providing a facile route towards their versatile applications in diverse fields.

10.5. CONCLUDING REMARKS

The electrochemistry and electrical properties of monolayer-protected nanoparticles are reviewed here. The key element of these novel nanomaterials is their molecular capacitor characters, which give rise to the electrochemical quantized charging behaviors, an unprecedented charge-transfer phenomenon. In addition, these discrete charging properties are maintained when the particles are assembled onto a substrate surface forming stable ensemble structures. More interestingly, the charging processes can be rectified by a combination of unique electrolytes and solvent media, paving the way toward the application of these nanoscale building blocks in the construction of electronic nanodevices/nanocircuits. In addition, their bulk electrical properties can be tailored readily by the nanoparticle molecular composition and chemical structures, leading to a transition from insulator to metal easily accessible.

REFERENCES

1. (a) C. P. Collier, T. Vossmeyer, and J. R. Heath, Nanocrystal superlattices, *Annu. Rev. Phys. Chem.* **49**, 371 (1998).
(b) Z. L. Wang, Transmission electron microscopy of shape-Controlled nanocrystals and their assemblies, *J. Phys. Chem. B* **104**, 1153 (2000).
2. G. Schmid, M. Bäumle, M. Geerkens, I. Heim, C. Osemann, and T. Sawitowski, Current and future applications of nanoclusters, *Chem. Soc. Rev.* **28**, 179 (1999).
3. (a) C. B. Murray, C. R. Kagan and M.G. Bawendi, Self-organization of CdSe nano-crystallites into three-dimensional quantum dots superlattices, *Science* **270**, 1335 (1995).
(b) H. Weller, Self-organized superlattices of nanoparticles, *Angew. Chem. Int. Ed. Engl.* **35**, 1079 (1996).
4. R. G. Freeman, K. C. Grabar, K. J. Allison, R. M. Bright, J. A. Davis, A. P. Guthrie, M. B. Hommer, M. A. Jackson, P. C. Smith, D. G. Walter, and M. J. Natan, Self-assembled metal colloid monolayers: An approach to SERS substrates, *Science* **267**, 1629 (1995).
5. R. P. Andres, T. Bein, M. Dorogi, S. Feng, J. I. Henderson, C. P. Kubiak, W. Mahoney, R. G. Osifchin, and R. Reinfenberger, "Coulomb staircase" at room temperature in a self-assembled molecular nanostructure, *Science* **272**, 1323 (1996).
6. J. H. Fendler, Self-assembled nanostructured materials, *Chem. Mater.* **8**, 1616 (1996).
7. D. L. Feldheim and C. D. Keating, Self-assembly of single electron transistors and related devices, *Chem. Rev.* **27**, 1 (1998).
8. (a) C. J. Loweth, W. B. Caldwell, X. Peng, A. P. Alivisatos, and P. G. Schultz, DNA-based assembly of gold nanocrystals, *Angew. Chem. Int. Ed. Engl.* **38**, 1808 (1999).
(b) T. A. Taton, R. C. Mucic, C. A. Mirkin, and R. L. Letsinger, The DNA-mediated formation of super-molecular mono- and multilayered nanoparticle structures, *J. Am. Chem. Soc.* **122**, 6305 (2000).
9. (a) M. Brust, M. Walker, D. Bethell, D. J. Schiffrin, and R. Whyman, Synthesis of thiol-derivatised gold nanoparticles in a two-phase liquid–liquid system, *J. Chem. Soc., Chem. Comm.*, 801 (1994).

(b) A. C. Templeton, W. P. Wuelfing, and R. W. Murray, Monolayer-protected cluster molecules, *Acc. Chem. Res.* **33**, 27 (2000).
(c) R. L. Whetten, M. N. Shafigullin, J. T. Khoury, T. G. Schaaff, I. Vezmar, M. M. Alvarez, and A. Wilkinson, Crystal structures of molecular gold nanocrystal arrays, *Acc. Chem. Res.* **32**, 397 (1999).

10. (a) R. S. Ingram, M. J. Hostetler, R. W. Murray, T. G. Schaaff, J. T. Khoury, R. L. Whetten, T. P. Bigioni, D. K. Guthrie, and P. N. First, 28 kDa Alkanethiolate-protected Au clusters give analogous solution electrochemistry and STM Coulomb staircases, *J. Am. Chem. Soc.* **119**, 9279 (1997).
(b) S. Chen, R. S. Ingram, M. J. Hostetler, J. J. Pietron, R. W. Murray, T. G. Schaaff, J. T. Khoury, M. M. Alvarez, and R. L. Whetten, Gold nanoelectrodes of varied size: Transition to molecule-like charging, *Science* **280**, 2098 (1998).
(c) J. F. Hicks, A. C. Templeton, S. Chen, K. M. Sheran, R. Jasti, R. W. Murray, J. Debord, T. G. Schaaff, and R. L. Whetten, The monolayer thickness dependence of quantized double-layer capacitances of monolayer-protected gold clusters, *Anal. Chem.* **71**, 3703 (1999)
(d) S. Chen, R. W. Murray, and S. W. Feldberg, Quantized capacitance charging of monolayer-protected Au clusters, *J. Phys. Chem. B* **102**, 9898 (1998).
11. (a) A. E. Hanna and M. Tinkam, Variation of the Coulomb staircase in a two-junction system by fractional electron charge, *Phys. Rev. B* **44**, 5919 (1991).
(b) M. Amman, R. Wilkins, E. Ben-Jacob, P. D. Maker, and R. C. Jaklevic, Analytic solution for the current–voltage characteristic of two mesoscopic tunnel junctions coupled in series, *Phys. Rev. B* **43**, 1146 (1991).
12. (a) T. Sato, and H. Ahmed, Observation of a Coulomb staircase in electron transport through a molecularly linked chain of gold colloidal particles, *Appl. Phys. Lett.* **70**, 2759 (1997).
(b) T. Sato, H. Ahmed, D. Brown, and B. F. G. Johnson, Single electron transistor using a molecularly linked gold colloidal particle chain, *J. Appl. Phys.* **82**, 696 (1997).
(c) L. Guo, E. Leobandung and S. Y. Chou, A Silicon single-electron transistor memory operating at room temperature, *Science* **275**, 649 (1997).
(d) W. Lu, A. J. Rimberg, K. D. Maranowski, and A. C. Gossard, Single-electron transistor strongly coupled to an electrostatically defined quantum dot, *Appl. Phys. Lett.* **77**, 2746 (2000).
13. M. J. Hostetler, J. E. Wingate, C.-J. Zhong, J. E. Harris, R. W. Vachet, M. R. Clark, J. D. Londono, S. J. Green, J. J. Stokes, G. D. Wignall, G. L. Glish, M. D. Porter, N. D. Evans, and R. W. Murray, Alkanethiolate gold cluster molecules with core diameters from 1.5 to 5.2 nm: Core and monolayer properties as a function of core size, *Langmuir* **14**, 17 (1998).
14. S. Chen and R. W. Murray, Arenethiolate monolayer-protected gold clusters, *Langmuir* **15**, 682 (1999).
15. S. Chen, A. C. Templeton, and R. W. Murray, Monolayer-protected cluster growth dynamics, *Langmuir* **16**, 3543 (2000).
16. S. L. Horswell, C. J. Kiely, I. A. O'Neil, and D. J. Schiffrin, Alkyl isocyanide-derivatized platinum nanoparticles, *J. Am. Chem. Soc.* **121**, 5573 (1999).
17. S. Chen, K. Huang, and J. A. Stearns, Alkanethiolate-protected palladium nanoparticles, *Chem. Mater.* **12**, 540 (2000).
18. C. P. Collier, R. J. Saykally, J. J. Shiang, S. E. Henrichs, and J. R. Heath, Reversible tuning of silver quantum dot monolayers through the metal–insulator transition, *Science* **277**, 1978 (1997).
19. S. Chen, and J. M. Sommers, Alkanethiolate-protected copper nanoparticles: spectroscopy, electrochemistry and solid-state morphological evolution, *J. Phys. Chem. B* **105**, 8816 (2001).
20. M. J. Hostetler, C.-J. Zhong, B. K. H. Yen, J. Anderegg, S. M. Gross, N. D. Evans, M. D. Porter, and R. W. Murray, Stable, monolayer-protected metal alloy J. clusters, *J. Am. Chem. Soc.* **120**, 9396 (1998).
21. T. G. Schaaff, G. Knight, M. N. Shafigullin, R. F. Borkman, and R. L. Whetten, Isolation and selected properties of a 10.4 kDa gold : glutathione cluster compound, *J. Phys. Chem. B* **102**, 10643 (1998).
22. T. G. Schaaff, M. N. Shafigullin, J. T. Khoury, I. Vezmar, R. L. Whetten, W. G. Cullen, and P. N. First, Isolation of smaller nanocrystal Au molecules: Robust quantum effects in optical spectra, *J. Phys. Chem. B* **101**, 7885 (1997).

23. N. Z. Clarke, C. Waters, K. A. Johnson, J. Satherley, and D. J. Schiffrin, Size-dependent solubility of thiol-derivatized gold nanoparticles in supercritical ethane, *Langmuir* **17**, 6048 (2001).
24. (a) S. Chen, L. A. Truax, and J. M. Sommers, Alkanethiolate-protected PbS nanoparticles: Synthesis, spectroscopic and electrochemical studies, *Chem. Mater.* **12**, 3864 (2000).
(b) S. K. Haram, B. M. Quinn, and A. J. Bard, Electrochemistry of CdS nanoparticles: A correlation between optical and electrochemical band gaps, *J. Am. Chem. Soc.* **123**, 8860 (2001).
25. D. I. Gittins, D. Bethell, R. J. Nichols, and D. J. Schiffrin, Diode-like electron transfer across nanostructured films containing a redox ligand, *J. Mater. Chem.* **10**, 79 (2000).
26. S. Chen, and R. W. Murray, Electrochemical quantized capacitance charging of surface ensembles of gold nanoparticles, *J. Phys. Chem. B* **103**, 9996 (1999).
27. (a) S. Chen, Self-assembling of monolayer-protected gold nanoparticles, *J. Phys. Chem. B* **104**, 663 (2000).
(b) S. Chen, Nanoparticle assemblies: "Rectified" quantized charging in aqueous media, *J. Am. Chem. Soc.* **122**, 7420 (2000).
(c) S. Chen, and R. Pei, Ion-induced rectification of nanoparticle quantized capacitance charging in aqueous solutions, *J. Am. Chem. Soc.* **123**, 10607 (2001).
28. S. Chen, and K. Huang, Electrochemical studies of water-soluble palladium nanoparticles, *J. Cluster Sci.* **11**, 405 (2001).
29. F. P. Zamborini, J. F. Hicks, and R. W. Murray, Quantized double layer charging of nanoparticle films assembled using carboxylate/(Cu^{2+} or Zn^{2+})/carboxylate bridges, *J. Am. Chem. Soc.* **122**, 4515 (2000).
30. S. Chen, R. Pei, T. Zhao, D. J. Dyer, Gold nanoparticle assemblies by metal ion-pyridine complexation and their rectified quantized charging in aqueous solutions, *J. Phys. Chem. B* **106**, 1903 (2002).
31. J. J. Pietron, J. F. Hicks, and R. W. Murray, Using electrons stored on quantized capacitors in electron transfer reactions, *J. Am. Chem. Soc.* **121**, 5565 (1999).
32. A. J. Bard and L. R. Faulkner, *Electrochemical Methods*, 2nd edition, (John Wiley & Sons, New York, 2001).
33. (a) H. D. B. Jenkins, and K. P. Thakur, Reappraisal of thermochemical radii for complex ions, *J. Chem. Educ.* **56**, 576 (1979).
(b) N. Matsushita, H. Kitagawa, and T. Mitani, Counter-ion radius dependence of the mixed-valence state in MX chain platinum complexes, Synth. Met. **71**, 1933 (1995).
(c) TMA^+ is more hydrophobic than TEA^+ despite its smaller ionic size, reflected in a lower solubility of its salts in water.
34. (a) R. E. Holmlin, R. Haag, M. L. Chabinyc, R. F. Ismagilov, A. E. Cohen, A. Terfort, M. A. Rampi, and G. M. Whitesides, Electron transport through thin organic films in metal-insulator–metal junctions based on self-assembled monolayers, *J. Am. Chem. Soc.* **123**, 5075 (2001).
(b) K. Slowinski, R. V. Chamberlain, C. J. Miller, and M. Majda, Through-bond and chain-to-chain coupling. Two pathways in electron tunneling through liquid alkanethiol monolayers on mercury electrodes, *J. Am. Chem. Soc.* **119**, 11910 (1997).
(c) J. F. Smalley, S. W. Feldberg, C. E. D. Chidsey, M. R. Linford, M. D. Newton, and Y. Liu, The kinetics of electron transfer through ferrocene-terminated alkanethiol monolayers on gold, *J. Phys. Chem.* **99**, 13141 (1995).
(d) K. Weber, L. Hockett, and S. E. Creager, Long-range electronic coupling between ferrocene and gold in alkanethiolate-based monolayers on electrodes, *J. Phys. Chem. B* **101**, 8286 (1997).
35. S. E. Creager, and T. T. Wooster, A new way of using ac voltammetry to study redox kinetics in electroactive monolayers, *J. Anal. Chem.* **70**, 4257 (1998).
36. (a) W. P. Wuelfing, S. J. Green, J. J. Pietron, D. E. Cliffel, and R. W. Murray, Electronic conductivity of solid-state, mixed-valent, monolayer-protected Au clusters, *J. Am. Chem. Soc.* **122**, 11465 (2000).
(b) J. F. Hicks, F. P. Zamborini, A. J. Osisek, and R. W. Murray, The dynamics of electron self-exchange between nanoparticles, *J. Am. Chem. Soc.* **123**, 7048 (2001).
37. D. C. S. Tse, R. L. McCreery, and R. N. Adams, Potential oxidative pathways of brain catecholamines, *J. Med. Chem.* **19**, 37 (1976).

38. R. H. Terrill, T. A. Postlethwaite, C. Chen, C. Poon, A. Terzis, A. Chen, J. E. Hutchison, M. R. Clark, G. Wignall, J. D. Londono, R. Superfine, M. Falvo, C. S. Johnson, Jr., E. T. Samulski, and R. W. Murray, Monolayers in three dimensions: NMR, SAXS, thermal and electron hopping studies of alkanethiol stabilized gold clusters, *J. Am. Chem. Soc.* **117**, 12537 (1995).
39. A. W. Snow, and H. Wohltjen, Size-induced metal to semiconductor transition in a stabilized gold cluster ensemble, *Chem. Mater.* **10**, 947 (1998).
40. W.-Y. Lee, N. J. Hostetler, R. W. Murray, M. Majda, Electron hopping and electronic conductivity in monolayers of alkanethiol-stabilized gold nano-clusters at the air/water interface, *Isr. J. Chem.* **37**, 213 (1997).
41. J. R. Heath, C. M. Knobler, and D. V. Leff, Pressure/temperature phase diagrams and superlattices of organically functionalized metal nanocrystal monolayers: The influence of particle size, size distribution and surface passivant, *J. Phys. Chem. B* **101**, 189 (1997).
42. M. J. Hostetler, A. C. Templeton, and R. W. Murray, Dynamics of place-exchange reactions on monolayer-protected gold cluster molecules, *Langmuir* **15**, 3782 (1999).
43. (a) S. J. Green, J. J. Stokes, M. J. Hostetler, J. J. Pietron, and R. W. Murray, Three-dimensional monolayers: Nanometer-sized electrodes of alkanethiolate-stabilized gold cluster molecules, *J. Phys. Chem. B* **101**, 2663 (1997).
(b) S. J. Green, J. J. Pietron, J. J. Stokes, M. J. Hostetler, H. Vu, W. P. Wuelfing, and R. W. Murray, Three-dimensional monolayers: Voltammetry of alkanethiolate-stabilized gold cluster molecules, *Langmuir* **14**, 5612 (1998).
44. R. S. Ingram, and R. W. Murray, Electroactive three-dimensional monolayers: Anthraquinone-functionalized alkanethiolate-stabilized gold clusters, *Langmuir* **14**, 4115 (1998).
45. D. T. Miles and R. W. Murray, Redox and double-layer charging of phenothiazine functionalized monolayer-protected clusters, *Anal. Chem.* **73**, 921 (2001).
46. A. C. Templeton, M. J. Hostetler, E. K. Warmoth, S. Chen, C. M. Hartshorn, V. M. Krishnamurthy, M. D. E. Forbes, and R. W. Murray, Gateway reactions to diverse, polyfunctional monolayer-protected gold clusters, *J. Am. Chem. Soc.* **120**, 4845 (1998).

Index

MIX
Papier aus verantwortungsvollen Quellen
Paper from responsible sources
FSC® C105338

If you have any concerns about our products,
you can contact us on
ProductSafety@springernature.com

In case Publisher is established outside the EU,
the EU authorized representative is:
Springer Nature Customer Service Center GmbH
Europaplatz 3, 69115 Heidelberg, Germany

Printed by Libri Plureos GmbH
in Hamburg, Germany